Ulrich Kohlenbach   Pablo Barce
Ruy de Queiroz (Eds.)

T0238931

Logic, Language, Information, and Computation

21st International Workshop, WoLLIC 2014
Valparaíso, Chile, September 1-4, 2014
Proceedings

Springer

Volume Editors

Ulrich Kohlenbach
TU Darmstadt
Department of Mathematics
Darmstadt, Germany
E-mail: kohlenbach@mathematik.tu-darmstadt.de

Pablo Barceló
Universidad de Chile
Department of Computer Science
Santiago, Chile
E-mail: pbarcelo@dcc.uchile.cl

Ruy de Queiroz
Universidade Federal de Pernambuco (UFPE)
Centro de Informática
Recife, PE, Brazil
E-mail: ruy@cin.ufpe.br

ISSN 0302-9743   e-ISSN 1611-3349
ISBN 978-3-662-44144-2   e-ISBN 978-3-662-44145-9
DOI 10.1007/978-3-662-44145-9
Springer Heidelberg New York Dordrecht London

Library of Congress Control Number: 2014943009

LNCS Sublibrary: SL 1 – Theoretical Computer Science and General Issues

© Springer-Verlag Berlin Heidelberg 2014
This work is subject to copyright. All rights are reserved by the Publisher, whether the whole or part of
the material is concerned, specifically the rights of translation, reprinting, reuse of illustrations, recitation,
broadcasting, reproduction on microfilms or in any other physical way, and transmission or information
storage and retrieval, electronic adaptation, computer software, or by similar or dissimilar methodology
now known or hereafter developed. Exempted from this legal reservation are brief excerpts in connection
with reviews or scholarly analysis or material supplied specifically for the purpose of being entered and
executed on a computer system, for exclusive use by the purchaser of the work. Duplication of this publication
or parts thereof is permitted only under the provisions of the Copyright Law of the Publisher's location,
in ist current version, and permission for use must always be obtained from Springer. Permissions for use
may be obtained through RightsLink at the Copyright Clearance Center. Violations are liable to prosecution
under the respective Copyright Law.
The use of general descriptive names, registered names, trademarks, service marks, etc. in this publication
does not imply, even in the absence of a specific statement, that such names are exempt from the relevant
protective laws and regulations and therefore free for general use.
While the advice and information in this book are believed to be true and accurate at the date of publication,
neither the authors nor the editors nor the publisher can accept any legal responsibility for any errors or
omissions that may be made. The publisher makes no warranty, express or implied, with respect to the
material contained herein.

Typesetting: Camera-ready by author, data conversion by Scientific Publishing Services, Chennai, India

Printed on acid-free paper

Springer is part of Springer Science+Business Media (www.springer.com)

Preface

This volume contains the papers presented at the 21st Workshop on Logic, Language, Information and Computation (WoLLIC 2014) held during September 1–4, 2014, at the Department of Informatics, Universidad Técnica Federico Santa Maria in Valparaiso, Chile.

The WoLLIC series of workshops started in 1994 with the aim of fostering interdisciplinary research in pure and applied logic. The idea is to have a forum that is large enough for the number of possible interactions between logic and the sciences related to information and computation, and yet is small enough to allow for concrete and useful interaction among participants.

There were 34 submissions of which five were withdrawn. Each of the remaining 29 submissions was reviewed by at least three Program Committee (PC) members who were assisted by 25 external reviewers. The committee decided to accept 15 papers. We very much like to thank all PC members and external reviewers for the work they put into reviewing the submissions. The help provided by the EasyChair system created by Andrei Vorokonkov is hardly to be overestimated.

The program also included six invited lectures by Verónica Becher (Universidad de Buenos Aires), Juha Kontinen (University of Helsinki), Aarne Ranta (University of Gothenburg), Kazushige Terui (Kyoto University), Luca Vigano (King's College London), and Thomas Wilke (Christian-Albrechts-Universität zu Kiel).

We would like to thank the entire Organizing Committee (Anjolina G. de Oliveira, Juan Reutter, and Cristian Riveros) for making WoLLIC 2014 a success. Finally, we would like to acknowledge the generous financial support provided by NIC (Chile) and the Pontificia Universidad Católica de Chile as well as the scientific sponsorship of the following organizations: Interest Group in Pure and Applied Logics (IGPL), The Association for Logic, Language and Information (FoLLI), Association for Symbolic Logic (ASL), European Association for Theoretical Computer Science (EATCS), European Association for Computer Science Logic (EACSL), Sociedade Brasileira de Computação (SBC), and Sociedade Brasileira de Lógica (SBL).

By the time this volume was to be delivered to the publishers, we heard the sad news of the passing away of Prof Grigory Mints (Stanford University). Grisha, as he was usually known by colleagues and students alike, was an enthusiastic and very active member of WoLLIC community, having also acted in past instances of the meeting as an invited speaker, member of steering committee, member of Programme Committee, chair of Programme Committee, guest editor of proceedings as well as of a special issue, and chair of the organizing committee. Grisha will be missed for all his intellectual and personal leadership qualities. This volume is dedicated to his memory.

May 2014

Ulrich Kohlenbach
Pablo Barceló
Ruy de Queiroz

Organization

Program Chair

Ulrich Kohlenbach   Technische Universität Darmstadt, Germany

Program Committee

Natasha Alechina	University of Nottingham, UK
Eric Allender	Rutgers University, USA
Marcelo Arenas	PUC, Chile
Steve Awodey	Carnegie Mellon University, USA
Stefano Berardi	Università di Torino, Italy
Julian Bradfield	University of Edinburgh, UK
Xavier Caicedo	Universidad de los Andes, Colombia
Olivier Danvy	Aarhus University, Denmark
Marcus Kracht	University of Bielefeld, Germany
Klaus Meer	Brandenburg University of Technology Cottbus, Germany
George Metcalfe	University of Bern, Switzerland
Dale Miller	Ecole Polytechnique Paris, France
Russell Miller	CUNY, USA
Sara Negri	University of Helsinki, Finland
Grigory Olkhovikov	Ural State University, Russia
Nicole Schweikardt	Goethe-University, Frankfurt am Main, Germany
Sebastiaan Terwijn	Radboud University Nijmegen, The Netherlands
Hans van Ditmarsch	Loria Nancy, France
Michiel van Lambalgen	University of Amsterdam, The Netherlands

Steering Committee

Samson Abramsky	Oxford University, UK
Johan van Benthem	University of Amsterdam, The Netherlands
Anuj Dawar	Cambridge University, UK
Joe Halpern	Cornell University, USA
Wilfrid Hodges	Queen Mary and Westfield College, UK
Daniel Leivant	Indiana University, USA
Leonid Libkin	Edinburgh University, UK
Angus Macintyre	Queen Mary and Westfield College, UK

Grigori Mints (†)   Stanford University, USA
Luke Ong   Oxford University, UK
Hiroakira Ono   Japan Advanced Institute of Science and
                                   Technology, Japan
Ruy de Queiroz   Universidade Federal de Pernambuco, Brasil

Additional Reviewers

Baader, Franz   Kuznets, Roman
Barceló, Pablo   Lehtinen, Karoliina
Bova, Simone   Montagna, Franco
Castiglioni, José Luis   Nieves, Juan Carlos
Dal Lago, Ugo   Pérez, Jorge
Goudsmit, Jeroen   Renne, Bryan
Grädel, Erich   Sano, Katsuhiko
Heijltjes, Willem   Urzyczyn, Pawel
Iemhoff, Rosalie   Van Eijck, Jan
Kamide, Norihiro   Van Gool, Sam
Kashefi, Elham   Volpe, Marco
Knoks, Aleks   von Plato, Jan
Kuyper, Rutger

Invited Talks
(Abstracts)

On Normal Numbers

Verónica Becher

Universidad de Buenos Aires and CONICET
Argentina
vbecher@dc.uba.ar

Normality is a basic form of randomness. A real number is *simply normal* to a given base if each digit occurs in the expansion with the same limit frequency. A real number is *normal* to a given base if each *block* of digits of equal length occurs in the expansion with the same limit frequency. And a real is *absolutely normal* if it is normal to all bases. This definition was introduced by Émile Borel more than one hundred years ago, but still not much is known about normal numbers. One of the famous open problems is whether the usual mathematical constants, as π, e and $\sqrt{2}$, are simply normal to any base.

In this talk I will summarize some recent results on normal numbers that answer the following questions:

- How does simple normality (respectively normality) to one base relates to simple normality (respectively normality) to other bases?
- How does normality relates to compressibility on different automata?
- How can we efficiently compute the expansion of absolutely normal numbers?
- How can we construct absolutely normal Liouville numbers?

The proofs integrate logical, combinatorial and number-theoretic tools.

This research has been done partly with Yann Bugeaud, Olivier Carton, Pablo Heiber and Theodore Slaman.

Dependence Logic

Juha Kontinen

University of Helsinki, Department of Mathematics and Statistics, Finland
juha.kontinen@helsinki.fi

Abstract. Dependence logic, introduced by Jouko Väänänen in 2007, is a new logic incorporating the concept of dependence into first-order logic. In the past few years, the team semantics of dependence logic has grown into a new framework in which various notions of dependence and independence can be formalized and studied. We review recent results on dependence logic and its applications.

Syntax and Semantics for Translation

Aarne Ranta

Department of Computer Science and Engineering
Chalmers University of Technology and University of Gothenburg

Abstract. Translation is expected to preserve the semantics of the source text and produce correct syntax in the target language. Obvious as this is for human translators, machine translation usually involves shortcuts that compromise both of these requirements. In this talk, we will take a look at what is needed to fulfil them. In particular, we will see how formal grammars should be written in order to help translation preserve semantics. The syntax and semantics that result are in many ways different from grammars that are written in a monolingual perspective. The question has its roots in the old ideas of a Universal Grammar, and has in modern times been suggested by Curry. Translation-oriented grammars were put to use in the Rosetta system at Philips in the 1980's, and are used today in the Grammatical Framework (GF). This talk will start from the basic concepts of syntax, semantics, and translation, and end up discussing some recent developments in GF.

References

1. Angelov, K., Bringert, B., Ranta, A.: Speech-enabled hybrid multilingual translation for mobile devices. In: EACL 2014, System Demonstration, pp. 41–44 (2014)
2. Curry, H.B.: Some logical aspects of grammatical structure. In: Jakobson, R. (ed.) Structure of Language and its Mathematical Aspects: Proceedings of the Twelfth Symposium in Applied Mathematics, pp. 56–68. American Mathematical Society (1961)
3. Ranta, A.: Grammatical Framework: Programming with Multilingual Grammars. CSLI Publications, Stanford (2011)
4. Ranta, A.: Machine Translation and Type Theory. In: Dybjer, P., Lindströom, S., Palmgren, E., Sundholm, G. (eds.) Epistemology versus Ontology. Essays on the Philosophy and Foundations of Mathematics in Honour of Per Martin-Löf, pp. 281–312. Springer, Heidelberg (2012)
5. Rosetta, M.T.: Compositional Translation. Springer, Heidelberg (1994)

Intersection Types for Normalization and Verification

Kazushige Terui

RIMS, Kyoto University,
Kitashirakawa-Oiwakecho, Sakyo-ku, Kyoto 606-8502, Japan

One of the basic principles in typed lambda calculi is that typable lambda terms are normalizable. Since the converse direction does not hold for simply typed lambda calculus, people have been studying its extensions. This gave birth to the *intersection type systems*, that exactly characterize various classes of lambda terms, such as strongly/weakly normalizable terms and solvable ones (see e.g. [6] for a survey).

There is another, more recent trend: intersection types are not only useful for extending simple types but also for *refining* them [4]. One thus obtains finer information on simply typed terms by assigning intersection types. This in particular leads to the concept of *normalization by typing*, that turns out to be quite efficient in some situations [5]. Moreover, intersection types are invariant under $\beta\eta$-equivalence (when assigned to simply typed terms), so that they constitute a denotational semantics (the *Scott model of linear logic* [1]), that provides a seemingly more direct interpretation of lambda terms than the traditional filter model. Finally, intersection types also work in an infinitary setting, where terms may represent infinite trees and types play the role of automata. This leads to a model checking framework for higher order recursion schemes via intersection types [2, 3].

The purpose of this talk is to outline the recent development of intersection types described above. In particular, we explain how an efficient evaluation algorithm is obtained by combining normalization by typing, β-reduction and Krivine's abstract machine, to result in the following complexity characterization. Consider simply typed lambda terms of boolean type $o \to o \to o$ and of order r. Then the problem of deciding whether a given term evaluates to "true" is complete for n-EXPTIME if $r = 2n + 2$, and complete for n-EXPSPACE if $r = 2n + 3$ [5].

References

1. Ehrhard, T.: Collapsing non-idempotent intersection types. In: Proceedings of 26th CSL, pp. 259–273 (2012)
2. Kobayashi, N.: Types and higher-order recursion schemes for verification of higher-order programs. In: Proceedings of 36th POPL, pp. 416–428 (2009)

3. Kobayashi, N., Luke Ong, C.-H.: A type system equivalent to the modal mu-calculus model checking of higher-order recursion schemes. In: Proceedings of 24th LICS, pp. 179–188 (2009)
4. Salvati, S.: On the membership problem for non-linear abstract categorial grammars. Journal of Logic, Language and Information 19(2), 163–183 (2010)
5. Terui, K.: Semantic evaluation, intersection types and complexity of simply typed lambda calculus. In: Proceedings of 23rd RTA, pp. 323–338 (2012)
6. van Bakel, S.: Intersection type assignment systems. Theoretical Computer Science 151(2), 385–435 (1995)

Backward Deterministic Büchi Automata

Thomas Wilke*

Department of Computer Science, Kiel University
thomas.wilke@email.uni-kiel.de

Finite words are symmetric: they have a first and a last letter. Opposed to this, ω-words are not symmetric: they have a first letter, but no last letter. This is why forward and backward deterministic finite-state ω-automata are fundamentally different. For instance, a theorem by Olivier Carton and Max Michel [1] states that for every regular ω-language there exists a reverse deterministic Büchi automaton recognizing this language, whereas this is not true for ordinary forward deterministic Büchi automata. The talk gives an overview of the theory of backward deterministic ω-automata, focusing on Büchi automata.

Reference

1. Carton, O., Michel, M.: Unambiguous Büchi automata. Theor. Comput. Sci. 297(1-3), 37–81 (2003)

---

* Work supported by DFG.

Quantum State Transformations and Branching Distributed Temporal Logic[*]

Luca Viganò[1], Marco Volpe[2], and Margherita Zorzi[2]

[1] Department of Informatics, King's College London, UK
[2] Dipartimento di Informatica, Università di Verona, Italy

Abstract. The Distributed Temporal Logic DTL allows one to reason about temporal properties of a distributed system from the local point of view of the system's agents, which are assumed to execute independently and to interact by means of event sharing. In this paper, we introduce the Quantum Branching Distributed Temporal Logic QBDTL, a variant of DTL able to represent quantum state transformations in an abstract, qualitative way. In QBDTL, each agent represents a distinct quantum bit (the unit of quantum information theory), which evolves by means of quantum transformations and possibly interacts with other agents, and n-ary quantum operators act as communication/synchronization points between agents. We endow QBDTL with a DTL-style semantics, which fits the intrinsically distributed nature of quantum computing, we formalize a labeled deduction system for QBDTL, and we prove the soundness of this deduction system with respect to the given semantics. Finally, we discuss possible extensions of our system in order to reason about entanglement phenomena.

[*] The work presented in this paper was partially supported by the EU FP7 Marie Curie PIRSES-GA-2012-318986 project "GeTFun: Generalizing Truth-Functionality". Part of this work was carried out while Luca Viganò was at the Dipartimento di Informatica, Università di Verona, Italy.

Table of Contents

Quantum State Transformations
and Branching Distributed Temporal Logic[*]
(Invited Paper)

Luca Viganò[1], Marco Volpe[2], and Margherita Zorzi[2]

[1] Department of Informatics, King's College London, UK
[2] Dipartimento di Informatica, Università di Verona, Italy

Abstract. The Distributed Temporal Logic DTL allows one to reason about temporal properties of a distributed system from the local point of view of the system's agents, which are assumed to execute independently and to interact by means of event sharing. In this paper, we introduce the Quantum Branching Distributed Temporal Logic QBDTL, a variant of DTL able to represent quantum state transformations in an abstract, qualitative way. In QBDTL, each agent represents a distinct quantum bit (the unit of quantum information theory), which evolves by means of quantum transformations and possibly interacts with other agents, and n-ary quantum operators act as communication/synchronization points between agents. We endow QBDTL with a DTL-style semantics, which fits the intrinsically distributed nature of quantum computing, we formalize a labeled deduction system for QBDTL, and we prove the soundness of this deduction system with respect to the given semantics. Finally, we discuss possible extensions of our system in order to reason about entanglement phenomena.

1 Introduction

Background and Motivation. The *Distributed Temporal Logic DTL* [12, 5, 6] allows one to reason about temporal properties of a distributed system from the local point of view of the system's agents: each asynchronous agent executes independently, evolves linearly along a time-line built upon some local events, and can interact with the other agents by means of event sharing. Distribution is implicit and properties of an entire system are formulated in terms of the local properties of the system's agents and their interaction. DTL's semantics was inspired by a conflict-free version of Winskel's *event structures* (see, e.g., [26]), enriched with information about sequential agents.

DTL has been initially proposed as a logic for specifying and reasoning about distributed information [12], but it has also been used in the context of security protocol analysis to reason about the interplay between protocol models and security properties [6]. In this paper, we show that, after a proper extension of the logic's syntax and semantics, DTL is also able to formally model quantum state transformations in an abstract, qualitative way.

---

[*] The work presented in this paper was partially supported by the EU FP7 Marie Curie PIRSES-GA-2012-318986 project "GeTFun: Generalizing Truth-Functionality". Part of this work was carried out while Luca Viganò was at the Dipartimento di Informatica, Università di Verona, Italy.

U. Kohlenbach et al. (Eds.): WoLLIC 2014, LNCS 8652, pp. 1–19, 2014.
© Springer-Verlag Berlin Heidelberg 2014

Quantum computing is one of the most promising research fields of computer science as well as a concrete future technology (see [22] for a useful introduction to the basic notions of quantum computing as we here only very briefly summarize the notions that are relevant to our work in this paper). However, at least from the point of view of theoretical computer science, a number of foundational aspects are still underdeveloped: quantum complexity, quantum computability, quantum programming theory (and its logical account), quantum cryptography and security are all active but open research areas, which still require the development of ad hoc formal methods. These issues are complex to face since the physical model quantum computing is based on is sophisticated and all basic definitions and formal tools have to be reformulated in a non-standard way.

To illustrate this, and our contributions in this paper, in more detail, let us focus our attention on quantum data, in particular on the unit of quantum information, the *quantum bit* or *qubit*, for short. The qubit is the quantum counterpart of the classical bit and, mathematically, it is simply a normalized vector of the Hilbert Space $\mathbb{C}^2$. Qubits can assume both classical values 0 and 1 (as the classical bit) and all their *superpositional values*, i.e., linear combinations such as $\alpha|0\rangle + \beta|1\rangle$, where $\alpha, \beta \in \mathbb{C}$ are called *amplitudes*, $|\alpha|^2 + |\beta|^2 = 1$ and $|c\rangle$, for $c \in \{0, 1\}$, is the so called *Dirac Notation*, which is simply a denotation of basis states (which corresponds to the classical values a bit can assume).

Intuitively, whereas a classical bit can only be 0 or 1, a quantum bit can assume both the value 0 and the value 1 (with a certain associated probability) at the same time. It is possible to modify a quantum bit in two ways:

- by means of a suitable class of algebraic operators called *unitary transformations* (that are also called *quantum gates* and are a class of algebraic operators enjoying some good properties, which represent the pure quantum computational steps) or
- by *measuring* it, i.e., probabilistically reducing it to 0 or 1.

In this paper, we deal only with unitary transformations, leaving measurement for future work.

The definition of a qubit can, of course, be generalized: a *quantum register* or *quantum state* is the representation of a system of n qubits (mathematically, it is a normalized vector of the Hilbert space $\mathbb{C}^{2^n}$). As for the single qubit, a quantum state can be modified by means of unitary algebraic operators.

Abstracting from any notion of control and considering only pure quantum transformations (i.e., unitary evolution of quantum states as computational steps), it seems to be interesting to provide a logical account of such a computation. The question then is: what is a logical approach suitable to represent quantum state evolution?

Contributions. The main contribution of this paper is the formalization of a logic and of an associated deduction system that allows one to formally represent and reason about unitary transformations of quantum states from a temporal multi-agent system perspective. More specifically, we view our contributions as two-fold.

First, we define the *Quantum Branching Distributed Temporal Logic* QBDTL, a significant variant of DTL that we introduce here to represent quantum state transformations in an abstract, *qualitative* way. In QBDTL, we abstract from the value of the qubits:

we are not interested in encoding into our system syntactical and semantical information about amplitudes or basis values 0 and 1 (in this way, we avoid any *quantitative* information) and we focus instead on the way qubits evolve by means of unitary transformations. Following DTL's central notion, in **QBDTL** we do not only consider globally quantum states but also, and in particular, the single unit of information, i.e., we maintain the local perspective of the qubit in the quantum computation.

In other words, in **QBDTL** each agent represents a distinct qubit, which is the object/subject of computation and which evolves in time by means of quantum transformations and possibly interacts with other agents/qubits.

There is a crucial difference between our **QBDTL** and the original DTL formulation. DTL is based on linear time life-cycles for agents. In **QBDTL**, we go beyond linearity and consider branching time since we want to be as general as possible: at each step of the temporal evolution of an agent/qubit, the accessibility relation between worlds in the subtended Kripke-style model aims to capture each possible unitary transformation that can be applied to the qubit. A world (a state in the temporal life-cycle of an agent) represents (an abstraction of) a 1-qubit quantum state. n-ary quantum operators, which act simultaneously on more than one qubit (such as control operators, which play a crucial role in quantum computing), act as communication/synchronization points between agents/qubits.

Second, we give a deduction system $\mathcal{N}$(**QBDTL**) for **QBDTL**. In order to deal with all the semantical notions—temporal, quantum and synchronization information—, we follow the style of *labeled deduction* [15, 24, 25], a framework for giving uniform presentations of different non-classical logics, where labels allow one to explicitly encode in the syntax additional information, of a semantic or proof-theoretical nature, that is otherwise implicit in the logic one wants to capture.

In addition to the works on DTL, and in particular the labeled tableaux system given in [5], our starting points for $\mathcal{N}$(**QBDTL**) are the labeled natural deduction system for the logic *UB* (i.e., the until-free fragment of *CTL*) given in [10] and the approach developed in [19, 20], where a labeled modal deduction system with specific modalities able to describe quantum state transformations is given. Fittingly, in $\mathcal{N}$(**QBDTL**), we consider composed labels (i, x, q) that represent an agent/qubit i, a time instant x, and the quantum information q in the underlying semantics. A further class of labels is used to represent paths in the life-cycles of the agents.

The rules of $\mathcal{N}$(**QBDTL**) can then be divided into rules that formalize the *local* temporal evolution of an agent/qubit, and synchronization rules that are, in a sense, *global* as they lift the reasoning from the local perspective of the agent to the *distributed* perspective induced by agent's synchronizations.

It is important to observe that our **QBDTL** is not a quantum logic. Since the work of Birkhoff and von Neumann [9], various logics have been investigated as a means to formalize reasoning about propositions taking into account the principles of quantum theory, e.g., [11]. In general, it is possible to view quantum logic as a logical axiomatization of quantum theory, which provides an adequate foundation for a theory of reversible quantum processes, e.g., [21, 1–4, 13, 14]. Research has focused also on automated reasoning (e.g., model checking for quantum systems as considered in [16]) and on formal analysis of quantum protocols (e.g., [18]). Our work moves from quite a different point of view, which, to reiterate, is the wish to provide a deduction system able to represent and reason about unitary transformations of quantum states from a temporal multi-agent system perspective and, as will become clear below, thereby provide a basis to reason about other, more complex properties of quantum states such as entanglement.

Organization. After a preliminary discussion about aims and motivations of our approach (Section 2), in Section 3 we introduce the logic QBDTL and a DTL-style semantics. In Section 4 we define the natural deduction system $\mathcal{N}(\text{QBDTL})$, providing some example derivations, and in Section 5 we state and prove the Soundness Theorem (of $\mathcal{N}(\text{QBDTL})$ with respect to the semantics). Section 6 is devoted to discussions about our ongoing and future works.

2   Why Branching Temporal Logic and Synchronization?

In this section, we describe how it is possible to use temporal logic and synchronization rules (the core of the DTL approach) to reason in a simple way about quantum state transformations, whenever one is not interested in the encoding of the mathematical object that represents a quantum state (i.e., a vector in a suitable Hilbert Space) but in the evolution itself as a sequence of transformations and in a notion of synchronization between different quantum bits.

Modal logics are a flexible instrument to describe qualitatively state transformations as they allow one to put the emphasis on the underlying "transition system"—the set of possible worlds of the Kripke semantics and the properties of the accessibility relations between them, which model the dynamical behavior of the system—rather than on the concrete meaning of the internal structures of possible worlds. This intuition was followed in [19, 20], where two pure modal systems were introduced and studied. In such systems, a world represents the abstraction of a quantum state and modal operators reflect general properties of quantum state transformations, since the subtended models are $S5$-models. The accessibility relation between worlds is therefore an equivalence relation, i.e., it enjoys reflexivity, symmetry and transitivity. This captures, in an abstract way, key properties of unary quantum operators: roughly speaking, reflexivity says that the class of the unitary operators includes the identity transformation; symmetry captures reversibility (it is always possible to reverse a quantum transformation, since the inverse operator is easily definable and is unitary); finally, transitivity models algebraic compositionality, i.e., the composition of two or more unitary operators is always a unitary operator [22].

The main difference between the modal systems proposed in [19, 20] and QBDTL is that whereas in the former case a world represents the abstraction of an *arbitrary* quantum state (i.e., a state that describes an arbitrary number n of qubits), in the case of QBDTL we focus on the single qubit and on its transformation by means of *unary* quantum operators and on a notion of local formula built upon a local language. Moreover, we move from a modal to a temporal system: in some sense we "unfold" the accessibility relation between worlds obtaining, for each agent, a tree-like structure that represents the agent's local life-cycle. In this way, we "link" the subtended branching temporal model to the abstract transition system induced by all the unary quantum transformations possibly occurring in each world, which are uniformly modeled in the semantics and in the deduction system by an equivalence relation. Reflexivity, symmetry and transitivity can be plainly expressed in **QBDTL**: for example, symmetry can be abstractly captured by the labeled formula $(i, x, q) : p \supset \exists \bigcirc \exists \bigcirc p$, where p is a propositional symbol, $\supset$ is implication and $\exists \bigcirc A$ expresses that the formula A is true at the next time instant in some possible future.

A licit question at this stage is what is the meaning of the set of propositional symbols **QBDTL** formulas are built upon. We maintain an abstract definition of the set (we simply say that is a set of syntactic objects), following the style of DTL and also in the spirit of modal/temporal logic as we discussed above. Then, working with labeled expressions like $(i, x, q) : A$, where the formula A is built by temporal operators, synchronization and propositional symbols, it is not actually crucial to say what propositional symbols stand for.[1] Still, it is important to consider what modal/temporal formulas, possible worlds and the accessibility relation stand for.

One could even choose to instantiate the set of propositional symbols to capture quantitative information about quantum states or general properties that permit one to reason about them. We provide here a simple example (partially related to the examples that we will provide later in Fig. 5). A possible choice is to fix a set of atomic propositions representing mathematical descriptions of the qubit, i.e., a normalized vector in $\mathbb{C}^2$. In other words, given a qubit $a = \alpha|0\rangle + \beta|1\rangle$, the encoding $\lceil a \rceil$ of this mathematical description is an atomic proposition. Let s_i stand for a label (i, x, q), take p as $\lceil a \rceil$ and consider the labeled formula $s_i : p \supset \exists \Box p$ (whose derivation will be given in Fig. 5 and where $\exists \Box p$ expresses that p is true at every time instant in some possible future). This labeled formula can be intuitively interpreted as follows: a (potentially infinite) sequence of identity unitary transformations does not change the mathematical description of the qubit.

Let p still be the encoding $\lceil a \rceil$ of a state $a = \alpha|0\rangle + \beta|1\rangle$ and let us consider again the labeled formula $(i, x, q) : p \supset \exists \bigcirc \exists \bigcirc p$, which fits a peculiar feature of quantum computation, i.e., reversibility. This labeled formula says that: if p holds for i in some state x, then there exists a temporal path such that, in two steps, i reaches a new state in which p still holds (i.e., the mathematical description of such a state is again $\alpha|0\rangle + \beta|1\rangle$). This models the fact that if one transforms a qubit state by means of a unitary operator U, then one can obtain again the same state by applying the adjoint U^* of U, where, in the class of unitary operators, the adjoint corresponds to the inverse U^{-1}, and algebraically, one has $U^*(U(\alpha|0\rangle + \beta|1\rangle)) = U(U^*(\alpha|0\rangle + \beta|1\rangle)) = \alpha|0\rangle + \beta|1\rangle$, i.e., $U^*U = UU^* = I$, where I is the identity operator. Looking for a concrete example, we can take $\alpha = \frac{1}{\sqrt{3}}$ and $\beta = \frac{\sqrt{2}}{\sqrt{3}}$ and instantiate U to X, the complementation gate, which corresponds to an exchange between amplitudes of basis states. Among the temporal states reachable from x there exists, in particular, the successor state in which $\overline{p}$ and $\exists \bigcirc p$ hold, where $\overline{p} = \lceil \frac{\sqrt{2}}{\sqrt{3}}|0\rangle + \frac{1}{\sqrt{3}}|1\rangle \rceil$.

---

[1] In analogy, note, e.g., that temporal logics developed to deal with concurrent systems do not possess any concurrent feature.

In quantum computing it is useful to compose small states in order to obtain bigger quantum states (this operation has a precise algebraic meaning, see [22]). Collecting agents, one can model quantum systems of n qubits. In some sense, we can see a quantum state of n qubits as a global state built upon the local states of the single qubits. Each qubit evolves independently but, in a realistic perspective, different qubits do not always evolve asynchronously, and so sometimes they interact, by means of n-ary quantum gates. This is modeled, in our system, by means of ad hoc "tools", properly adapted from DTL: by a special construct in the local language (an operator © named *calling*), it is possible to express the fact that an agent/qubit i synchronizes with another agent/qubit j. This choice has a precise quantum meaning. In quantum computing, one can of course globally modify a set of n qubits by means of n-ary algebraic operators. We view n-ary quantum gates as *synchronization* points between states of different life-cycles, i.e., between states of different qubits. The inputs of an n-ary quantum gate may each have previously been subject to a sequence of other transformations, i.e., in DTL terms, a sequence of events, and the gate itself then can be seen as a transformation event that is shared by the inputs. In this paper, we model this synchronization mechanism abstractly (since, as we said, we model unitary transformations by an equivalence relation), but it is possible to plan a concrete research direction based on the further development of this interpretation of n-ary gates as synchronization mechanisms. See Section 6 for a more detailed discussion of our ongoing and future works.

3   The Logic QBDTL

We introduce the Quantum Branching Distributed Temporal Logic QBDTL by presenting its syntax and semantics.

3.1   Syntax

Given a finite set $\mathsf{Id} = \{i, j, \ldots\}$ of *agent identifiers* and a set $\mathsf{Prop} = \{p, p_1, p_2, \ldots\}$ of *atomic propositions* (which characterize the current local states of the agents), we define the *local language* of an agent $i \in \mathsf{Id}$ by the following grammar:

$$\mathcal{L}_i ::= p \mid \perp \mid \mathcal{L}_i \supset \mathcal{L}_i \mid \exists \bigcirc \mathcal{L}_i \mid \exists \Box \mathcal{L}_i \mid \forall \Box \mathcal{L}_i \mid ©_j \mathcal{L}_j \,,$$

where $p \in \mathsf{Prop}$ and $j \in \mathsf{Id}$ with $i \neq j$. *Local formulas*, as their names suggest, hold locally for the different agents. $\perp$ is *falsum* and $\supset$ is implication. As in DTL, the communication formula $©_j A$ means that agent i has just communicated (i.e., synchronized) with agent j, for whom A holds. We follow here the Peircean branching temporal logic *UB* [7] and only consider the temporal operators that are obtained as a combination of one single linear-time operator immediately preceded by one single path quantifier. More specifically, we consider here the Peircean operators

- $\exists \bigcirc$ (as we noted previously, $\exists \bigcirc A$ expresses that the formula A in the scope of this operator is true at the next time instant in some possible future),
- $\exists \Box$ ("it is true at every time instant in some possible future") and
- $\forall \Box$ ("it is true at every time instant in every possible future").

For simplicity, in this work we do not consider the temporal operator until, although such an extension would not be problematic. Moreover, as usual, other connectives and temporal operators can be defined as abbreviations.

The *global language* of QBDTL is defined by the grammar:

$$\mathcal{L} ::= @_{i_1}\mathcal{L}_{i_1} \mid \ldots \mid @_{i_n}\mathcal{L}_{i_n},$$

where $i_1, \ldots, i_n \in \mathsf{Id}$. The global formula $@_{i_k}A$ means that A holds for agent i_k.

3.2 Semantics

The models of QBDTL are inspired by those of DTL and built upon a form of Winskel's event structures (cf. [26], where also the relationship to other concurrency models is discussed). There is, however, a fundamental difference with respect to the semantics that has (actually, with respect to the slightly different semantics that in the literature have) been given for DTL, which is based on distributed families of linear life-cycles local to each agent, i.e., countable, discrete and totally ordered local events. Since our logic QBDTL is inherently branching, we need to define its semantics accordingly, and we thus modify DTL's semantics as follows.

Given an agent $i \in \mathsf{Id}$, a *branching local life-cycle* of i is an ω-tree, i.e., a pair $\lambda_i = \langle \mathsf{Ev}_i, <_i \rangle$, where Ev_i is the set of *local events of* i and $<_i \subseteq \mathsf{Ev}_i \times \mathsf{Ev}_i$ is a binary relation such that:

(i) $<_i$ is transitive and irreflexive;

(ii) for each $e \in \mathsf{Ev}_i$, the set $\{e' \in \mathsf{Ev}_i \mid e' <_i e\}$ is linearly ordered by $<_i$;

(iii) there is a $<_i$-smallest element 0_i called the *root* of λ_i;

(iv) each maximal linearly $<_i$-ordered subset of Ev_i is order-isomorphic to the natural numbers.

We write $e \to_i e'$ to denote the fact that e' is an immediate local successor of e, i.e., $e <_i e'$ and there is no e'' such that $e <_i e'' <_i e'$. A $\to_i$-*path* is a sequence of local events $(e_0, \ldots, e_n)$ such that $e_k \to_i e_{k+1}$ for $0 \le k \le n - 1$. An e-*branch b of* i is an infinite $\to_i$-path $b = (e_0, e_1, \ldots)$ such that $e = e_0$ and we write $\to_i^b$ to denote the restriction of $\to_i$ to b, i.e., $e' \to_i^b e''$ iff $e' = e_k$ and $e'' = e_{k+1}$ for some k, and denote with $\mathcal{B}_i$ the set of all such $\to_i^b$. Further, we denote with $\to_i^{b*}$ the reflexive and transitive closure of $\to_i^b$.

A *local state* is a finite set $\xi \in \mathsf{Ev}_i$ down-closed for local causality, i.e., if $e <_i e'$ and $e' \in \xi$ then also $e \in \xi$. In general, each non-empty local state ξ is reached by the occurrence of an event that we call $last(\xi)$, from the local state $\xi \setminus \{last(\xi)\}$. Given $e \in \mathsf{Ev}_i$, the set $e{\downarrow}i = \{e' \in \mathsf{Ev}_i | e' \le_i e\}$, where $\le_i$ denotes the reflexive closure of $<_i$, is always a local state. Moreover, if ξ is non-empty, then $last(\xi){\downarrow}i = \xi$.

A *branching distributed life-cycle* is a family of local life-cycles

$$\lambda = \{\lambda_i = \langle \mathsf{Ev}_i, <_i \rangle\}_{i \in \mathsf{Id}}$$

such that:

(i) $\le = (\bigcup_{i \in \mathsf{Id}} \le_i)^*$ defines a partial order of *global causality* on the set of events $\mathsf{Ev} = \bigcup_{i \in \mathsf{Id}} \mathsf{Ev}_i$;

(ii) if $e, e' \in \mathsf{Ev}_i$ and $e \le e'$ then $e \le_i e'$.

Condition *(i)* ensures that a distributed life-cycle respects global compatibility, i.e., there is no $e \in \mathsf{Ev}_i \cap \mathsf{Ev}_j$ such that $e <_i e'$ but $e' <_j e$, while condition *(ii)* ensures that synchronization $\leq$-relates two events of an agent i only if there exists a 0_i-branch in which both the events occur.

An *S5 Kripke frame* is a pair $\langle Q, \mathcal{U} \rangle$, where Q is a non-empty set of *qubit states* and $\mathcal{U}$ is a binary equivalence relation on Q, i.e., $\mathcal{U} : Q \to Q$ is reflexive, symmetric and transitive. An *S5 Kripke model* is a triple $\mathcal{M} = \langle Q, \mathcal{U}, \mathcal{V} \rangle$, where $\langle Q, \mathcal{U} \rangle$ is an *S5* Kripke frame and $\mathcal{V} : Q \to \mathcal{P}(\mathsf{Prop})$ is a valuation function assigning to each qubit state in Q a set of atomic propositions.

A **QBDTL** *model* is a triple $\mu = \langle \lambda, \mathcal{M}, \pi \rangle$, where $\lambda = \{\lambda_i\}_{i \in \mathsf{Id}}$ is a distributed life-cycle, $\mathcal{M} = \langle Q, \mathcal{U}, \mathcal{V} \rangle$ is an *S5* Kripke model and $\pi = \{\pi_i\}_{i \in \mathsf{Id}}$ is a family of local functions associating to each local state a qubit state in Q; for each $i \in \mathsf{Id}$ and set Ξ_i of local states of i, the function $\pi_i : \Xi_i \to Q$ is such that:

(i) if $\xi, \xi' \in \Xi_i$, $last(\xi) \to_i last(\xi')$, $\pi(\xi) = q$ and $\pi(\xi') = q'$, then $q\mathcal{U}q'$;

(ii) if $q, q' \in Q$, $q\mathcal{U}q'$ and $\pi(\xi) = q$, then there exists $\xi' \in \Xi_i$ such that $last(\xi) \to_i last(\xi')$ and $\pi(\xi') = q'$.

In what follows, we denote $\langle \lambda_i, \mathcal{M}, \pi_i \rangle$ by μ_i.

The *global satisfaction relation* is defined by:

$$\models^\mu @_i A \quad \text{iff} \quad \models^{\mu_i}_i A \quad \text{iff} \quad \models^{\mu_i, \xi}_i A \text{ for every } \xi \in \Xi_i,$$

where the *local satisfaction relation at a local state ξ of i* is defined by:

$$
\begin{aligned}
&\not\models^{\mu_i, \xi}_i \bot \\
&\models^{\mu_i, \xi}_i p && \text{iff} && p \in \mathcal{V}(\pi_i(\xi)), \text{ for } p \in \mathsf{Prop} \\
&\models^{\mu_i, \xi}_i A \supset B && \text{iff} && \models^{\mu_i, \xi}_i A \text{ implies } \models^{\mu_i, \xi}_i B \\
&\models^{\mu_i, \xi}_i \forall \Box A && \text{iff} && \text{for all } \xi', last(\xi) \leq_i last(\xi') \text{ implies } \models^{\mu_i, \xi'}_i A \\
&\models^{\mu_i, \xi}_i \exists \Box A && \text{iff} && \text{there exists a } last(\xi)\text{-branch } b \text{ such that for all } \xi', \\
& && && last(\xi) \to^{b*}_i last(\xi') \text{ implies } \models^{\mu_i, \xi'}_i A \\
&\models^{\mu_i, \xi}_i \exists \Diamond A && \text{iff} && \text{there exists } \xi' \text{ such that } last(\xi) \to_i last(\xi') \text{ and } \models^{\mu_i, \xi'}_i A \\
&\models^{\mu_i, \xi}_i \copyright_j A && \text{iff} && last(\xi) \in \mathsf{Ev}_j \text{ and } \models^{\mu_j, last(\xi) \downarrow j}_j A
\end{aligned}
$$

By extension, we define:

$$
\begin{aligned}
&\models^\mu \Gamma && \text{iff} && \models^\mu A \text{ for all } A \in \Gamma \\
&\Gamma \models^\mu A && \text{iff} && \models^\mu \Gamma \text{ implies } \models^\mu A \\
&\Gamma \models A && \text{iff} && \Gamma \models^\mu A \text{ for each } \mathbf{QBDTL} \text{ model } \mu
\end{aligned}
$$

4   A Deduction System for QBDTL

4.1   Syntax of the Labeled Logic

In order to formalize our labeled natural deduction system $\mathcal{N}(\mathbf{QBDTL})$, we extend the syntax and semantics of **QBDTL** by introducing four kinds of *labels* (that represent agents, states, quantum information and paths in the underlying semantics) and by defining labeled and relational formulas.

First of all, we use the agent identifiers in Id as labels. Further, we assume given two fixed denumerable sets of labels Lab_S and Lab_Q. Intuitively, the labels $x, y, z, \ldots$ in Lab_S refer to local states of an agent, whereas the labels $q, q', q_1, \ldots$ in Lab_Q refer to the quantum information concerning an agent.

A *labeled formula* is then a formula of the form

$$(i, x, q) : A,$$

where (i, x, q) is a *composed label* with $i \in \mathsf{Id}$, $x \in \mathsf{Lab}_S$ and $q \in \mathsf{Lab}_Q$, and A is a formula in the local language $\mathcal{L}_i$ of the agent i. Note that we do not use the operator @ inside labeled formulas as it is implicitly expressed by the first element of the composed label. For instance, in order to show that a global formula $@_i A$ is valid, we will prove that the labeled formula $(i, x, q) : A$, for arbitrary x and q, is derivable in our system.

In $\mathcal{N}(\mathsf{QBDTL})$, we also need formulas modeling the relation between the states referred by the labels. We thus assume given a further set of labels Lab_B, whose elements will be denoted by $\triangleleft, \triangleleft_1, \triangleleft_2, \ldots$, which intuitively refer to the successor relation between local states in the local life-cycle of an agent i along a given branch.

We define

$$\mathsf{Lab}_B^+ = \mathsf{Lab}_B \cup \{r(i, x, \bigstar A) \mid i \in \mathsf{Id}, x \in \mathsf{Lab}_S, \bigstar \in \{\square, \bigcirc\}, A \in \mathcal{L}_i\}.$$

The labels in $\mathsf{Lab}_B^+ \setminus \mathsf{Lab}_B$ will be used to refer to successor relations between local states along distinct branches. We will write $R, R_1, R_2, \ldots$ to denote generic elements of Lab_B^+ and we will use R^* to refer to the reflexive and transitive closure of R. Finally, we will use the symbol U to refer to the relation modeling unary quantum transformations and the symbol $\bowtie$ to denote that the local states of two agents are synchronized on a given event.

A *relational formula* is then a formula of the form

- $(i, x, q) \, R \, (i, y, q')$, or
- $(i, x, q) \, R^* \, (i, y, q')$, or
- $(i, x, q) \bowtie (j, y, q')$, or
- $q \, U \, q'$,

where $i, j \in \mathsf{Id}$, $x, y \in \mathsf{Lab}_S$, $R \in \mathsf{Lab}_B^+$, $q, q' \in \mathsf{Lab}_Q$. In the following, for simplicity, we will sometimes use metavariables of the form s_i, possibly superscripted, to refer to composed labels of the form (i, x, q).

4.2   Semantics of the Labeled Logic

In order to give a semantics for our labeled system, we need to define explicitly an interpretation of the labels. Given a QBDTL model μ, an *interpretation function* is a triple $\mathcal{I} = \langle \mathcal{I}_S, \mathcal{I}_Q, \mathcal{I}_B \rangle$, where:

- $\mathcal{I}_S = \{\mathcal{I}_S^i\}_{i \in \mathsf{Id}}$ is a set of functions such that $\mathcal{I}_S^i : \mathsf{Lab}_S \to \varXi_i$ for each $i \in \mathsf{Id}$;
- $\mathcal{I}_Q : \mathsf{Lab}_Q \to Q$;

- $I_B = \{I_B^i\}_{i \in \text{Id}}$ is a set of functions such that $I_B^i : \text{Lab}_B^+ \to \mathcal{B}_i$ for each $i \in \text{Id}$, and if $r(i, x, \bigstar A) \in \text{Lab}_B^+ \setminus \text{Lab}_B$, then:
  - $I_B^i(r(i, x, \bigstar A)) = \to_i^b$ for some $I_S^i(x)$-branch b;
  - if $\models^{\mu, I_S^i(x)} \exists \bigstar A$, then for all $\xi \in \Xi_i$:
    * if $\bigstar = \circ$, then $last(I_S^i(x)) \, I_B^i(r(i, x, \bigstar A)) \, last(\xi)$ implies $\models^{\mu, \xi} A$;
    * if $\bigstar = \square$, then $last(I_S^i(x)) \, I_B^i(r(i, x, \bigstar A))^* \, last(\xi)$ implies $\models^{\mu, \xi} A$.

The notion of interpretation allows us to extend the truth relation to labeled formulas, as well as define truth of relational formulas. Given a QBDTL model μ and an interpretation function $I = \langle I_S, I_Q, I_B \rangle$ on it, *truth for a labeled or relational formula γ is defined as follows:*

$\models^{\mu, I} (i, x, q) : A$	iff	$\mu_i, I_S^i(x) \models_i A$ and $\pi_i(I_S^i(x)) = I_Q(q)$
$\models^{\mu, I} (i, x, q) R (i, y, q')$	iff	$last(I_S^i(x)) \, I_B^i(R) \, last(I_S^i(y)), \pi_i(I_S^i(x)) = I_Q(q)$ and $\pi_i(I_S^i(y)) = I_Q(q')$
$\models^{\mu, I} (i, x, q) R^* (i, y, q')$	iff	$last(I_S^i(x)) \, I_B^i(R)^* \, last(I_S^i(y)), \pi_i(I_S^i(x)) = I_Q(q)$ and $\pi_i(I_S^i(y)) = I_Q(q')$
$\models^{\mu, I} (i, x, q) \bowtie (j, y, q')$	iff	$last(I_S^i(x)) = last(I_S^j(y)), \pi_i(I_S^i(x)) = I_Q(q)$ and $\pi_j(I_S^j(y)) = I_Q(q')$
$\models^{\mu, I} q \, U \, q'$	iff	$I_Q(q) \, \mathcal{U} \, I_Q(q')$

When $\models^{\mu, I} \gamma$, for γ a labeled or relational formula, we say that γ is *true* in μ according to I. By extension:

$\models^{\mu, I} \Gamma$	iff	$\models^{\mu, I} \gamma$ for all $\gamma \in \Gamma$
$\Gamma \models^{\mu, I} \gamma$	iff	$\models^{\mu, I} \Gamma$ implies $\models^{\mu, I} \gamma$
$\models^{\mu} \gamma$	iff	for every interpretation function I, $\models^{\mu, I} \gamma$
$\models^{\mu} \Gamma$	iff	for every interpretation function I, $\models^{\mu, I} \Gamma$
$\Gamma \models \gamma$	iff	for every QBDTL model $\mathcal{M}$ and interpretation function I, $\Gamma \models^{\mu, I} \gamma$

4.3   The Rules of $\mathcal{N}(\textsf{QBDTL})$

The rules of $\mathcal{N}(\textsf{QBDTL})$ are given in Fig. 1–4. We can classify them into four categories: (i) *local life-cycle rules* (inspired to the deduction system for the logic UB given in [10]), (ii) *distributed life-cycle rules* (reminiscent of the global labeled tableaux developed for DTL in [5]), (iii) *quantum transformations rules* (actually a fragment of the deduction systems studied in [20]) and (iv) *interaction rules*.

Local Life-Cycle Rules (Fig. 1). These rules all infer formulas "local" to an agent i, i.e., labeled with s_i. We can divide them further into rules for classical connectives ($\perp E$, $\supset I$ and $\supset E$), rules for temporal operators ($\forall \square I$, $\forall \square E$, $\exists \square I$, $\exists \square E$, $\exists \circ I$ and $\exists \circ E$), relational rules ($ser_\triangleleft$, ser_{sk}, $base_R$, $lin_\triangleleft$, $refl_R$, $trans_R$ and $comp_R$) and induction rules ($ind\forall$ and $ind\exists$).

Rules for classical connectives. The rule $\perp E$ is a labeled version of *reductio ad absurdum*, where we do not enforce Prawitz's side condition that $A \neq \perp$ and we do

$$\frac{\begin{array}{c}[s_i : A \supset \bot]\\ \vdots\\ s_j : \bot\end{array}}{s_i : A}\ \bot E \qquad \frac{\begin{array}{c}[s_i : A]\\ \vdots\\ s_i : B\end{array}}{s_i : A \supset B}\ \supset I \qquad \frac{s_i : A \supset B \quad s_i : A}{s_i : B}\ \supset E \qquad \frac{\begin{array}{c}[s_i \lhd^* s_i']\\ \vdots\\ s_i' : A\end{array}}{s_i : \forall\Box A}\ \forall\Box I \qquad \frac{s_i' : \forall\Box A \quad s_i' R^* s_i}{s_i : A}\ \forall\Box E$$

$$\frac{\begin{array}{c}[s_i R^* s_i']\\ \vdots\\ s_i' : A \quad s_i R s_i''\end{array}}{s_i : \exists\Box A}\ \exists\Box I \qquad \frac{(i,x,q) : \exists\Box A \quad (i,x,q)\, r(i,x,\Box A)^*\, s_i}{s_i : A}\ \exists\Box E$$

$$\frac{\begin{array}{c}[s_i R s_i']\\ \vdots\\ s_i' : A \quad s_i R s_i''\end{array}}{s_i : \exists\bigcirc A}\ \exists\bigcirc I \qquad \frac{(i,x,q) : \exists\bigcirc A \quad (i,x,q)\, r(i,x,\bigcirc A)\, s_i}{s_i : A}\ \exists\bigcirc E$$

$$\frac{\begin{array}{c}[s_j \lhd s_j']\\ \vdots\\ s_i : A\end{array}}{s_i : A}\ ser_{\lhd} \qquad \frac{\begin{array}{c}[(j,x,q)\, r(j,x,\bigstar B)\, s_j]\\ \vdots\\ s_i : A\end{array}}{s_i : A}\ ser_{sk} \qquad \frac{\begin{array}{c}\qquad [s_j R^* s_j']\\ \qquad \vdots\\ s_j R s_j' \quad s_i : A\end{array}}{s_i : A}\ base_R \qquad \frac{s_i \lhd s_i' \quad s_i \lhd s_i'' \quad s_i' : \alpha}{s_i'' : \alpha}\ lin_{\lhd}$$

$$\frac{\begin{array}{c}\qquad [s_j R^* s_j]\\ \qquad \vdots\\ s_j R s_j' \quad s_i : A\end{array}}{s_i : A}\ refl_R \qquad \frac{\begin{array}{c}\qquad\qquad [s_j R^* s_j'']\\ \qquad\qquad \vdots\\ s_j R^* s_j' \quad s_j' R^* s_j'' \quad s_i : A\end{array}}{s_i : A}\ trans_R \qquad \frac{\begin{array}{c}\qquad\qquad [s_j \lhd^* s_j'']\\ \qquad\qquad \vdots\\ s_j R_1^* s_j' \quad s_j' R_2^* s_j'' \quad s_i : A\end{array}}{s_i : A}\ comp_R$$

$$\frac{\begin{array}{c}\qquad\qquad [s_i' \lhd_1^* s_i'''] \quad [s_i''' \lhd_2 s_i''] \quad [s_i''' : A]\\ \qquad\qquad \vdots\\ s_i' : A \quad s_i' R^* s_i \qquad s_i'' : A\end{array}}{s_i : A}\ ind\forall$$

$$\frac{\begin{array}{c}\qquad\qquad [(i,x,q) \lhd^* (i,y,q')] \quad [(i,y,q')\, r(i,y,\bigcirc A)\, s_i'] \quad [(i,y,q') : A]\\ \qquad\qquad \vdots\\ (i,x,q) : A \quad (i,x,q)\, r(i,x,\Box A)^*\, s_i \qquad s_i' : A\end{array}}{s_i : A}\ ind\exists$$

In $\forall\Box I$, $\exists\Box I$ and $\exists\bigcirc I$, where $s_i' \equiv (i,x,q)$, the labels x and q are fresh. Moreover, in $\forall\Box I$, $\lhd$ is fresh.
In $ser_{\lhd}$, where $s_j' \equiv (j,x,q)$, the labels x, q and $\lhd$ are fresh.
In ser_{sk}, where $s_j \equiv (j,y,q')$, the labels y and q' are fresh.
In $comp_R$, $\lhd$ is fresh.
In $ind\forall$, where $s_i'' \equiv (i,x,q)$ and $s_i''' \equiv (i,y,q')$, the labels x, y, q, q', $\lhd_1$ and $\lhd_2$ are fresh.
In $ind\exists$, where $s_i' \equiv (i,z,q'')$, the labels y, z, q', q'' and $\lhd$ are fresh.

Fig. 1. The rules of $\mathcal{N}$(QBDTL): local life-cycle rules

$$\frac{s_j : A \quad s_i \bowtie s_j}{s_i : \copyright_j A} \; \copyright I \qquad \qquad \begin{array}{c} [s_i \bowtie s_j][s_j : A] \\ \vdots \\ \dfrac{s_i : \copyright_j A \qquad s_k : A}{s_k : A} \; \copyright E \end{array}$$

$$\begin{array}{c} [s_k \bowtie s_j] \\ \vdots \\ \dfrac{s_j \bowtie s_k \quad s_i : A}{s_i : A} \; symm_\bowtie \end{array} \qquad \begin{array}{c} [s_j \bowtie s_l] \\ \vdots \\ \dfrac{s_j \bowtie s_k \quad s_k \bowtie s_l \quad s_i : A}{s_i : A} \; trans_\bowtie \end{array}$$

$$\begin{array}{c} [s_i \triangleleft^* s_i'] \\ \vdots \\ \dfrac{s_i \bowtie s_j \quad s_j R^* s_j' \quad s_j' \bowtie s_i' \quad s_k : A}{s_k : A} \; comp_\bowtie \end{array}$$

In $\copyright I$ and $\copyright E$, $i \neq j$. In $\copyright E$, where $s_j \equiv (j, x, q)$, the labels x and q are fresh. In $comp_\bowtie$, $\triangleleft$ is fresh.

Fig. 2. The rules of $\mathcal{N}$(QBDTL): distributed life-cycle rules

$$\begin{array}{c} [q \, U \, q] \\ \vdots \\ \dfrac{s_i : A}{s_i : A} \; refl_U \end{array} \qquad \begin{array}{c} [q' \, U \, q] \\ \vdots \\ \dfrac{q \, U \, q' \quad s_i : A}{s_i : A} \; symm_U \end{array}$$

$$\begin{array}{c} [q \, U \, q''] \\ \vdots \\ \dfrac{q \, U \, q' \quad q' \, U \, q'' \quad s_i : A}{s_i : A} \; trans_U \end{array} \qquad \dfrac{(i, x, q) : p \quad \gamma(j, y, q)}{(j, y, q) : p} \; prop$$

In *prop*, $\gamma(j, y, q)$ is a (labeled or relational) formula where (j, y, q) occurs and $p \in$ Prop is an atomic proposition.

Fig. 3. The rules of $\mathcal{N}$(QBDTL): quantum transformation rules

not constrain the "world" in which we derive a contradiction to be the same as in the assumption. The rules $\supset I$ and $\supset E$ are the labeled version of the standard [23] natural deduction rules for implication introduction and elimination.

Rules for temporal operators. The rules for the introduction and the elimination of $\forall\square$, $\exists\square$ and $\exists\bigcirc$ follow the same structure as the rules for introduction and elimination of $\square$ in labeled systems for modal logics. Let us consider $\forall\square I$; the idea is that the meaning of $s_i : \forall\square A$ is given by the metalevel implication $s_i \triangleleft^* s_i' \Longrightarrow s_i' : A$ for an arbitrary path denoted by the relation $\triangleleft$ and an arbitrary s_i' $\triangleleft^*$-accessible from s_i. The arbitrariness, i.e., the *freshness*, of both the path denoted by $\triangleleft$ and s_i' is ensured by the side-conditions

$$[(i,x,q) \triangleleft (i,y,q')] \qquad\qquad\qquad [q \ U \ q']$$
$$\vdots \qquad\qquad\qquad\qquad\qquad\qquad \vdots$$

$$\frac{q \ U \ q' \quad \gamma(i,x,q) \qquad s_j : A}{s_j : A} \ U{\Rightarrow}R \qquad \frac{(i,x,q) \triangleleft (i,y,q') \quad s_j : A}{s_j : A} \ R{\Rightarrow}U$$

In $U{\Rightarrow}R$, $\gamma(i,x,q)$ is a (labeled or relational) formula where (i,x,q) occurs. Moreover, y is fresh.

Fig. 4. The rules of $\mathcal{N}$(QBDTL): interaction rules

of the rule, e.g., s_i must be different from s_i and not occur in any assumption on which $s_i' : A$ depends other than the discharged assumption $s_i \triangleleft^* s_i'$.

Introductions of $\exists\square$ and $\exists\bigcirc$ follow the same principle, but relax the freshness condition on the label denoting the relation, thus allowing us to reason on a single specific path. Note that in this case a further premise (s_iRs_i'') is required: such a premise works as a "witness", in the sense that it ensures that the relation R considered is indeed a relation passing through the state s_i.

For what concerns the elimination rules, the intuition behind $\forall\square E$ is that if $\forall\square A$ holds in a state s_i' and s_i is accessible from s_i' (along some path), then it is possible to conclude that A holds in s_i. The case of $\exists\square E$ and $\exists\bigcirc E$ is similar but complicated by the fact that the universal linear-time operator ($\square$ or $\bigcirc$) is preceded by an existential path quantifier ($\exists$), which prevents us from inferring the conclusion for a successor along an arbitrary relation. Our solution is based on the idea (originally proposed in [10]) of using Skolem functions as names for particular relations, e.g., $r(i,x,\square A)$ denotes a relation passing at x and such that if $\exists\square A$ holds in x, then A holds at each successor of x along $r(x,\square A)$.

Relational rules. Relational rules allow for modeling properties of the accessibility relations.[2] The rule $base_R$ expresses the fact that for each relation R, R^* contains R; i.e., $base_R$ says that if (i) s_j is such that there is some R-accessible s_j' and (ii) from the assumption that s_j' is also R^*-accessible from s_j we can infer some labeled formula $s_i : A$ (where s_i might be different from s_j and s_j'), then we can discharge the assumption $s_jR^*s_j'$ and conclude that indeed $s_i : A$ holds. $refl_R$ and $trans_R$ model reflexivity and transitivity of each relation, respectively, whereas $comp_R$ states that it is possible to compose two relations, i.e., if $s_jR_1^*s_j'$ and $s_j'R_2^*s_j''$, then there exists a third relation $\triangleleft^*$ such that $s_j \triangleleft^* s_j''$. We also have two rules capturing two different aspects of the seriality of the relations. $ser_\triangleleft$ captures the fact that, given a state s_j, there is at least a relation passing through s_j and a successor along that relation. ser_{sk} says that, given a state s_j and a Skolem function $r(j,x,\bigstar B)$, there exists a successor of s_j along that relation.

Induction rules. Finally, we have two rules that model the induction principle underlying the relation between R and R^*. This modeling of the induction principle is inspired to the one proposed in [10] and it is reminiscent of deduction systems for Peano

---

[2] Note that in these rules we use relational formulas as auxiliary formulas in order to derive labeled formulas. Rules treating relational formulas as full-fledged first class formulas, which can be assumed and derived in the rules, could also be defined in the style of [25].

Arthimetic. An example of use of the rule $ind\exists$ can be found in Fig. 5, as we discuss below.

Distributed Life-Cycle Rules (Fig. 2). The rules for communication (ⓒI and ⓒE) follow quite closely the semantics. Consider, e.g., ⓒI: if agent i in state s_i synchronizes with agent j in state s_j, and A holds for j in that state, then i just communicated with agent j. The rules for synchronization are also quite intuitive, except maybe $comp_\bowtie$. Intuitively, $comp_\bowtie$ models a notion of compatibility between different synchronizations that involves the same agents and reflects condition (*ii*) in the definition of branching distributed life-cycle.

Quantum Transformations Rules (Fig. 3). The rules $refl_U$, $symm_U$, $trans_U$ formalize, quite straightforwardly, the reflexivity, symmetry and transitivity of the U relation, in order to uniformly model the class of algebraic unitary operators as an equivalence relation. This captures, in an abstract way, key properties of quantum operators. Roughly speaking: reflexivity says that the class of the unitary operators includes the identity transformation; symmetry captures reversibility (it is always possible to reverse a quantum transformation, since the inverse operator is easily definable and is unitary [22]); finally, transitivity models algebraic compositionality, i.e., the composition of two or more unitary operators is always a unitary operator.

The rule *prop* says that the third element in a composed label fully captures the quantum information contained in a state: thus if two composed labels (i, x, q) and (j, y, q) share the same q, each atomic proposition holding in (i, x, q) must also hold in (j, y, q).

Interaction Rules (Fig. 4). The rules $U \Rightarrow R$ and $R \Rightarrow U$ model the interaction between U and R and express respectively the conditions (i) and (ii) in the definition of function π_i of **QBDTL** models. More specifically, $U \Rightarrow R$ says that if qUq' and the label (i, x, q) occurs in the labeled or relational formula $\gamma(j, x, q)$, then (i, x, q) has a ◄-successor (i, y, q'); this means that the local state labeled by y is an immediate successor of the state labeled by x in local life-cycle of the agent i, along a given branch. The rule $R \Rightarrow U$ captures the fact that if (i, y, q') is a ◄-successor of (i, x, q) then also the quantum labels q and q' have to be related by U.

4.4   Derivations

Given the rules in Fig. 1–4, the notions of *derivation*, *conclusion*, *open* and *discharged assumption* are the standard ones for natural deduction systems (see, e.g., [17], pp. 127-129). We write

$$\Gamma \vdash_{\mathcal{N}(\text{QBDTL})} (i, x, q) : A$$

to say that there exists a derivation of $(i, x, q) : A$ in the system $\mathcal{N}(\text{QBDTL})$ whose open assumptions are all contained in the set of (labeled and relational) formulas Γ. A derivation of $(i, x, q) : A$ in $\mathcal{N}(\text{QBDTL})$ where all the assumptions are discharged is a *proof* of $(i, x, q) : A$ in $\mathcal{N}(\text{QBDTL})$ and we then say that $(i, x, q) : A$ is a *theorem* of $\mathcal{N}(\text{QBDTL})$ and write $\vdash_{\mathcal{N}(\text{QBDTL})} (i, x, q) : A$.

$$
\cfrac{
  \cfrac{
    [q'Uq']^5 \quad [(i,y,q'):p]^4
  }{
    \cfrac{(i,y,q'):\exists\Box p}{(i,y,q'):\exists\Box p}\;refl_U^5
  }
}{
  \cfrac{[s_i:p]^1 \quad [s_i\,r(i,x,\Box p)^*\,s_i^a]^3}{s_i^a:p}
}
\qquad
\cfrac{
  \cfrac{
    \cfrac{[(i,y,q'):p]^4 \quad [(i,y,q')\lhd(i,z,q')]^6}{(i,z,q'):p}\;prop
  }{
    s_i^d:p
  }\;lin_\lhd
}{
  \cfrac{(i,y,q'):\exists\Box p}{(i,y,q'):\exists\Box p}\;U\!\Rightarrow\! R^6
}
\qquad [(i,y,q')\lhd(i,z,q')]^6 \;\;\exists\Box I^7
$$

$$
\cfrac{
  \cfrac{s_i^c:p}{\;}\;ind\exists^4
  \quad
  [(i,y,q')\,r(i,y,\bigcirc p)s_i^c]^4 \;\;\exists\Box E
}{
  \cfrac{
    \cfrac{s_i:\exists\Box p}{s_i:\exists\Box p}\;ser_{sk}^2
  }{
    s_i:p\supset\exists\Box p
  }\;\supset I^1
}
\quad [s_i\,r(i,x,\Box p)s_i^b]^2 \;\;\exists\Box I^3
$$

$$
\cfrac{
  \cfrac{
    \cfrac{[s_i:\forall\Box p]^1 \quad [s_i\lhd_1^* s_i']^7}{s_i':p}\;\forall\Box E
  }{
    s_j':\bigcirc_i p
  }\;comp_\bowtie^7
  \quad [s_j'\bowtie s_i']^6 \;\;\bigcirc I
}{
  \cfrac{
    \cfrac{s_j':\bigcirc_i p}{s_j':\bigcirc_i\top\supset\bigcirc_i p}\;\supset I^5
  }{
    s_j:\forall\Box(\bigcirc_i\top\supset\bigcirc_i p)
  }\;\forall I^4
}
\qquad
[s_i\bowtie s_j]^3 \quad [s_j\lhd^* s_j']^4 \quad [s_j'\bowtie s_i']^6
$$

$$
[s_j':\bigcirc_i\top]^5 \qquad [s_i:\bigcirc_j\top]^2
$$

$$
\cfrac{
  \cfrac{s_i:\bigcirc_j\forall\Box(\bigcirc_i\top\supset\bigcirc_i p)}{\;}\;\bigcirc E^3
  \quad [s_i\bowtie s_j]^3 \;\;\bigcirc I
}{
  \cfrac{
    \cfrac{s_i:\bigcirc_j\top\supset\bigcirc_j\forall\Box(\bigcirc_i\top\supset\bigcirc_i p)}{\;}\;\supset I^2
  }{
    s_i:\forall\Box p\supset(\bigcirc_j\top\supset\bigcirc_j\forall\Box(\bigcirc_i\top\supset\bigcirc_i p))
  }\;\supset I^1
}
$$

Fig. 5. Example derivations

Fig. 5 contains two examples of derivations (actually, proofs). The first is based on the fact that it is always possible to apply the identity transformation to a qubit. It follows that if a qubit is in a state where an atomic proposition p holds, then there exists a path along which p always holds.

The formula derived in the second example describes a property of the synchronization between qubits and can be read as a consequence of condition (*ii*) in the definition of a distributed life-cycle. If the qubit i is in a state from which a proposition p always holds in the future, then if i synchronizes with j, i.e., the two qubits are combined by means of some n-ary quantum operator, and after that, j synchronizes with i again, we end up in a state of i where p still holds. Note that in this derivation we use the *verum* $\top$ as an abbreviation for $\bot\supset\bot$.

5 Soundness

$\mathcal{N}$(QBDTL) is sound with respect to the given semantics.

Theorem 1 (Soundness). *For every set Γ of labeled and relational formulas and every labeled formula $(i,x,q):A$, it holds that $\Gamma \vdash_{\mathcal{N}(\text{QBDTL})} (i,x,q):A \;\Rightarrow\; \Gamma \models (i,x,q):A$.*

This theorem can be shown by adapting standard proof techniques for labeled natural deduction systems [25]. The proof proceeds by induction on the structure of the derivation of $(i, x, q) : A$. The base case is when $(i, x, q) : A \in \Gamma$ and is trivial. There is one step case for every rule (where, for what concerns local life-cycle rules, we refer to [10], whose treatment can be quite easily adapted to work here). We show a few representative step cases.

Consider the case when the last rule applied is *prop*:

$$\frac{\begin{array}{cc} \Pi_1 & \Pi_2 \\ (i, x, q) : p & \gamma(j, y, q) \end{array}}{(j, y, q) : p} \ prop$$

where Π_1 is a proof of $(i, x, q) : p$ from hypotheses in Γ_1 and Π_2 is a proof of $\gamma(j, y, q)$ from hypotheses in Γ_2, for some sets Γ_1, Γ_2 of formulas. By the induction hypothesis, for each model $\mu = \langle \lambda, M, \pi \rangle$ and interpretation function I, if $\models^{\mu, I} \Gamma_1$ then $\models^{\mu, I} (i, x, q) : p$ and if $\models^{\mu, I} \Gamma_2$ then $\models^{\mu, I} \gamma(j, y, q)$. We consider an I and a μ such that $\models^{\mu, I} \Gamma = \Gamma_1 \cup \Gamma_2$, and show that $\models^{\mu, I} (j, y, q) : p$. As a consequence of the induction hypothesis, we get: (i) $\models_i^{\mu, I_S^i(x)} p$; (ii) $\pi_i(I_S^i(x)) = I_\varrho(q)$; and (iii) $\pi_j(I_S^j(y)) = I_\varrho(q)$. It follows from (i) that $p \in \mathcal{V}(\pi_i(I_S^i(x)))$, i.e., by (ii), $p \in \mathcal{V}(I_\varrho(q))$ and, by (iii), $p \in \mathcal{V}(\pi_j(I_S^j(y)))$. By definition, we have $\models^{\mu, I_S^j(y)} p$, from which we infer $\models^{\mu, I} (j, y, q) : p$.

Now consider the case of an application of $©I$:

$$\frac{\begin{array}{cc} \Pi \\ (j, y, q') : A & (i, x, q) \bowtie (j, y, q') \end{array}}{(i, x, q) : ©_j A} \ ©I$$

where Π is a proof of $(j, y, q') : A$ from hypotheses in Γ_1. By the induction hypothesis, we have $\Gamma_1 \models (j, y, q') : A$. We want to show that $\Gamma = \Gamma_1 \cup \{(i, x, q) \bowtie (j, y, q')\} \models (i, x, q) : ©_j A$. Let us consider an arbitrary QBDTL model $\mu = \langle \lambda, M, \pi \rangle$ and an interpretation function I, and assume that $\models^{\mu, I} \Gamma$ holds. This implies $last(I_S^i(x)) = last(I_S^i(y))$ and $\pi_i(I_S^i(x)) = I_\varrho(q)$. By the induction hypothesis, we also obtain $\models^{\mu, I} (j, y, q') : A$, which yields $\models_j^{\mu_j, I_S^i(y)} A$. By the definition of local satisfaction relation, we infer $\models_i^{\mu_i, I_S^i(x)} ©_j A$ and then $\models^{\mu, I} (i, x, q) : ©_j A$. Since μ and I are arbitrary, we can conclude $\Gamma \models (i, x, q) : ©_j A$.

Finally, consider the case of an application of $R \Rightarrow U$:

$$\frac{\begin{array}{cc} & [q'Uq''] \\ & \Pi \\ (i, y, q') \triangleleft (i, z, q'') & (j, x, q) : A \end{array}}{(j, x, q) : A} \ R \Rightarrow U$$

where Π is a proof of $(j, x, q) : A$ from hypotheses in $\Gamma_1 \cup \{q'Uq''\}$ for some set Γ_1 of formulas. By the induction hypothesis, we have $\Gamma_1 \cup \{q'Uq''\} \models (j, x, q) : A$. We want to show that $\Gamma = \Gamma_1 \cup \{(i, y, q') \triangleleft (i, z, q'')\} \models (j, x, q) : A$. Let us consider arbitrary $\mu = \langle \lambda, M, \pi \rangle$ and I, and assume that $\models^{\mu, I} \Gamma$ holds. This implies $\models^{\mu, I} (i, y, q') \triangleleft (i, z, q'')$, from which we infer: $last(I_S^i(y)) I_B^i(\triangleleft) last(I_S^i(z)); \pi_i(I_S^i(y)) = I_\varrho(q')$; and $\pi_i(I_S^i(z)) = I_\varrho(q'')$. By condition (i) in the definition of a QBDTL model, this yields $I_Q(q')\mathcal{U}I_Q(q'')$ and thus $\models^{\mu,I} q'Uq''$. By the induction hypothesis, we obtain $\models^{\mu,I} (j,x,q):A$. Since μ and I are arbitrary, we can conclude $\Gamma \models (j,x,q):A$.

6 Concluding Remarks

We have proved that the system $\mathcal{N}$(QBDTL) is sound with respect to the given semantics. We expect $\mathcal{N}$(QBDTL) to be also complete, since it is "built" by composing subsystems that are complete with respect to the semantics of the sublogics that they capture [10, 5, 20], with the addition of rules tightly related to the interactions between those subsystems. A thorough proof, however, requires a non-trivial use of refining techniques to get appropriate models from those obtained by a standard canonical-model construction, similarly to what happens with related temporal logics. We have thus left it for future work.

We are also working at extending QBDTL in order to deal with peculiar properties of quantum states such as *entanglement*. Roughly speaking, in physics, an *entangled* state is a quantum state where two or more qubits behave as connected, independently of their real physical distance. As a consequence, operations on an entangled qubit can (possibly) have side-effects on other entangled qubits. This phenomenon (that does not have a classical counterpart) plays a major role in quantum computing (see, e.g., the teleportation protocol [22]).

In this paper, we have modeled quantum state transformations from an abstract perspective: in QBDTL, no reference to a specific quantum computation or to a notion of input/output of values is required. This allowed us to design a manageable high-level formalization oriented to modeling the behavior of quantum systems, but it is probably not enough if one wants to capture more complex properties such as entanglement. This does not mean that one has to completely convert the qualitative approach into a quantitative one (following the "philosophy" of quantum logic, cf. the discussion in the introduction). We believe that a distributed logic is a promising tool not only for the simple description of quantum states, but also to model the correct amount of quantitative information needed to capture properties like entanglement. In some sense, we aim at integrating into the QBDTL high-level perspective, able to model the "control" of quantum computation (which treats qubits and quantum gates as black-boxes), more detailed information about quantum data, so that it is possible to "look inside" the qubits and specifically model the quantitative behavior of some interesting unitary operators.

In QBDTL, we are as general as possible with respect to the application of transformations: in a local-life cycle the subtended temporal transition tree represents at each step all the possible unary gates that can be applied to the current state, while the synchronization mechanism between agents models all possible n-ary operators. It is well known that one can fix a complete computational basis (finite or infinite) of unitary operators and represent other operators in terms of the elements of such a basis. An infinite complete basis can be defined by taking all unary operators and a particular binary quantum gate called *controlled-not* (or *cnot*). Intuitively, the cnot acts as follows: it takes in input two distinct qubits and complements the *target* qubit (the second one) if the *control* qubit (the first one) is different from 0; otherwise, it does not perform any action. Noticeably, when the control qubit assumes some specific superpositional value (e.g., $\frac{1}{\sqrt{2}}|0\rangle + \frac{1}{\sqrt{2}}|1\rangle$), the output of the cnot is an entangled state. This suggests that restricting our perspective about arbitrary n-ary gates as synchronization operators to a single binary gate, the cnot, and lifting syntax and semantics to capture its behavior would provide us with all the ingredients needed to model entanglement. Following this standpoint, one can now view synchronizations exactly as control operators: a target qubit has to synchronize (by sharing an event) with the control qubit in order to perform its own, controlled evolution. Moreover, we observe that the notion of synchronization, in presence of entanglement, assumes a non-local (with respect to time) meaning: a synchronization that creates entanglement does not only represent the sharing of local events, but it also influences the subsequent events in the local life cycle of the involved agents. We thus aim to make the connection between agent synchronization and (possible) entanglement of qubits explicit.

Finally, we are considering the explicit modeling inside QBDTL of measurement steps, which can be seen as a further class of transformations. We believe that these extensions will also enable the use of our approach for the analysis of quantum security protocols, which are based on entanglement phenomena [8], along the lines of what has been done with respect to classical security protocols by using DTL [6].

References

1. Abramsky, S., Duncan, R.: A categorical quantum logic. Math. Structures Comput. Sci. 16(3), 469–489 (2006)
2. Baltag, A., Smets, S.: The logic of quantum programs. In: Proceedings of the 2nd QPL (2004)
3. Baltag, A., Smets, S.: LQP: the dynamic logic of quantum information. Math. Structures Comput. Sci. 16(3), 491–525 (2006)
4. Baltag, A., Smets, S.: Quantum logic as a dynamic logic. Synthese 179(2), 285–306 (2011)
5. Basin, D., Caleiro, C., Ramos, J., Viganò, L.: Labelled tableaux for distributed temporal logic. J. Log. and Comput. 19(6), 1245–1279 (2009)
6. Basin, D., Caleiro, C., Ramos, J., Viganò, L.: Distributed Temporal Logic for the Analysis of Security Protocol Models. Theor. Comput. Sci. 412(31), 4007–4043 (2011)
7. Ben-Ari, M., Manna, Z., Pnueli, A.: The temporal logic of branching time. In: Proceedings of POPL. ACM Press (1981)
8. Bennett, C.H., Brassard, G.: Quantum cryptography: Public key distribution and coin tossing. In: Proceedings of IEEE International Conference on Computers, Systems, and Signal Processing, pp. 175–179 (1984)
9. Birkhoff, G., von Neumann, J.: The logic of quantum mechanics. Ann. of Math. (2) 37(4), 823–843 (1936)
10. Caleiro, C., Viganò, L., Volpe, M.: A Labeled Deduction System for the Logic UB. In: Proceedings of TIME. IEEE CS Press (2013)
11. Dalla Chiara, M.L.: Quantum logic. In: Handbook of Philosophical Logic III, pp. 427–469. Reidel (1986)
12. Ehrich, H.-D., Caleiro, C.: Specifying communication in distributed information systems. Acta Informatica 36, 591–616 (2000)
13. Engesser, K., Gabbay, D.M., Lehmann, D.: A New Approach to Quantum Logic. College Publications (2007)
14. Engesser, K., Gabbay, D.M., Lehmann, D. (eds.): Handbook of Quantum Logic and Quantum Structures. Elsevier (2009)
15. Gabbay, D.M.: Labelled Deductive Systems, vol. 1. Clarendon Press (1996)

16. Gay, S.J., Nagarajan, R., Papanikolaou, N.: QMC: A Model Checker for Quantum Systems. In: Gupta, A., Malik, S. (eds.) CAV 2008. LNCS, vol. 5123, pp. 543–547. Springer, Heidelberg (2008)
17. Girard, J.-Y.: Proof theory and logical complexity, vol. 1. Bibliopolis (1987)
18. Kouvaros, P., Lomuscio, A.: Automatic verification of parameterised multi-agent systems. In: Proceedings of AAMAS (2013)
19. Masini, A., Viganò, L., Zorzi, M.: A qualitative modal representation of quantum state transformations. In: Proceedings of the 38th ISMVL. IEEE CS Press (2008)
20. Masini, A., Viganò, L., Zorzi, M.: Modal Deduction Systems for Quantum State Transformations. Multiple-Valued Logic and Soft Computing 17(5-6), 475–519 (2011)
21. Mittelstaedt, P.: The modal logic of quantum logic. J. Philos. Logic 8(4), 479–504 (1979)
22. Nielsen, M., Chuang, I.: Quantum computation and quantum information. Cambridge University Press (2000)
23. Prawitz, D.: Natural Deduction: a Proof-Theoretical Study. Almquist and Wiskell (1965)
24. Simpson, A.K.: The Proof Theory and Semantics of Intuitionistic Modal Logic. PhD thesis, School of Informatics, University of Edinburgh (1994)
25. Viganò, L.: Labelled Non-Classical Logics. Kluwer Academic Publishers (2000)
26. Winskel, G., Nielsen, M.: Event structures. In: Handbook of Logic in Computer Science IV. Oxford University Press (1995)

Definability for Downward and Vertical XPath on Data Trees

Sergio Abriola[1], María Emilia Descotte[1], and Santiago Figueira[1,2]

[1] University of Buenos Aires, Argentina
[2] CONICET, Argentina

Abstract. We study the expressive power of the downward and vertical fragments of XPath equipped with (in)equality tests over data trees. We give necessary and sufficient conditions for a class of pointed data trees to be definable by a set of formulas or by a single formula of each of the studied logics. To do so, we introduce a notion of saturation, and show that over saturated data trees bisimulation coincides with logical equivalence.

Keywords: XPath, data tree, bisimulation, definability, first-order logic, ultraproduct, saturation.

1 Introduction

The abstraction of an XML document is a data tree, i.e. a tree whose every node contains a tag or label (such as *LastName*) from a finite domain, and a data value (such as *Smith*) from an infinite domain. XPath is the most widely used query language for XML documents; it is an open standard and constitutes a World Wide Web Consortium (W3C) Recommendation [5]. XPath has syntactic operators to navigate the tree using the 'child', 'parent', 'sibling', etc. accessibility relations, and can make tests on intermediate nodes. Core-XPath [9] is the fragment of XPath 1.0 containing only the navigational behavior of XPath. It can express properties of the underlying tree structure of the XML document, such as *"the root of the tree has a child labeled a and a child labeled b"*, but it cannot express conditions on the actual data contained in the attributes, such as *"the root of the tree has two children with same tag a but different data value"*. However, Core-Data-XPath [3], here called XPath$_=$, can. Indeed, XPath$_=$ is the extension of Core-XPath with (in)equality tests between attributes of elements in an XML document.

In a recent paper [8], the expressive power of XPath$_=$ was studied, from a logical and modal model theoretical point of view. A notion of bisimulation is introduced for some fragments of XPath$_=$, and a van Benthem like characterization theorem is shown for some of them. In this work we show a definability theorem, which answers the basic question of when a class of data trees is definable by a set of formulas, or by a single formula, over two fragments of XPath$_=$: the *downward* fragment (which only has the 'child' accessibility relation) and the *vertical* fragment (which has both 'child' and 'parent' axes).

U. Kohlenbach et al. (Eds.): WoLLIC 2014, LNCS 8652, pp. 20–35, 2014.
© Springer-Verlag Berlin Heidelberg 2014

Our main result is the analog of the classic first-order definability theorem (see, e.g.[4, Cor. 6.1.16]), which can be stated as follows:

> A class of models K is definable by means of a set of first-order formulas if and only if K is closed under ultraproducts and isomorphisms, and the complement of K is closed under ultrapowers. Also K is definable by a single first-order formula if and only if both K and its complement are closed under ultraproducts and isomorphisms.

The above result was adapted to the context of many modal logics, where the notion of *isomorphism* is replaced by the weaker concept of *bisimulation* (the one which turns to be adequate for the chosen modal logic). Thus definability theorems were established for the basic modal logic [6], for temporal logics with *since* and *until* operators [11], for negation-free modal languages [12], etc. A global counterpart was studied in [7], and a general framework stating sufficient conditions for an arbitrary (modal) logic $\mathcal{L}$ to verify it was given in [1]. One of those requirements is that the models of $\mathcal{L}$ are closed under ultraproducts, which is true for the aforementioned logics, but not for XPath$_=$: models of XPath$_=$ are data trees, which may not remain connected under ultraproducts. Hence one cannot expect to use that framework in this case.

Though we take as motivation the current relevance of XML documents (which of course are finite) and the logics for reasoning over them, we do not restrict ourselves to the finite case. Indeed, an infinite set of formulas may force all its data tree models to be infinite. Hence, since we aim at working with arbitrary sets of formulas, we must consider arbitrary (i.e. finite or infinite) data trees.

Our definability theorems for XPath$_=$ themselves are shown using rather known techniques. The main contribution, however, is to devise and calibrate the adequate notions to be used in the XPath$_=$ scenario, and to study the subtle interaction between them:

- *Bisimulation*: already introduced in [8], it is the counterpart of *isomorphisms* in the classical theorem for first-order logic. In [8] it is shown that if two (possibly infinite) data trees are bisimilar then they are logically equivalent (that is, they are not distinguishable by an XPath$_=$ formula) but that the converse is not true in general.
- *Saturation*: we define and study the new notion of *XPath$_=$-saturation*. We show that for XPath$_=$-saturated data trees being bisimilar is the same as being logically equivalent. It is also shown that a 2-saturated data tree (regarded as a first-order structure) is already XPath$_=$-saturated.
- *Utlraproducts*: contrary to other adaptations of the classical first-order definability theorem to modal logics, in our case we have to adjust also the notion of *ultraproduct*, and so we work with a variant of it called *quasi-ultraproduct*. The reason is that we must not abandon the universe of data trees, as these are the only allowed models of XPath$_=$.

There are many works in the literature studying the expressive power of Core-XPath (see e.g. [10,13,14]). All these consider the navigational fragment

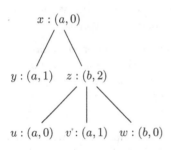

- $\varphi_1 = \langle\downarrow[a]\rangle \wedge \langle\downarrow[b]\rangle$, $[\![\varphi_1]\!]^{\mathcal{T}} = \{x, z\}$
  ("nodes with a child labeled a and a child labeled b")
- $\varphi_2 = \langle\downarrow[a] \neq \downarrow[a]\rangle$, $[\![\varphi_2]\!]^{\mathcal{T}} = \{z\}$
  ("nodes with two children with same tag a but different data value")
- $\varphi_3 = \langle\varepsilon \neq \uparrow[\langle\varepsilon \neq \uparrow\rangle]\rangle$, $[\![\varphi_3]\!] = \{u, v, w\}$
  ("nodes with a data value different from the one of his parent, who, in turn, has a data value different from his parent")
- $\varphi_4 = \langle\varepsilon \neq \downarrow\downarrow[\varphi_3]\rangle$, $[\![\varphi_4]\!] = \{x\}$
  ("nodes with a downward path of length 2, with all distinct data values")

Fig. 1. A data tree $\mathcal{T} \in \mathit{Trees}(\{a, b\} \times \mathbb{N})$ and the meaning of some XPath$^{\updownarrow}_{=}$-formulas

of XPath. A first step towards the study of the expressive power of XPath when equipped with (in)equality test over data trees, is the recent paper [8]. We aim to shed more light in this direction.

The paper is organized as follows, In §2 we introduce the formal syntax and semantics of the downward and vertical fragments of XPath$_{=}$, together with notions of bisimulations from [8]. Suitable notions of saturation for both fragments are given in §3, where it is also shown that for saturated trees bisimilarity coincides with logical equivalence. In §4 we explain the connection between XPath$_{=}$ and first-order logic, and we introduce the idea of quasi-ultraproducts for the downward and vertical fragments. In §5 we state the theorems on definability, and we close in §6 with a few words about future research and show some applications of the definability results.

2   Preliminaries

Data trees. Let $\mathit{Trees}(A)$ be the set of ordered and unranked (finite or infinite) trees over an arbitrary alphabet A. We say that $\mathcal{T}$ is a **data tree** if it is a tree from $\mathit{Trees}(\mathbb{A} \times \mathbb{D})$, where $\mathbb{A}$ is a finite set of **labels** and $\mathbb{D}$ is an infinite set of **data values** (see Figure 1 for an example). A data tree is **finitely branching** if every node has finitely many children. For any given data tree $\mathcal{T}$, we denote by T its set of nodes. We use letters x, y, z, u, v, w as variables for nodes. Given a node $x \in T$ of $\mathcal{T}$, we write $label(x) \in \mathbb{A}$ to denote the node's label, and $data(x) \in \mathbb{D}$ to denote the node's data value.

Given two nodes $x, y \in T$ we write $x \rightarrow y$ if y is a child of x, and $x \xrightarrow{n} y$ if y is a descendant of x at distance n. In particular, $\xrightarrow{1}$ is the same as $\rightarrow$, and $\xrightarrow{0}$ is the identity relation. ($\xrightarrow{n} y$) denotes the sole ancestor of y at distance n (assuming it has one).

Vertical and Downward XPath with data tests. We work with a simplification of XPath, stripped of its syntactic sugar. We consider fragments of XPath that correspond to the navigational part of XPath 1.0 with data equality and inequality. XPath$_=$ is a two-sorted language, with **path expressions** (that we write α, β, γ) and **node expressions** (that we write φ, ψ, η). The **vertical XPath**, notated XPath$_=^{\updownarrow}$ is defined by mutual recursion as follows:

$$\alpha, \beta ::= o \mid [\varphi] \mid \alpha\beta \mid \alpha \cup \beta \qquad\qquad o \in \{\varepsilon, \uparrow, \downarrow\}$$

$$\varphi, \psi ::= a \mid \neg\varphi \mid \varphi \wedge \psi \mid \varphi \vee \psi \mid \langle\alpha\rangle \mid \langle\alpha = \beta\rangle \mid \langle\alpha \neq \beta\rangle \quad a \in \mathbb{A}$$

We call **downward XPath**, notated XPath$_=^{\downarrow}$, to the syntactic fragment which only uses the navigation axis $\downarrow$, but not $\uparrow$. An **XPath$_=^{\updownarrow}$-formula** [resp. **XPath$_=^{\downarrow}$-formula**] is simply a node expression of XPath$_=^{\updownarrow}$ [resp. XPath$_=^{\downarrow}$].

Semantics of XPath$_=^{\updownarrow}$ in a given data tree $\mathcal{T}$ are defined as follows:

$$[\![\downarrow]\!]^{\mathcal{T}} = \{(x,y) \mid x{\rightarrow}y\} \qquad [\![\neg\varphi]\!]^{\mathcal{T}} = T \setminus [\![\varphi]\!]^{\mathcal{T}} \qquad [\![a]\!]^{\mathcal{T}} = \{x \in T \mid label(x) = a\}$$

$$[\![\uparrow]\!]^{\mathcal{T}} = \{(x,y) \mid y{\rightarrow}x\} \qquad [\![\varphi \wedge \psi]\!]^{\mathcal{T}} = [\![\varphi]\!]^{\mathcal{T}} \cap [\![\psi]\!]^{\mathcal{T}} \qquad [\![[\varphi]]\!]^{\mathcal{T}} = \{(x,x) \mid x \in [\![\varphi]\!]^{\mathcal{T}}\}$$

$$[\![\varepsilon]\!]^{\mathcal{T}} = \{(x,x) \mid x \in T\} \qquad [\![\alpha \cup \beta]\!]^{\mathcal{T}} = [\![\alpha]\!]^{\mathcal{T}} \cup [\![\beta]\!]^{\mathcal{T}}$$

$$[\![\alpha\beta]\!]^{\mathcal{T}} = \{(x,z) \mid (\exists y \in T)\, (x,y) \in [\![\alpha]\!]^{\mathcal{T}}, (y,z) \in [\![\beta]\!]^{\mathcal{T}}\}$$

$$[\![\langle\alpha\rangle]\!]^{\mathcal{T}} = \{x \in T \mid (\exists y \in T)\, (x,y) \in [\![\alpha]\!]^{\mathcal{T}}\}$$

$$[\![\langle\alpha = \beta\rangle]\!]^{\mathcal{T}} = \{x \in T \mid (\exists y,z \in T)(x,y) \in [\![\alpha]\!]^{\mathcal{T}}, (x,z) \in [\![\beta]\!]^{\mathcal{T}}, data(y) = data(z)\}$$

$$[\![\langle\alpha \neq \beta\rangle]\!]^{\mathcal{T}} = \{x \in T \mid (\exists y,z \in T)(x,y) \in [\![\alpha]\!]^{\mathcal{T}}, (x,z) \in [\![\beta]\!]^{\mathcal{T}}, data(y) \neq data(z)\}$$

See Figure 1 for the semantics of some formulas. For a data tree $\mathcal{T}$ and $u \in T$, we write $\mathcal{T}, u \models \varphi$ to denote $u \in [\![\varphi]\!]^{\mathcal{T}}$, and we say that $\mathcal{T}, u$ satisfies φ or that φ is true at $\mathcal{T}, u$. Let $\mathrm{Th}_{\updownarrow}(\mathcal{T}, u)$ [resp. $\mathrm{Th}_{\downarrow}(\mathcal{T}, u)$] be the set of all XPath$_=^{\updownarrow}$-formulas [resp. XPath$_=^{\downarrow}$-formulas] true at $\mathcal{T}, u$. In terms of expressive power, it is easy to see that $\cup$ is unessential (see [8, §2.2]). We will henceforth assume that formulas do not contain union of path expressions.

Let $\mathcal{T}$ and $\mathcal{T}'$ be data trees, and let $u \in T$, $u' \in T'$. We say that $\mathcal{T}, u$ and $\mathcal{T}', u'$ are **equivalent for XPath$_=^{\updownarrow}$** [resp. **equivalent for XPath$_=^{\downarrow}$**] (notation: $\mathcal{T}, u \equiv^{\updownarrow} \mathcal{T}', u'$ [resp. $\mathcal{T}, u \equiv^{\downarrow} \mathcal{T}', u'$]) iff for all formulas $\varphi \in$ XPath$_=^{\updownarrow}$ [resp. $\varphi \in$ XPath$_=^{\downarrow}$], we have $\mathcal{T}, u \models \varphi$ iff $\mathcal{T}', u' \models \varphi$.

Bisimulations. In [8] the notions of downward and vertical bisimulations are introduced. We reproduce them here, as they are key concepts for our results.

We say that $u \in T$ and $u' \in T'$ are **bisimilar for XPath$_=^{\downarrow}$** (or $\downarrow$-**bisimilar**; notation: $\mathcal{T}, u \leftrightarrow^{\downarrow} \mathcal{T}', u'$) iff there is a relation $Z \subseteq T \times T'$ such that uZu' and for all $x \in T$ and $x' \in T'$ we have

- **Harmony:** If xZx' then $label(x) = label(x')$.
- **Zig:** If xZx', $x \xrightarrow{n} v$ and $x \xrightarrow{m} w$ then there are $v', w' \in T'$ such that $x' \xrightarrow{n} v'$, $x' \xrightarrow{m} w'$ and
  1. $data(v) = data(w) \Leftrightarrow data(v') = data(w')$,
  2. $(\xrightarrow{i} v)\, Z\, (\xrightarrow{i} v')$ for all $0 \le i < n$, and

3. $(\overset{i}{\to}w)\,Z\,(\overset{i}{\to}w')$ for all $0 \le i < m$.
  - **Zag:** If xZx', $x'\overset{n}{\to}v'$ and $x'\overset{m}{\to}w'$ then there are $v, w \in T$ such that $x\overset{n}{\to}v$, $x\overset{m}{\to}w$ and items 1, 2 and 3 above are verified.

We say that $u \in T$ and $u' \in T'$ are **bisimilar for XPath$_=^{\updownarrow}$** (or $\updownarrow$-bisimilar; notation: $\mathcal{T}, u \leftrightarrow^{\updownarrow} \mathcal{T}', u'$) iff there is a relation $Z \subseteq T \times T'$ such that uZu' and for all $x \in T$ and $x' \in T'$ we have

  - **Harmony:** If xZx' then $label(x) = label(x')$,
  - **Zig:** If xZx', $y\overset{n}{\to}x$ and $y\overset{m}{\to}z$ then there are $y', z' \in T'$ such that $y'\overset{n}{\to}x'$, $y'\overset{m}{\to}z'$, $data(z) = data(x) \Leftrightarrow data(z') = data(x')$, and zZz'.
  - **Zag:** If xZx', $y'\overset{n}{\to}x'$ and $y'\overset{m}{\to}z'$ then there are $y, z \in T$ such that $y\overset{n}{\to}x$, $y\overset{m}{\to}z$, $data(z) = data(x) \Leftrightarrow data(z') = data(x')$, and zZz'.

The main results establishing the connection between bisimulation and equivalence is the following:

Theorem 1 ([8]). *If $\mathcal{T}, u \leftrightarrow^{\downarrow} \mathcal{T}', u'$ then $\mathcal{T}, u \equiv^{\downarrow} \mathcal{T}', u'$, and if $\mathcal{T}, u \leftrightarrow^{\updownarrow} \mathcal{T}', u'$, then $\mathcal{T}, u \equiv^{\updownarrow} \mathcal{T}', u'$.*

3 Saturation

In [8] is is shown that the reverse implications of Theorem 1 hold over finitely branching trees. However, they do not hold in general. In this section we introduce notions of saturation for the downward and vertical fragments of XPath, and show that the reverse implications of Theorem 1 are true over saturated data trees.

Saturation for the downward fragment. Let $\langle \Sigma_1, \ldots, \Sigma_n \rangle$ and $\langle \Gamma_1, \ldots, \Gamma_m \rangle$ be tuples of sets of XPath$_=^{\downarrow}$-formulas. Given a data tree $\mathcal{T}$ and $u \in T$, we say that $\langle \Sigma_1, \ldots, \Sigma_n \rangle$ and $\langle \Gamma_1, \ldots, \Gamma_m \rangle$ are $=_{n,m}^{\downarrow}$-**satisfiable** [resp. $\neq_{n,m}^{\downarrow}$-**satisfiable**] at $\mathcal{T}, u$ if there exist $v_0 \to v_1 \to \cdots \to v_n \in T$ and $w_0 \to w_1 \to \cdots \to w_m \in T$ such that $u = v_0 = w_0$ and

1. for all $i \in \{1, \ldots, n\}$, $\mathcal{T}, v_i \models \Sigma_i$;
2. for all $j \in \{1, \ldots, m\}$, $\mathcal{T}, w_j \models \Gamma_j$; and
3. $data(v_n) = data(w_m)$ [resp. $data(v_n) \neq data(w_m)$].

We say that $\langle \Sigma_1, \ldots, \Sigma_n \rangle$ and $\langle \Gamma_1, \ldots, \Gamma_m \rangle$ are $=_{n,m}^{\downarrow}$-**finitely satisfiable** [resp. $\neq_{n,m}^{\downarrow}$-**finitely satisfiable**] at $\mathcal{T}, u$ if for every finite $\Sigma_i' \subseteq \Sigma_i$ and finite $\Gamma_j' \subseteq \Gamma_j$, we have that $\langle \Sigma_1', \ldots, \Sigma_n' \rangle$ and $\langle \Gamma_1', \ldots, \Gamma_m' \rangle$ are $=_{n,m}^{\downarrow}$-satisfiable [resp. $\neq_{n,m}^{\downarrow}$-satisfiable] at $\mathcal{T}, u$.

Definition 2. *We say that a data tree $\mathcal{T}$ is ↓-saturated if for every $n, m \in \mathbb{N}$, every pair of tuples $\langle \Sigma_1, \ldots, \Sigma_n \rangle$ and $\langle \Gamma_1, \ldots, \Gamma_m \rangle$ of sets of XPath$_=^{\downarrow}$-formulas, every $u \in T$, and $\star \in \{=, \neq\}$, the following is true:*

  if $\langle \Sigma_1, \ldots, \Sigma_n \rangle$ and $\langle \Gamma_1, \ldots, \Gamma_m \rangle$ are $\star_{n,m}^{\downarrow}$-finitely satisfiable at $\mathcal{T}, u$ then $\langle \Sigma_1, \ldots, \Sigma_n \rangle$ and $\langle \Gamma_1, \ldots, \Gamma_m \rangle$ are $\star_{n,m}^{\downarrow}$-satisfiable at $\mathcal{T}, u$.

Proposition 3. *Any finitely branching data tree is $\downarrow$-saturated.*

Proof. Suppose by contradiction that there is $u \in T$ and tuples $\langle \Sigma_1, \ldots, \Sigma_n \rangle$ and $\langle \Gamma_1, \ldots, \Gamma_m \rangle$ of sets of XPath$^{\downarrow}_{=}$-formulas which are finitely $=^{\downarrow}_{n,m}$-satisfiable at $\mathcal{T}, u$ but not $=^{\downarrow}_{n,m}$-satisfiable at $\mathcal{T}, u$ (the case for $\mathcal{T}$ being $\neq^{\downarrow}_{n,m}$-satisfiable is analogous). Let

$$P = \{(v, w) \in T^2 \mid u \xrightarrow{n} v \wedge u \xrightarrow{m} w \wedge data(v) = data(w)\}.$$

Observe that P is finite because $\mathcal{T}$ is finitely branching. It is clear that if $(v, w) \in P$, so that $u = v_0 \to v_1 \to \cdots \to v_n = v \in T$, and $u = w_0 \to w_1 \to \cdots \to w_m = w \in T$ then either

1. there is $i \in \{1, \ldots, n\}$ such that $\mathcal{T}, v_i \not\models \Sigma_i$, or
2. there is $j \in \{1, \ldots, m\}$ such that $\mathcal{T}, w_j \not\models \Gamma_j$.

We will define sets $(\Sigma_{i,v,w})_{1 \leq i \leq n}$ and $(\Gamma_{j,v,w})_{1 \leq j \leq m}$, each one of them with at most one element, as follows: If case 1 holds, assume i_0 is the least such number and define $\Sigma_{i_0,v,w}$ as $\{\rho\}$ for some formula $\rho \in \Sigma_{i_0}$ such that $\mathcal{T}, v_{i_0} \not\models \rho$, define $\Sigma_{i,v,w} = \emptyset$ for any $i \in \{1, \ldots, n\} \setminus \{i_0\}$, and define $\Gamma_{j,v,w} = \emptyset$ for any $j \in \{1, \ldots, m\}$. If case 1 does not hold then case 2 holds, so assume j_0 is the least such number and define $\Gamma_{j_0,v,w}$ as $\{\rho\}$ for some formula $\rho \in \Gamma_{j_0}$ such that $\mathcal{T}, w_{j_0} \not\models \rho$, define $\Gamma_{j,v,w} = \emptyset$ for any $j \in \{1, \ldots, m\} \setminus \{j_0\}$, and define $\Sigma_{i,v,w} = \emptyset$ for any $i \in \{1, \ldots, n\}$. Finally, define the finite sets $\Sigma'_i = \bigcup_{(v,w) \in P} \Sigma_{i,v,w}$ and $\Gamma'_j = \bigcup_{(v,w) \in P} \Gamma_{j,v,w}$. By construction, we have $\Sigma'_i \subseteq \Sigma_i$, $\Gamma'_j \subseteq \Gamma_j$ and $\langle \Sigma'_1, \ldots, \Sigma'_n \rangle$ and $\langle \Gamma'_1, \ldots, \Gamma'_m \rangle$ are not $=^{\downarrow}_{n,m}$-satisfiable at $\mathcal{T}, u$ which is a contradiction.   □

Proposition 4. *Let $\mathcal{T}$ and $\mathcal{T}'$ be $\downarrow$-saturated data trees, and let $u \in T$ and $u' \in T'$. If $\mathcal{T}, u \equiv^{\downarrow} \mathcal{T}', u'$, then $\mathcal{T}, u \underleftrightarrow{\downarrow} \mathcal{T}', u'$.*

Proof. We show that Z, defined by xZx' iff $\mathcal{T}, x \equiv^{\downarrow} \mathcal{T}', x'$ is a $\downarrow$-bisimulation between $\mathcal{T}, u$ and $\mathcal{T}', u'$. Clearly uZu', and Harmony holds. We only need to show that Zig and Zag are satisfied. We see only Zig, as Zag is analogous.

Suppose xZx', $x = v_0 \to v_1 \to \cdots \to v_n$ and $x = w_0 \to w_1 \to \cdots \to w_m$ are paths on $\mathcal{T}$, and $data(v_n) = data(w_m)$ (the case $data(v_n) \neq data(w_m)$ is shown analogously). For $i \in \{1, \ldots, n\}$, let $\Sigma_i = \text{Th}_{\downarrow}(\mathcal{T}, v_i)$, and for $j \in \{1, \ldots, m\}$, let $\Gamma_j = \text{Th}_{\downarrow}(\mathcal{T}, w_j)$. Furthermore, let Σ'_i be a finite subset of Σ_i, and let Γ'_j be a finite subset of Γ_j. Define

$$\varphi = \langle \downarrow [\wedge \Sigma'_1] \downarrow \ldots \downarrow [\wedge \Sigma'_n] = \downarrow [\wedge \Gamma'_1] \downarrow \ldots \downarrow [\wedge \Gamma'_m] \rangle.$$

It is clear that $\mathcal{T}, x \models \varphi$, and since by definition of Z we have $\mathcal{T}, x \equiv^{\downarrow} \mathcal{T}', x'$, we conclude that $\mathcal{T}', x' \models \varphi$. Hence $\langle \Sigma'_1, \ldots, \Sigma'_n \rangle$ and $\langle \Gamma'_1, \ldots, \Gamma'_m \rangle$ are $=^{\downarrow}_{n,m}$-satisfiable at x'. This holds for *any* finite sets $\Sigma'_i \subseteq \Sigma_i$ and $\Gamma'_j \subseteq \Gamma_j$, and so $\langle \Sigma_1, \ldots, \Sigma_n \rangle$ and $\langle \Gamma_1, \ldots, \Gamma_m \rangle$ are $=^{\downarrow}_{n,m}$-finitely satisfiable at x', Since $\mathcal{T}'$ is $\downarrow$-saturated, $\langle \Sigma_1, \ldots, \Sigma_n \rangle$ and $\langle \Gamma_1, \ldots, \Gamma_m \rangle$ are $=^{\downarrow}_{n,m}$-satisfiable at $\mathcal{T}', x'$, so there are paths $x' = v'_0 \to v'_1 \to \cdots \to v'_n$ and $x' = w'_0 \to w'_1 \to \cdots \to w'_m$ on $\mathcal{T}'$ such that

   i. $data(v'_n) = data(w'_m)$;

  ii. for all $1 \leq i \leq n$, $\mathcal{T}', v'_i \models \text{Th}_\downarrow(\mathcal{T}, v_i)$, i.e. $\mathcal{T}, v_i \equiv^\downarrow \mathcal{T}', v'_i$; and

 iii. for all $1 \leq j \leq m$, $\mathcal{T}', w'_j \models \text{Th}_\downarrow(\mathcal{T}, w_j)$, i.e $\mathcal{T}, w_j \equiv^\downarrow \mathcal{T}', w'_j$.

By the definition of Z, conditions i, ii and iii above imply items 1, 2 and 3 of the Zig clause for $\downarrow$-bisimulation. □

Saturation for the vertical fragment. Given a data tree $\mathcal{T}$ and $u \in T$, we say that the set of XPath$^\updownarrow_=$-formulas Γ is $=^\updownarrow_{n,m}$-**satisfiable** [resp. $\neq^\updownarrow_{n,m}$-**satisfiable**] at $\mathcal{T}, u$ if there exist $v, w \in T$ such that $v \overset{n}{\to} u$, $v \overset{m}{\to} w$, $w \models \Gamma$ and $data(u) = data(w)$ [resp. $data(u) \neq data(w)$]. We say that Γ is $=^\updownarrow_{n,m}$-**finitely satisfiable** [resp. $\neq^\updownarrow_{n,m}$-**finitely satisfiable**] at $\mathcal{T}, u$ if for every finite $\Gamma' \subseteq \Gamma$, we have that Γ' is $=^\updownarrow_{n,m}$-satisfiable [resp. $=^\updownarrow_{n,m}$-satisfiable] at $\mathcal{T}, u$.

Definition 5. *We say that a data tree $\mathcal{T}$ is $\updownarrow$-saturated if for every set of XPath$^\updownarrow_=$-formulas Γ, every $u \in T$, every $n, m \in \mathbb{N}$, and $\star \in \{=, \neq\}$, the following is true:*

if Γ is $\star^\updownarrow_{n,m}$-finitely satisfiable at $\mathcal{T}, u$ then Γ is $\star^\updownarrow_{n,m}$-satisfiable at $\mathcal{T}, u$.

Proposition 6. *Let $\mathcal{T}$ and $\mathcal{T}'$ be $\updownarrow$-saturated data trees, and let $u \in T$ and $u' \in T'$. If $\mathcal{T}, u \equiv^\updownarrow \mathcal{T}', u'$, then $\mathcal{T}, u \underline{\leftrightarrow}^\updownarrow \mathcal{T}', u'$.*

Proof. We show that $Z \subseteq T \times T'$, defined by xZx' iff $\mathcal{T}, x \equiv^\updownarrow \mathcal{T}', x'$ is a $\updownarrow$-bisimulation between $\mathcal{T}, u$ and $\mathcal{T}', u'$. Clearly uZu', and Harmony also holds, so we only need to show that Zig and Zag are satisfied. We see only Zig, as Zag is analogous.

Suppose xZx', $y \overset{n}{\to} x$ and $y \overset{m}{\to} z$ are in $\mathcal{T}$, and $data(x) = data(z)$ (the case $data(x) \neq data(z)$ can be shown analogously). Let $\Gamma = \text{Th}_\updownarrow(\mathcal{T}, z)$, and let Γ' be a finite subset of Γ. Define

$$\varphi = \langle \varepsilon = \uparrow^n \downarrow^m [\wedge \Gamma'] \rangle.$$

It is clear that $\mathcal{T}, x \models \varphi$, and since by definition of Z we have $\mathcal{T}, x \equiv^\updownarrow \mathcal{T}', x'$, we conclude that $\mathcal{T}', x' \models \varphi$. Hence Γ' is $=^\updownarrow_{n,m}$-satisfiable at x'. This holds for *any* finite set $\Gamma' \subseteq \Gamma$, and so Γ is $=^\updownarrow_{n,m}$-finitely satisfiable at x'. Since $\mathcal{T}'$ is $\updownarrow$-saturated, Γ is $=^\updownarrow_{n,m}$-satisfiable at x', and thus there are $y' \overset{n}{\to} x'$ and $y' \overset{m}{\to} z'$ on $\mathcal{T}'$ such that $data(x') = data(z')$ and $\mathcal{T}', z' \models \text{Th}_\updownarrow(\mathcal{T}, z)$, i.e $\mathcal{T}, z \equiv^\updownarrow \mathcal{T}', z'$. By the definition of Z, we have zZz' and hence the Zig clause for $\updownarrow$-bisimulation is verified. □

4   Weak Data Trees and Quasi-ultraproducts

We fix the signature σ with binary relations $\rightsquigarrow$ and $\sim$, and a unary predicate P_a for each $a \in \mathbb{A}$. Any data tree $\mathcal{T}$ can be seen as a first-order σ-structure, where

$$\rightsquigarrow^\mathcal{T} = \{(x, y) \in T^2 \mid x \to y \text{ in } \mathcal{T}\};$$

$$\sim^{\mathcal{T}} = \{(x, y) \in T^2 \mid data(x) = data(y)\};$$
$$P_a^{\mathcal{T}} = \{x \in T \mid label(x) = a\}.$$

If $\varphi(x)$ is a first-order formula with a free variable x, we use $\mathcal{T} \models \varphi[a]$, as usual, to denote that that φ is true in $\mathcal{T}$ under the valuation which maps x to $a \in T$. In [8] it is shown a truth preserving translation Tr_x mapping XPath$_=^{\updownarrow}$-formulas into first-order σ-formulas with one free variable x. By *truth preserving* we mean that for $\varphi \in \mathrm{XPath}_=^{\updownarrow}$ we have $\mathcal{T}, u \models \varphi$ iff $\mathcal{T} \models \mathrm{Tr}_x(\varphi)[u]$.

For reasons that will become clearer later on, we will need to work with σ-structures which are slightly more general than data trees.

Definition 7. *A σ-structure $\mathcal{T}$ is a **weak data tree** if $\sim$ is an equivalence relation; there is exactly one node r with no u such that $u \rightsquigarrow r$ (r is called* root *of $\mathcal{T}$); for all nodes $x \neq r$ there is exactly one y such that $y \rightsquigarrow x$; and for each $n \geq 0$ the relation $\rightsquigarrow$ has no cycles of length n.*

Observe that a weak data tree need not be connected, and that the class of weak data trees is elementary, i.e. definable by a set of first-order σ-sentences (with equality). For a weak data tree $\mathcal{T}$ and $u \in T$, let $\mathcal{T}|u$ denote the substructure of $\mathcal{T}$ induced by $\{v \in T \mid u \rightsquigarrow^* v\}$. Observe that in this case $\mathcal{T}|u$ is a data tree.

The following proposition shows the 'local' aspect of XPath$_=^{\downarrow}$ and XPath$_=^{\updownarrow}$. It is stated in terms of first-order because models are weak data trees.

Proposition 8. *Let $\mathcal{T}$ be a weak data tree and let $r \rightsquigarrow^* u$ in $\mathcal{T}$.*

1. *If $\varphi \in \mathrm{XPath}_=^{\downarrow}$-formula then $\mathcal{T} \models \mathrm{Tr}_x(\varphi)[u]$ iff $\mathcal{T}|r \models \mathrm{Tr}_x(\varphi)[u]$.*
2. *If r is the root of $\mathcal{T}$ and $\varphi \in \mathrm{XPath}_=^{\updownarrow}$ then $\mathcal{T} \models \mathrm{Tr}_x(\varphi)[u]$ iff $\mathcal{T}|r \models \mathrm{Tr}_x(\varphi)[u]$.*

Observe that the condition of r being the root in the second item is needed. Suppose for example we are on the data tree with only 2 nodes, the root r and its child u, with same data value. Consider now $\varphi = \langle \varepsilon = \uparrow \rangle$. Clearly $\mathcal{T} \models \mathrm{Tr}_x(\varphi)[u]$, but $\mathcal{T}|u \not\models \mathrm{Tr}_x(\varphi)[u]$.

If $\mathcal{M}$ is a first-order σ-structure and $A \subseteq M$, we denote by σ_A the language obtained by adding to σ constant symbols for each $a \in A$. $\mathcal{M}$ can be seen as a σ_A structure by interpreting the new symbols in the obvious way. Let $\mathrm{Th}_A(\mathcal{M})$ be the set of all σ_A-sentences true in $\mathcal{M}$. Let κ be a cardinal. We recall the definition of κ-saturated first-order structures. We say that the σ-structure $\mathcal{M}$ is κ-saturated if for all $A \subseteq M$ and all n, if $|A| < \kappa$ and $\Gamma(x_1, \ldots, x_n)$ is a set of σ_A-formulas with free variables among $x_1, \ldots, x_n$ such that $\Gamma(x_1, \ldots, x_n) \cup \mathrm{Th}_A(\mathcal{M})$ is satisfiable, then $\Gamma(x_1, \ldots, x_n)$ is realized in $\mathcal{M}$.

We now show that 2-saturated data trees are already both downward and vertical saturated. For technical reasons we state these results in the more general setting of weak data trees.

Proposition 9. *Let $\mathcal{T}$ be a 2-saturated weak data tree and $r \in T$.*

1. *$\mathcal{T}|r$ is a $\downarrow$-saturated data tree.*
2. *If r is the root of T then $\mathcal{T}|r$ is a $\updownarrow$-saturated data tree.*

Proof. Let $\mathcal{T}' = \mathcal{T}|r$ and let $u \in T'$. For item 1, let $\langle \Sigma_1, \ldots, \Sigma_n \rangle$ and $\langle \Gamma_1, \ldots, \Gamma_m \rangle$ be tuples of sets of XPath$_{=}^{\downarrow}$-formulas. Suppose $\langle \Sigma_1, \ldots, \Sigma_n \rangle$ and $\langle \Gamma_1, \ldots, \Gamma_m \rangle$ are $=_{n,m}^{\downarrow}$-finitely satisfiable at $\mathcal{T}', u$ (the case for $\neq_{n,m}^{\downarrow}$-finitely satisfiable is analogous). We show that $\langle \Sigma_1, \ldots, \Sigma_n \rangle$ and $\langle \Gamma_1, \ldots, \Gamma_m \rangle$ are $=_{n,m}^{\downarrow}$-satisfiable at $\mathcal{T}', u$. Consider the following first-order $\sigma_{\{u\}}$-formula with free variables $\bar{x} = x_1, \ldots, x_n$ and $\bar{y} = y_1, \ldots, y_m$:

$$\varphi(\bar{x}, \bar{y}) = u \rightsquigarrow x_1 \wedge \bigwedge_{i=1}^{n-1} x_i \rightsquigarrow x_{i+1} \wedge u \rightsquigarrow y_1 \wedge \bigwedge_{j=1}^{m-1} y_j \rightsquigarrow y_{j+1} \wedge x_n \sim y_m.$$

Define the following set of first-order $\sigma_{\{u\}}$-formulas:

$$\Delta(\bar{x}, \bar{y}) = \{\varphi(\bar{x}, \bar{y})\} \cup \bigcup_{i=1}^{n} \mathrm{Tr}_{x_i}(\Sigma_i) \cup \bigcup_{j=1}^{m} \mathrm{Tr}_{y_j}(\Gamma_j).$$

Let $\Delta'(\bar{x}, \bar{y})$ be a finite subset of $\Delta(\bar{x}, \bar{y})$. Since $\langle \Sigma_1, \ldots, \Sigma_n \rangle$ and $\langle \Gamma_1, \ldots, \Gamma_m \rangle$ are $=_{n,m}^{\downarrow}$-finitely satisfiable at $\mathcal{T}', u$, then $\Delta'(\bar{x}, \bar{y})$ is satisfiable and, by item 1 of Proposition 8, consistent with $\mathrm{Th}_{\{u\}}(\mathcal{T})$. By compactness, $\Delta(\bar{x}, \bar{y})$ is satisfiable and consistent with $\mathrm{Th}_{\{u\}}(\mathcal{T})$. By 2-saturation, we conclude that $\Delta(\bar{x}, \bar{y})$ is realizable in $\mathcal{T}$, say at $\bar{v} = v_1, \ldots, v_n$ and $\bar{w} = w_1, \ldots, w_m$. Thus we have:

i. $u \rightsquigarrow v_1 \rightsquigarrow \cdots \rightsquigarrow v_n$ and $u \rightsquigarrow w_1 \rightsquigarrow \cdots \rightsquigarrow w_m$ in $\mathcal{T}$, and hence in $\mathcal{T}'$;

ii. for all $i \in \{1, \ldots, n\}$, $\mathcal{T} \models \mathrm{Tr}_{x_i}(\Sigma_i)[v_i]$, and for all $j \in \{1, \ldots, m\}$, $\mathcal{T} \models \mathrm{Tr}_{y_j}(\Gamma_j)[w_j]$; by item 1 of Proposition 8 this implies that $\mathcal{T}' \models \mathrm{Tr}_{x_i}(\Sigma_i)[v_i]$ and $\mathcal{T}' \models \mathrm{Tr}_{y_j}(\Gamma_j)[w_j]$;

iii. $v_n \sim w_m$ in $\mathcal{T}$, and hence in $\mathcal{T}'$.

Since Tr is truth preserving, we have that for all $i \in \{1, \ldots, n\}$, $\mathcal{T}', v_i \models \Sigma_i$, and for all $j \in \{1, \ldots, m\}$, $\mathcal{T}', w_i \models \Gamma_i$. Together with i and iii we conclude that $\langle \Sigma_1, \ldots, \Sigma_n \rangle$ and $\langle \Gamma_1, \ldots, \Gamma_m \rangle$ are $=_{n,m}^{\downarrow}$-satisfiable at $\mathcal{T}', u$.

For item 2, let Γ be a set of XPath$_{=}^{\updownarrow}$-formulas. Suppose Γ is $=_{n,m}^{\updownarrow}$-finitely satisfiable at $\mathcal{T}', u$ (the case for $\neq_{n,m}^{\updownarrow}$-finitely satisfiable is analogous). We show that Γ are $=_{n,m}^{\updownarrow}$-satisfiable at $\mathcal{T}', u$.

Consider the following first-order $\sigma_{\{u\}}$-formula with free variable y:

$$\varphi(y) = (\exists x_0 \ldots \exists x_n)(\exists y_0 \ldots \exists y_m)[x_n = u \wedge y = y_m \wedge x_0 = y_0 \wedge$$
$$\bigwedge_{i=0}^{n-1} x_i \rightsquigarrow x_{i+1} \wedge \bigwedge_{j=0}^{m-1} y_j \rightsquigarrow y_{j+1} \wedge x_n \sim y_m]$$

Define the following set of first-order $\sigma_{\{u\}}$-formulas: $\Delta(y) = \{\varphi(y)\} \cup \mathrm{Tr}_y(\Gamma)$. Let $\Delta'(y)$ be a finite subset of $\Delta(y)$. Since Γ is $=_{n,m}^{\updownarrow}$-finitely satisfiable at $\mathcal{T}', u$, then $\Delta'(y)$ is satisfiable and, by item 2 of Proposition 8, consistent with $\mathrm{Th}_{\{u\}}(\mathcal{T})$. By compactness, $\Delta(y)$ is satisfiable and consistent with $\mathrm{Th}_{\{u\}}(\mathcal{T})$. By 2-saturation, we conclude that $\Delta(y)$ is realizable in $\mathcal{T}$, say at w. Thus we have:

iv. There is $v \in T$ such that $v \xrightarrow{n} u$ and $v \xrightarrow{m} w$ in $\mathcal{T}$ and hence in $\mathcal{T}'$.

v. $\mathcal{T} \models \mathrm{Tr}_y(\Gamma)[w]$; by item 2 of Proposition 8 this implies that $\mathcal{T}' \models \mathrm{Tr}_y(\Gamma)[w]$;

vi. $u \sim w$ in $\mathcal{T}$, and hence in $\mathcal{T}'$.

Since Tr is truth preserving, we have that $\mathcal{T}', w \models \Gamma$. Together with iv and vi we conclude that Γ is $=_{n,m}^{\updownarrow}$-satisfiable at $\mathcal{T}', u$. $\square$

In what follows, we introduce the notion of quasi-ultraproduct, a variant of the usual notion of first-order model theory, which will be needed for the definability theorems.

Let $I \neq \emptyset$, let U be an ultrafilter over I and let $(\mathcal{T}_i)_{i \in I}$ be a family of data trees. As usual, we denote with $\prod_U \mathcal{T}_i$ the ultraproduct of $(\mathcal{T}_i)_{i \in I}$ modulo U. Observe that by the fundamental theorem of ultraproducts (see e.g. [4, Thm. 4.1.9]), $\prod_U \mathcal{T}_i$ is a weak data tree σ-structure —though it may not be a data tree because it may be disconnected, as it is shown next:

Example 10. For $i \in \mathbb{N}$, let $\mathcal{T}_i$ as any data tree of height at least n, and let u_i as any node of T_i at distance n from the root of $\mathcal{T}_i$. Let $\varphi_n(x)$ be the first-order property "x is at distance at least n from the root". It is clear that $\mathcal{T}_m \models \varphi_n[u_m]$ for every $m \geq n$. Let u^* be the ultralimit of $(u_i)_{i \in I}$ modulo U. Since $\{m \mid m \geq n\} \in U$ for any non-principal U, we conclude that $\prod_U \mathcal{T}_i \models \varphi_n[u^*]$ for every n, and so u^* is disconnected from the root of $\prod_U \mathcal{T}_i$.

Let $(\mathcal{T}_i, u_i)_{i \in I}$ be a family of pointed data trees. The ultraproduct of such *pointed* data trees is defined, as usual, by $(\prod_U \mathcal{T}_i, u^*)$, where u^* is the ultralimit of $(u_i)_{i \in I}$ modulo U.

Definition 11. *Suppose $(\mathcal{T}_i, u_i)_{i \in I}$ is a family of pointed data trees, r_i is the root of $\mathcal{T}_i$, U is an ultrafilter over I, $\mathcal{T}^* = \prod_U \mathcal{T}_i$, and u^* and r^* are the ultralimits of $(u_i)_{i \in I}$ and $(r_i)_{i \in I}$ modulo U respectively.*

1. *The $\downarrow$-**quasi ultraproduct** of $(\mathcal{T}_i, u_i)_{i \in I}$ modulo U is the pointed data tree $(\mathcal{T}^* | u^*, u^*)$.*
2. *The $\updownarrow$-**quasi ultraproduct** of $(\mathcal{T}_i, u_i)_{i \in I}$ modulo U is the pair $(\mathcal{T}^* | r^*, u^*)$.*

Observe that both $\mathcal{T}^* | u^*$ and $\mathcal{T}^* | r^*$ are data trees. However, while u^* is in the domain of the former, it may not be in the domain of the latter (cf. Example 10). Hence, in general, pointed data trees are not closed under $\updownarrow$-quasi ultraproduct. Let $k \geq 0$, let $\mathcal{T}$ be a data tree and let $u \in \mathcal{T}$. We say that $(\mathcal{T}, u)$ is a k-**bounded pointed data tree** if u is at distance at most k from the root of $\mathcal{T}$. In particular, if r is the root of $\mathcal{T}$ (as it is often the case) then $(\mathcal{T}, r)$ is a 0-bounded pointed data tree. The following proposition states that k-bounded data trees are closed under $\updownarrow$-quasi ultraproducts.

Proposition 12. *Let $(\mathcal{T}_i, u_i)_{i \in I}$ be a family of k-bounded pointed data trees. Then the $\updownarrow$-quasi ultraproduct of $(\mathcal{T}_i, u_i)_{i \in I}$ is a k-bounded pointed data tree.*

Proof. Let $(\mathcal{T}^{\updownarrow}, u^*)$ be the $\updownarrow$-quasi ultraproduct of $(\mathcal{T}_i, u_i)_{i \in I}$ modulo U. By definition it is clear that $\mathcal{T}^{\updownarrow}$ is a data tree. To see that $u^* \in T^{\updownarrow}$, let

$$\varphi(x) = (\exists r) \, [\neg(\exists y) y \rightsquigarrow r \land [r = x \lor r \rightsquigarrow x \lor$$

$$\bigvee_{1 \le i < k} (\exists z_1 \dots \exists z_i)[r \rightsquigarrow z_1 \wedge z_{i-1} \rightsquigarrow x \wedge \bigwedge_{1 \le j < i-1} z_j \rightsquigarrow z_{j+1}]]],$$

which is a first-order formula for "r is the root and x is at distance at most k from r". Since for every $i \in I$ we have $\mathcal{T}_i \models \varphi[u_i]$, we conclude that $\mathcal{T}^{\updownarrow} \models \varphi[u^*]$ and hence u^* is at distance at most k from the root of $\mathcal{T}^{\updownarrow}$.   □

As a particular case one has the notion of $\downarrow$-**quasi ultrapower** and $\updownarrow$-**quasi ultrapower** of a family of pointed data trees. Observe that if $(\mathcal{T}^{\updownarrow}, u^*)$ is the $\updownarrow$-quasi ultrapower of $(\mathcal{T}, u)_{i \in I}$ then u^* belongs to the domain of $\mathcal{T}^{\updownarrow}$ and so $(\mathcal{T}^{\updownarrow}, u^*)$ is a pointed data tree.

5   Definability

In this section we state the main results. If K is a class of pointed data trees, we denote its complement by $\overline{K}$. We begin with the downward fragment.

Lemma 13. *Let $(\mathcal{T}, u)$ and $(\mathcal{T}', u')$ be two pointed data trees such that $\mathcal{T}, u \equiv^{\downarrow} \mathcal{T}', u'$. Then there exist $\downarrow$-quasi ultrapowers $(\mathcal{T}^{\downarrow}, u^*)$ and $(\mathcal{T}'^{\downarrow}, u'^*)$ of $(\mathcal{T}, u)$ and $(\mathcal{T}', u')$ respectively such that $(\mathcal{T}^{\downarrow}, u^*) \underline{\leftrightarrow}^{\downarrow} (\mathcal{T}'^{\downarrow}, u'^*)$*

Proof. It is known that there is a suitable ultrafilter U such that $\prod_U \mathcal{T}$ and $\prod_U \mathcal{T}'$ are ω-saturated (see e.g. [2, Lem. 2.7.3]). By item 1 Proposition 9, $\mathcal{T}^{\downarrow} = (\prod_U \mathcal{T})|u^*$ and $\mathcal{T}'^{\downarrow} = (\prod_U \mathcal{T}')|u'^*$ are $\downarrow$-saturated data trees. By hypothesis $\mathcal{T}, u \equiv^{\downarrow} \mathcal{T}', u'$, and hence $\mathcal{T}^{\downarrow}, u^* \equiv^{\downarrow} \mathcal{T}'^{\downarrow}, u'^*$. Finally, by Proposition 4, $\mathcal{T}^{\downarrow}, u^* \underline{\leftrightarrow}^{\downarrow} \mathcal{T}'^{\downarrow}, u'^*$.   □

Lemma 14. *Let K be a class of pointed data trees and let Σ be a set of $XPath_{=}^{\downarrow}$-formulas finitely satisfiable in K. Then Σ is satisfiable in some $\downarrow$-quasi ultraproduct of pointed data trees in K.*

Proof. Let $I = \{\Sigma_0 \subset \Sigma \mid \Sigma_0 \text{ is finite}\}$ and for each $\varphi \in \Sigma$, let $\hat{\varphi} = \{i \in I \mid \varphi \in i\}$. Then the set $E = \{\hat{\varphi} \mid \varphi \in \Sigma\}$ has the finite intersection property: $\{\varphi_1, \dots, \varphi_n\} \in \hat{\varphi}_1 \cap \dots \cap \hat{\varphi}_n$. By the Ultrafilter Theorem (see [4, Prop. 4.1.3]) E can be extended to an ultrafilter U over I.

Since Σ is finitely satisfiable in K, for each $i \in I$ there is $(\mathcal{T}_i, u_i) \in K$ such that $\mathcal{T}_i, u_i \models i$. Let $(\mathcal{T}^{\downarrow}, u^*)$ be the $\downarrow$-quasi ultraproduct of $(\mathcal{T}_i, u_i)_{i \in I}$ modulo U. We show that $\mathcal{T}^{\downarrow}, u^* \models \Sigma$: let $\varphi \in \Sigma$. Then $\hat{\varphi} \in E \subseteq U$ and $\hat{\varphi} \subset \{i \in I \mid \mathcal{T}_i, u_i \models \varphi\}$. Hence $\{i \in I \mid \mathcal{T}_i, u_i \models \varphi\} \in U$, which implies that $\prod_U \mathcal{T}_i \models \text{Tr}_x(\varphi)[u^*]$, where u^* is the ultralimit of $(u_i)_{i \in I}$. Since $\mathcal{T}^{\downarrow} = (\prod_U \mathcal{T}_i)|u^*$, by item 1 of Proposition 8 we conclude that $\mathcal{T}^{\downarrow}, u^* \models \varphi$.   □

Theorem 15. *Let K be a class of pointed data trees. Then K is definable by a set of $XPath_{=}^{\downarrow}$-formulas iff K is closed under $\downarrow$-bisimulations and $\downarrow$-quasi ultraproducts, and $\overline{K}$ is closed under $\downarrow$-quasi ultrapowers.*

Proof. For ($\Rightarrow$), suppose that K is definable by a set of XPath$_=^\downarrow$-formulas. By Theorem 1 it is clear that K is closed under $\downarrow$-bisimulations. By the fundamental theorem of ultraproducts together with item 1 of Proposition 8 it is clear that K is closed under $\downarrow$-quasi ultraproducts. It is also clear that the fundamental theorem of ultraproducts and the fact that any XPath$_=^\downarrow$-formula is expressible in first-order imply that $\mathcal{T}, u \equiv^\downarrow \mathcal{T}^\downarrow, u^*$ for any $(\mathcal{T}^\downarrow, u^*)$ $\downarrow$-quasi ultrapower modulo U, and therefore that $\overline{K}$ is closed under $\downarrow$-quasi ultrapowers.

For ($\Leftarrow$), suppose K is closed under bisimulations and $\downarrow$-quasi ultraproducts, and $\overline{K}$ is closed under $\downarrow$-quasi ultrapowers. We show that $\Gamma = \bigcap_{(\mathcal{T},u) \in K} \mathrm{Th}_\downarrow(\mathcal{T}, u)$ defines K. It is clear that if $(\mathcal{T}, u) \in K$ then $\mathcal{T}, u \models \Gamma$.

Now suppose that $\mathcal{T}, u \models \Gamma$ and consider $\Sigma = \mathrm{Th}_\downarrow(\mathcal{T}, u)$. Let Δ be a finite subset of Σ, and assume that Δ is not satisfiable in K. Then $\neg \wedge \Delta$ is true in every pointed data tree of K, so $\neg \wedge \Delta \in \Gamma$. Therefore $\mathcal{T}, u \models \neg \wedge \Delta$ which is a contradiction because $\Delta \subseteq \Sigma$. This shows that Σ is finitely satisfiable in K.

By Lemma 14, Σ is satisfiable in some $\downarrow$-quasi ultraproduct of pointed data trees in K, and since K is closed under $\downarrow$-quasi ultraproducts, Σ is satisfiable in K. Then there exists $(\mathcal{T}', u') \in K$ such that $\mathcal{T}', u' \models \Sigma$ and therefore $\mathcal{T}, u \equiv^\downarrow \mathcal{T}', u'$. By Lemma 13, there exist $\downarrow$-quasi ultrapowers $(\mathcal{T}^\downarrow, u^*)$ and $(\mathcal{T}'^\downarrow, u'^*)$ of $(\mathcal{T}, u)$ and $(\mathcal{T}', u')$ respectively such that $(\mathcal{T}^\downarrow, u^*) \leftrightarrow^\downarrow (\mathcal{T}'^\downarrow, u'^*)$. Since K is closed under $\downarrow$-bisimulations, $(\mathcal{T}^\downarrow, u^*) \in K$. Suppose $(\mathcal{T}, u) \in \overline{K}$. Since K is closed under $\downarrow$-quasi ultrapowers, $(\mathcal{T}^\downarrow, u^*) \in \overline{K}$, and this is a contradiction. Hence we conclude $(\mathcal{T}, u) \in K$. $\qquad\square$

Theorem 16. *Let K be a class of pointed data trees. Then K is definable by an XPath$_=^\downarrow$-formula iff both K and $\overline{K}$ are closed under $\downarrow$-bisimulations and $\downarrow$-quasi ultraproducts.*

Proof. For ($\Rightarrow$) suppose that K is definable by an XPath$_=^\downarrow$-formula. By Theorem 1 it is clear that K and $\overline{K}$ are closed under bisimulations. By the fundamental theorem of ultraproducts together with item 1 of Proposition 8 it is clear that K and $\overline{K}$ are closed under $\downarrow$-quasi ultraproducts.

For ($\Leftarrow$) suppose K and $\overline{K}$ are closed under bisimulations and $\downarrow$-quasi ultraproducts. Then, by Theorem 15, there exist sets Γ_1 and Γ_2 of XPath$_=^\downarrow$-formulas defining K and $\overline{K}$ respectively. Consider the set of XPath$_=^\downarrow$-formulas $\Gamma_1 \cup \Gamma_2$. This set is clearly inconsistent and so, by compactness, there are finite sets Δ_1 and Δ_2 such that $\Delta_i \subseteq \Gamma_i$ ($i = 1, 2$) and

$$\mathcal{T}, u \models \wedge \Delta_1 \rightarrow \neg \wedge \Delta_2 \qquad (1)$$

for any pointed data tree $(\mathcal{T}, u)$. We show that $\varphi = \wedge \Delta_1$ defines K. On the one hand, it is clear that if $(\mathcal{T}, u) \in K$ then $\mathcal{T}, u \models \varphi$. On the other hand, suppose that $\mathcal{T}, u \models \varphi$. From (1) we conclude $\mathcal{T}, u \models \neg \wedge \Delta_2$ and so $\mathcal{T}, u \not\models \Gamma_2$. Then $(\mathcal{T}, u) \notin \overline{K}$ as we wanted to prove. $\qquad\square$

In [8, §3.1.1] a restricted version of $\downarrow$-bisimulations, called ℓ-bisimulation, is introduced. It is shown to coincide with the notion of ℓ-equivalence, which informally means *indistinguishable by XPath$_=^\downarrow$ formulas that cannot "see" beyond ℓ*

'child'-steps from the current point of evaluation. Like Theorem 16, the following result characterizes when a class of pointed data trees is definable by a single XPath$_=^\downarrow$-formula. However, instead of using the rather abstract notion of $\downarrow$-quasi ultraproducts, it uses the perhaps more natural notion of ℓ-bisimulation.

Theorem 17. Let K be a class of pointed data trees. Then K is definable by a formula of XPath$_=^\downarrow$ iff K is closed by ℓ-bisimulations for XPath$_=^\downarrow$ for some ℓ.

Proof. ($\Rightarrow$) is a direct consequence of Theorem 1. Let us see ($\Leftarrow$). We know [8, Cor. 3.2] that $\{\mathcal{T}', u' \mid \mathcal{T}, u \equiv_\ell^\downarrow \mathcal{T}', u'\}$ is definable by an XPath$_=^\downarrow$-formula $\chi_{\ell, \mathcal{T}, u}$ of downward depth $\leq \ell$. We show that

$$\varphi = \bigvee_{(\mathcal{T}, u) \in K} \chi_{\ell, \mathcal{T}, u}$$

defines K. In [8, Prop. 3.1] it is shown that $\equiv_\ell^\downarrow$ has finite index, and therefore the above disjunction is equivalent to a finite one. On the one hand, if $\mathcal{T}', u' \in K$ then it is clear that $\mathcal{T}', u' \models \chi_{\ell, \mathcal{T}', u'}$ and so $\mathcal{T}', u' \models \varphi$. On the other hand, we have $\mathcal{T}', u' \models \varphi$ iff there is $(\mathcal{T}, u) \in K$ such that $\mathcal{T}', u' \models \chi_{\ell, \mathcal{T}, u}$ iff there is $(\mathcal{T}, u) \in K$ such that $\mathcal{T}, u \underline{\leftrightarrow}_\ell^\downarrow \mathcal{T}', u'$. Hence since K is closed under $\underline{\leftrightarrow}_\ell^\downarrow$, if $\mathcal{T}', u' \models \varphi$ we have $\mathcal{T}', u' \in K$. □

We turn to the vertical fragment.

Lemma 18. Let $(\mathcal{T}, u)$ and $(\mathcal{T}', u')$ be two pointed data trees such that $\mathcal{T}, u \equiv^\updownarrow \mathcal{T}', u'$. Then there exist $\updownarrow$-quasi ultrapowers $(\mathcal{T}^\updownarrow, u^*)$ and $(\mathcal{T}'^\updownarrow, u'^*)$ of $(\mathcal{T}, u)$ and $(\mathcal{T}', u')$ respectively such that $(\mathcal{T}^\updownarrow, u^*) \underline{\leftrightarrow}^\updownarrow (\mathcal{T}'^\updownarrow, u'^*)$

Proof. The proof is analogous to the proof of Lemma 13 but using item 2 instead of item 1 of Proposition 9 and Proposition 6 instead of Proposition 4. □

Lemma 19. Let K be a class of k-bounded pointed data trees and let Σ be a set of XPath$_=^\updownarrow$-formulas finitely satisfiable in K. Then Σ is satisfiable in some $\updownarrow$-quasi ultraproduct of pointed data trees in K.

Proof. The proof is analogous to the proof of Lemma 14 but taking $\updownarrow$-quasi ultraproducts instead of $\downarrow$-quasi ultraproducts and using item 2 instead of item 1 of Proposition 8. To apply this Proposition, one has to note that $u^* \in \mathcal{T}^\updownarrow$ since the $\mathcal{T}_i, u_i$ are k-bounded pointed. □

In the next two theorems, the universe of pointed data trees is restricted to those which are k-bounded (for any fixed k). Therefore, the operations of closure and complement must be taken with respect to this universe.

Theorem 20. *Over k-bounded pointed data trees: K is definable by a set of XPath$_=^{\updownarrow}$-formulas iff K is closed under $\updownarrow$-bisimulations and $\updownarrow$-quasi ultraproducts, and $\overline{K}$ is closed under $\updownarrow$-quasi ultrapowers.*

Proof. The proof is analogous to the proof of Theorem 15 but replacing pointed data trees for k-bounded pointed data trees and every occurrence of $\downarrow$ for $\updownarrow$. Also, for ($\Rightarrow$), one has to use item 2 instead of item 1 of Proposition 8 and for ($\Leftarrow$), Lemmas 19 and 18 instead of Lemmas 14 and 13.   □

Theorem 21. *Over k-bounded pointed data trees: K is definable by an XPath$_=^{\updownarrow}$-formula iff both K and $\overline{K}$ are closed under $\updownarrow$-bisimulations and $\updownarrow$-quasi ultraproducts.*

As in Theorem 17, one can also restate Theorem 21 in terms of (r, s, k)-bisimulations for XPath$_=^{\updownarrow}$ (see [8, §3.2.3] for a definition).

Theorem 22. *Let K be a class of pointed data trees. Then K is definable by a formula of XPath$_=^{\updownarrow}$ iff K is closed by (r, s, k)-bisimulations for XPath$_=^{\updownarrow}$ for some r, s, k.*

6   Future Research and Applications

In this work we introduced new tools for showing definability results for the downward and vertical fragments of XPath with (in)equality tests over data trees. The general road to prove these theorems themselves is somewhat similar to the one used for the basic modal logic BML (namely, a detour to first-order), but the new concepts (and their interactions) needed to be used in the context of XPath$_=$ are more sophisticated. The notions of $\downarrow$-saturation and $\updownarrow$-saturation are more refined than the usual notions of BML, as they need to take care of the (in)equality tests over the data. Another difference with respect to the models of BML, namely Krike models, is that models of XPath$_=$ are trees (in particular, connected) and so they are not closed under ultraproducts. Thus the notions of $\downarrow$-quasi and $\updownarrow$-quasi ultraproducts arise. These are variants of the classical first-order ultraproducts, and they are, of course, absent in the BML framework.

Our development may be useful for showing other basic model theoretical results such as separation or interpolation of XPath$_=^{\downarrow}$ and XPath$_=^{\updownarrow}$. We plan to study those and other properties using the tools introduced in this work.

An interesting question is what can be said about other fragments of XPath$_=$ such as XPath$_=^{\downarrow\downarrow^*}$ ('child' and 'descendant' axes) or XPath$_=^{\updownarrow\uparrow^*}$ ('child', 'parent', 'descendant' and 'ancestor' axes). As it is mentioned in [8, §5], the bisimulation notions of these two fragments correspond to those for XPath$_=^{\downarrow}$ and XPath$_=^{\updownarrow}$ respectively. However, in the case of XPath$_=^{\downarrow\downarrow^*}$ and XPath$_=^{\updownarrow\uparrow^*}$, the connection to first-order logic is not clear, as we cannot express *transitive closure*.

We finish with some applications:

Example 23. Let K be the class of pointed data trees $(\mathcal{T}, u)$ where u is the root of $\mathcal{T}$ and $\mathcal{T}$ has some node labeled a. On the one hand, K is definable by a first-order σ-sentence. On the other, K is closed under $\text{XPath}^{\updownarrow}_=$-bisimulations but not closed under $\updownarrow$-quasi ultraproducts: for $i \in \mathbb{N}$ define $\mathcal{T}_i$ as any tree of height i whose only node labeled a is at distance i from the root, and define u_i as the root of $\mathcal{T}_i$. By an argument similar to the one used in Example 10 one can show that if $(\mathcal{T}^{\updownarrow}, u^*)$ is any $\updownarrow$-quasi ultraproduct of $(\mathcal{T}_i, u_i)_{i \in \mathbb{N}}$ then no node of $\mathcal{T}^{\updownarrow}$ has label a. By Theorem 20, K is not definable by a set of $\text{XPath}^{\updownarrow}_=$-formulas.

Example 24. Let $dist_3(x)$ be the property stating that there are nodes y, z so that $x {\to} y {\to} z$ and x, y, z have pairwise distinct data values. It can be checked that the $\text{XPath}^{\updownarrow}_=$-formula φ_4 from Figure 1 expresses $dist_3(x)$. Let K be the class of pointed data trees $(\mathcal{T}, u)$, where u is the root of $\mathcal{T}$, and for all $v \in T$ we have $dist_3(v)$. On the one hand, K is definable by the set of $\text{XPath}^{\updownarrow}_=$-formulas $\{\neg\langle\downarrow^n [\neg\varphi_4]\rangle \mid n \geq 0\}$. On the other, for $i \in \mathbb{N}$, let $(\mathcal{T}_i, u_i)$ be any pointed data tree not in K, of height at least $i + 1$, where u_i is the root of $\mathcal{T}_i$, and such that for all $v \in T_i$ at distance at most i from u_i we have $dist_3(v)$. Let $(\mathcal{T}^{\updownarrow}, u^*)$ be any $\updownarrow$-quasi ultraproduct of $(\mathcal{T}_i, u_i)_{i \in \mathbb{N}}$. One can see that all nodes of $v \in T^{\updownarrow}$ satisfy $dist_3(v)$, and so $(\mathcal{T}^{\updownarrow}, u^*) \in K$. Therefore $\overline{K}$ is not closed under $\updownarrow$-quasi ultraproducts and by Theorem 21, K is not definable by an $\text{XPath}^{\updownarrow}_=$-formula. The reader can verify that K is not closed under $\downarrow$-bisimulations (see [8, Prop. 7.5]) and hence, by Theorem 15, K is not definable by a set of $\text{XPath}^{\downarrow}_=$-formulas.

Acknowledgements. This work was partially supported by grant ANPCyT-PICT-2011-0365, UBACyT 20020110100025, the FP7-PEOPLE-2011-IRSES Project MEALS and the Laboratoire International Associé INFINIS.

References

1. Areces, C., Carreiro, F., Figueira, S.: Characterization, definability and separation via saturated models. In: Theoretical Computer Science (to appear, 2014)
2. Blackburn, P., de Rijke, M., Venema, Y.: Modal Logic. Cambridge Tracts in Theoretical Computer Science, vol. 53. Cambridge University Press (2001)
3. Bojańczyk, M., Muscholl, A., Schwentick, T., Segoufin, L.: Two-variable logic on data trees and XML reasoning. Journal of the ACM 56(3), 1–48 (2009)
4. Chang, C.C., Keisler, H.J.: Model theory. Studies in logic and the foundations of mathematics. North-Holland (1990)
5. Clark, J., DeRose, S.: XML path language (XPath). Website (1999), W3C Recommendation, http://www.w3.org/TR/xpath
6. De Rijke, M.: Modal model theory. Annals of Pure and Applied Logic (1995)
7. De Rijke, M., Sturm, H.: Global definability in basic modal logic. Essays on Non-Classical Logic 1, 111 (2001)
8. Figueira, D., Figueira, S., Areces, C.: Basic model theory of XPath on data trees. In: ICDT, pp. 50–60 (2014)
9. Gottlob, G., Koch, C., Pichler, R.: Efficient algorithms for processing XPath queries. ACM Transactions on Database Systems 30(2), 444–491 (2005)
10. Gyssens, M., Paredaens, J., Van Gucht, D., Fletcher, G.H.L.: Structural characterizations of the semantics of XPath as navigation tool on a document. In: PODS, pp. 318–327. ACM (2006)

11. Kurtonina, N., de Rijke, M.: Bisimulations for temporal logic. Journal of Logic, Language and Information 6, 403–425 (1997)
12. Kurtonina, N., de Rijke, M.: Simulating without negation. Journal of Logic and Computation 7, 503–524 (1997)
13. Marx, M., de Rijke, B.: Semantic characterizations of navigational XPath. SIGMOD Record 34(2), 41–46 (2005)
14. ten Cate, B.: The expressivity of XPath with transitive closure. In: Vansummeren, S. (ed.) PODS, pp. 328–337. ACM (2006)

On the Value of Variables

Beniamino Accattoli and Claudio Sacerdoti Coen

Department of Computer Science and Engineering,
University of Bologna, Italy

Abstract. Call-by-value and call-by-need λ-calculi are defined using the distinguished syntactic category of values. In theoretical studies, values are variables and abstractions. In more practical works, values are usually defined simply as abstractions. This paper shows that practical values lead to a more efficient process of substitution—for both call-by-value and call-by-need—once the usual hypothesis for implementations hold (terms are closed, reduction does not go under abstraction, and substitution is done in micro steps, replacing one variable occurrence at the time). Namely, the number of substitution steps becomes linear in the number of β-redexes, while theoretical values only provide a quadratic bound.

1 Introduction

The theory and the practice of functional programming languages are sometimes far apart. For instance, the theory is based on the λ-calculus, where terms may have free variables, reduction is non-deterministic (but confluent), and can take place everywhere in the term. In practice—*i.e.* in the implementation of functional languages—only closed λ-terms are considered, reduction is deterministic, and weak, *i.e.* it does not take place under abstraction.

Theoretical and Practical Values. Plotkin's call-by-value λ-calculus [1] is a theoretical object of study introduced to model a concrete case, Landin's SECD machine [2]. In such a calculus there is a primitive notion of *value* and β-redexes can fire only when the argument is a value. For Plotkin—and for most of the huge theoretical literature that followed—values are *variables* and *abstractions*; let us call them *theoretical values*. However, most call-by-value abstract machines (or imperative extensions of Plotkin's calculus [3]) employ a notion of *practical value* that includes abstractions and excludes variables. For instance, Paolini and Ronchi della Rocca's book [4] on the *parametric λ-calculus*, a generalization of Plotkin's calculus based on a parametric notion of value, requires that the given notion of value is theoretical (*i.e.* that it includes variables), while Pierce's book [5], driven by programming and implementations, uses practical values. Under the usual practical hypotheses—terms are closed, reduction does not go under abstraction—the difference between the two notions of value is not *extensionally* observable, as it does not affect the result of evaluation.

In this paper we close the gap between theory and practice, providing a theoretical justification for practical values. We show that the difference between the

U. Kohlenbach et al. (Eds.): WoLLIC 2014, LNCS 8652, pp. 36–50, 2014.
© Springer-Verlag Berlin Heidelberg 2014

two notions of value is *intensionally* observable: the practical variant leads to a more efficient implementation of substitution, where efficiency is measured in relation to the number of β-redexes. To state and prove our claim it is necessary to switch to a refinement of the call-by-value λ-calculus where the usual *small-step semantics* is decomposed in a *micro-step semantics*, in which substitution acts on a variable occurrence at the time, *i.e.* with the granularity of abstract machines (or of that of *substructural* operational semantics [6]).

The Linear Substitution Calculus. Our framework is the Linear Substitution Calculus (LSC) [7,8,9], a calculus with explicit substitutions that is in between theory and practice. It is theoretically well-founded, as it arises from graphical and logical studies on the λ-calculus (of which it is a refinement), and practically useful, as it faithfully models most environment-based abstract machines [10], and—remarkably—the number of evaluation steps in the LSC is a reasonable measure of the time complexity of a λ-term [8,11]. One of its key features is its simplicity: it can model an abstract machine using only two rules, corresponding to multiplicative and exponential cut-elimination in linear logic. The first rule, the multiplicative one $\multimap_m$, deals with β-redexes, replacing them with an explicit substitution. The second rule, the exponential one $\multimap_e$, replaces a single occurrence of a variable with the content of its associated explicit substitution, mimicking the mechanism at work in abstract machines.

Call-by-Name. Call-by-name does not rely on values, or, equivalently, everything, including variables, is a value. Using the call-by-name LSC, in [8] it is shown that the number of substitutions steps $(\multimap_e)$ is quadratic in the number of β steps $(\multimap_m)$. The worst cases, *i.e.* those reaching the quadratic bound, are given by sequences where between any two multiplicative steps (corresponding to β-redexes) there is a chain of substitution steps of length linear in the number of preceding multiplicative steps.

Call-by-Value. In the call-by-value LSC, if values are *theoretical* then the chains of substitution steps at work in call-by-name case are still possible, and so the bound is quadratic. On the other hand, we show that it is enough to remove variables from values—therefore switching to *practical* values—to avoid these expensive chains and obtain a *globally linear* relationship between the number of substitution steps $(\multimap_e)$ and the number of β steps $(\multimap_m)$. The proof of the bound is particularly simple and, curiously, it holds only under the assumption that evaluation terminates.

Call-by-Need. We then deal with call-by-need evaluation, that is usually defined using practical values [12,13,14,15,16] and that can be modularly expressed in the LSC. As for call-by-value, theoretical values induce a quadratic bound, while practical values provide a linear bound. The proof, however, is inherently different. It is technically more involved and it does not require the termination assumption.

New Speed-Up. Summing up, the two contributions of the paper are the linear bounds for call-by-value and call-by-need. These evaluation strategies are usually considered to speed up call-by-name evaluation because they reduce redexes in arguments *before* the arguments are substituted, implementing a form of sharing. Our results show that they also provide a subtler and deeper speed-up with respect to call-by-name: there are terms that take the same number k of β-steps to evaluate to normal form in call-by-name/value, and yet their micro-step evaluation takes $O(k^2)$ steps in call-by-name and $O(k)$ steps in call-by-value/need.

Justifying Practical Values. One of the motivations of this work is to find a theoretical justification for practical values, that escape usual argument based on logic or rewriting. Indeed, while both call-by-value and call-by-need have a logical foundation in the so-called *boring* translation of λ-calculus into linear logic [17,18], such translation wraps both variables and abstractions inside the ! modality—the connective allowing non-linear behaviour—thus enabling the substitution of both. At the rewriting level, the strategies implemented by abstract machines can be justified as being standard strategies, in the sense of the standardization theorem. Now, the strategies with practical values are not standard in the wider calculi with theoretical values, so that the switch to practical values cannot be justified that way. Our results provide an alternative explanation, based on the relative complexity of the substitution process.

Abstract Machines. Let us conclude pointing out a companion paper [10], where for the LSC calculi considered here and several abstract machines from the literature, we show that the number of execution steps of the abstract machine is *linear* in the number of steps in the calculus. Via that work, our bounds apply to concrete implementation models.

Related Work. The only similar work we are aware of is Dal Lago and Martini's [19], where it is shown that evaluation in the call-by-value λ-calculus (corresponding to our $\multimap_m$) and evaluation in a related graph-rewriting formalism (playing the role of the LSC, and accounting for $\multimap_m$ and $\multimap_e$) are linearly related (and so $\multimap_e$ is linear in $\multimap_m$). They do not discuss the difference between theoretical and practical values, however they employ practical values at the graphical level, exactly as our results prescribe.

2 Call-by-Name Analysis

Terms and Contexts. The language of the *linear substitution calculus*, that will be shared by all the calculi treated in the paper, is generated by the following grammar:

$$t, u, w, r ::= x \mid \lambda x.t \mid tu \mid t[x \leftarrow u]$$

The constructor $t[x \leftarrow u]$ is called an *explicit substitution* (of u for x in t). Both $\lambda x.t$ and $t[x \leftarrow u]$ bind x in t, with the usual notion of α-equivalence and of free/bound variable (occurrence).

An *initial term* is a closed term (*i.e.* without free variables) with no explicit substitutions.

The operational semantics is defined using contexts, *i.e.* terms with one occurrence of the hole $\langle \cdot \rangle$, an additional constant. For call-by-name (shortened CBN), evaluation contexts are defined by the following grammar:

$$H ::= \langle \cdot \rangle \mid Ht \mid H[x \leftarrow t]$$

The *plugging* $H\langle t \rangle$ (resp. $H\langle H' \rangle$) of a term t (resp. context H') in a context H is defined as $\langle t \rangle := t$ (resp. $\langle H' \rangle := H'$), $(Ht)\langle u \rangle := H\langle u \rangle t$ (resp. $(Ht)\langle H' \rangle := H\langle H' \rangle t$), and so on. *Substitution contexts* are defined by $L ::= \langle \cdot \rangle \mid L[x \leftarrow t]$.

Rewriting Rules. As usual, the rewriting rules are obtained by first defining the rewriting rules at top level, and then taking their closure by evaluation contexts. A peculiar aspect of the LSC is that contexts are also used to define the rules at top level. Such a use of contexts is how locality on proof nets (the graphical language for linear logic proofs) is reflected on terms. For CBN, the rewriting relation is $\multimap := \multimap_m \cup \multimap_e$, where $\multimap_m$ and $\multimap_e$ are given by:

RULE AT TOP LEVEL	CONTEXTUAL CLOSURE
$L\langle \lambda x.t \rangle u \mapsto_m L\langle t[x \leftarrow u] \rangle$	$H\langle t \rangle \multimap_m H\langle u \rangle$ iff $t \mapsto_m u$
$H\langle x \rangle[x \leftarrow u] \mapsto_e H\langle u \rangle[x \leftarrow u]$	$H\langle t \rangle \multimap_e H\langle u \rangle$ iff $t \mapsto_e u$

We silently work modulo α-equivalence to avoid variable capture in the rewriting rules, and in $\mapsto_e$ we assume that the context H does not capture the variable x nor the free variables of u.

In the literature, $\multimap$ is known as *weak linear head reduction*. The rule $\multimap_m$, turning (generalized) β-redexes into explicit substitutions, corresponds to the *multiplicative* case of cut-elimination in proof nets, while $\multimap_e$, implementing substitution in micro steps, corresponds to the exponential case.

Exponential vs Multiplicative Analysis. For CBN, the relationship between $\multimap_m$ and $\multimap_e$ is already well-known from the literature [8,11]. Given a derivation $d : t \multimap^* u$ let us note $|d|_e$ and $|d|_m$ the number of exponential and multiplicative steps, respectively. Then:

Theorem 1 (Quadratic Bound [8]). *Let $d : t \multimap^* u$ be a CBN derivation from an initial term t. Then $|d|_e = O(|d|_m^2)$ (and so $|d| = O(|d|_m^2)$).*

In [11] this result is generalized and its proof is axiomatized. In fact, it holds for any strategy having the two following abstract properties of $\multimap$ (using the notation of the theorem):

1. *Trace*: the number $|u|_{[\,]}$ of explicit substitutions in u is exactly $|d|_m$.
2. *Syntactic Bound*: the length of a sequence of $\multimap_e$ steps from u is $\leq |u|_{[\,]}$.

Their proof for $\multimap$ can be found in [8] or—in a more general form—in [11]. Then the bound can be proved easily.

Proof. Note that $\multimap_{\mathsf{m}}$ terminates, as the number of constructors decreases. The syntactic bound property gives termination of $\multimap_{\mathsf{e}}$. Then d has the shape:

$$t = w_1 \multimap_{\mathsf{m}}^{a_1} r_1 \multimap_{\mathsf{e}}^{b_1} w_2 \multimap_{\mathsf{m}}^{a_2} r_2 \multimap_{\mathsf{e}}^{b_2} \ldots w_k \multimap_{\mathsf{m}}^{a_k} r_k \multimap_{\mathsf{e}}^{b_k} u.$$

By the syntactic bound property, we obtain $b_i \leq |r_i|_{[\,]}$. By the trace property we obtain $|r_i|_{[\,]} = \sum_{j=1}^{i} a_j$, and so $b_i \leq \sum_{j=1}^{i} a_j$. Then:

$$|d|_{\mathsf{e}} = \sum_{i=1}^{k} b_i \leq \sum_{i=1}^{k} \sum_{j=1}^{i} a_j.$$

Note that $\sum_{j=1}^{i} a_j \leq \sum_{j=1}^{k} a_j = |d|_{\mathsf{m}}$ and $k \leq |d|_{\mathsf{m}}$. So

$$|d|_{\mathsf{e}} \leq \sum_{i=1}^{k} \sum_{j=1}^{i} a_j \leq \sum_{i=1}^{k} |d|_{\mathsf{m}} \leq |d|_{\mathsf{m}}^2. \qquad \square$$

The bound is tight, as it is reached for instance by $\delta\delta$ (where $\delta = \lambda x.(xx)$). In particular, its evaluation has subsequences of variable renamings of the form:

$$
\begin{aligned}
&(x_n x_n)[x_n \leftarrow x_{n-1}] \ldots [x_2 \leftarrow x_1][x_1 \leftarrow \delta] && \multimap_{\mathsf{e}} \\
&(x_{n-1} x_n)[x_n \leftarrow x_{n-1}] \ldots [x_2 \leftarrow x_1][x_1 \leftarrow \delta] && \multimap_{\mathsf{e}} \\
&\ldots \\
&(x_1 x_n)[x_n \leftarrow x_{n-1}] \ldots [x_2 \leftarrow x_1][x_1 \leftarrow \delta] && \multimap_{\mathsf{e}} \\
&(\delta x_n)[x_n \leftarrow x_{n-1}] \ldots [x_2 \leftarrow x_1][x_1 \leftarrow \delta] && \multimap_{\mathsf{m}} \\
&(x_{n+1} x_{n+1})[x_{n+1} \leftarrow x_n][x_n \leftarrow x_{n-1}] \ldots [x_2 \leftarrow x_1][x_1 \leftarrow \delta] \multimap_{\mathsf{e}} \ldots
\end{aligned}
\tag{1}
$$

where it takes n renaming steps to obtain a multiplicative redex, that in turn generates a new sequence of $n + 1$ renamings, and so on. In other words, these sequences meet the bound in the syntactic bound property.

Let us point out that the bound is reached also by some normalizing terms. Consider $\tau\tau n$ where $\tau = \lambda x.\lambda n.(n(\lambda y.y)(xx))$ and n is any Scott's numeral [20], defined by $[\![0]\!] = \lambda x.\lambda y.x$ and $[\![n + 1]\!] = \lambda x.\lambda y.y[\![n]\!]$. Evaluating the term takes $(n + 1)(n + 4)/2$ exponential steps but only $4(n + 1)$ multiplicative steps.

The trace and syntactic bound properties can be proved also for call-by-value and call-by-need variants of the calculus, obtaining a quadratic bound. But the next sections will show that for the variants of these strategies that employ practical values a finer analysis is possible, leading to a linear bound. These two results are new, and surprising in various ways:

1. *Variables*: for the linear bound it is crucial that values do not include variables. For instance, if variables are values $\delta\delta$ has exactly the same reductions in the three evaluation scheme considered, matching the quadratic bound. What is surprising is that it is enough to remove variables from values to decrease the asymptotic complexity of substitution.
2. *New Speed-Up*: the terms of the form $\tau\tau n$ mentioned before take the same number k of β-steps to evaluate to normal form in call-by-name and call-by-value, and yet their micro-step evaluation takes $O(k^2)$ steps in call-by-name and $O(k)$ steps in call-by-value.

3. *Linear Logic*: from a linear logic perspective the bound is quite unexpected. The exponentials (*i.e.* the substitutions), responsible for duplications, are expected to capture most of the computing time, while the multiplicatives are somehow negligible in terms of cost. One may suspect that the number of steps is not a good complexity measure, as substitution may be very costly to implement. But it is not the case here, as our exponential steps can be implemented in time linear in the size of the initial term (because of the properties of the micro-step evaluation strategy we consider), and can thus be taken as a realistic measure of complexity, see [8,11].

3   Call-by-Value Analysis

For call-by-value (CBV), the underlying language is the same as for call-by-name, but we distinguish (practical) *values*, noted v, that are given only by abstractions, and *answers* $L\langle v \rangle$, given by a value in a substitution context (see Sect. 2). Evaluation contexts for CBV, implementing left-to-right CBV, are defined as:

$$V ::= \langle \cdot \rangle \mid Vt \mid L\langle v \rangle V \mid V[x \leftarrow t]$$

In CBV, it can be easily shown that a closed term either diverges or produces an answer (but this property will not play a role in our analysis), and moreover the definiens of substitutions are also answers.

Rewriting Rules. We re-define $\multimap_m$ and $\multimap_e$ as follows:

RULE AT TOP LEVEL	CONTEXTUAL CLOSURE
$L\langle \lambda x.t \rangle L'\langle v \rangle \mapsto_m L\langle t[x \leftarrow L'\langle v \rangle] \rangle$	$V\langle t \rangle \multimap_m V\langle u \rangle$ iff $t \mapsto_m u$
$V\langle x \rangle [x \leftarrow L\langle v \rangle] \mapsto_e L\langle V\langle v \rangle [x \leftarrow v] \rangle$	$V\langle t \rangle \multimap_e V\langle u \rangle$ iff $t \mapsto_e u$

As for call-by-name, we silently work modulo α-equivalence and in $\mapsto_e$ the context V does not capture x nor the free variables of v. We also still use the notation $\multimap := \multimap_m \cup \multimap_e$.

Let us revisit the $\delta\delta$ example of Sect. 2, used to show that the quadratic bound is strict for CBN. Using CBV and theoretical values one obtains the same evaluation sequence. Practical values, instead, give:

$$
\begin{aligned}
\delta\delta \multimap_m \; &(x_1 x_1)[x_1 \leftarrow \delta] &&\multimap_e \\
&(\delta x_1)[x_1 \leftarrow \delta] &&\multimap_e \\
&(\delta\delta)[x_1 \leftarrow \delta] &&\multimap_m \\
&(x_2 x_2)[x_2 \leftarrow \delta][x_1 \leftarrow \delta] &&\multimap_e \qquad\qquad (2)\\
&(\delta x_2)[x_2 \leftarrow \delta][x_1 \leftarrow \delta] &&\multimap_e \\
&(\delta\delta)[x_2 \leftarrow \delta][x_1 \leftarrow \delta] &&\multimap_m \\
&(x_3 x_3)[x_3 \leftarrow \delta][x_2 \leftarrow \delta][x_1 \leftarrow \delta] &&\multimap_e \ldots
\end{aligned}
$$

Where it is easily seen that for any $d : \delta\delta \multimap^* t$ we have the linear relationship $|d|_e \leq 2 \cdot |d|_m$. This fact suggests that any CBV derivation d verifies $|d|_e = O(|d|_m)$. Curiously, this is not true in general. In particular, in CBV a chain of substitution steps can be arbitrarily longer than the number of previous multiplicative steps. Let us give an example. Let t^n stay for t applied to itself n times, associating to the right, i.e. $t^n := t(t(t(t \dots)))$ n times, and set $I := \lambda y.y$. We have

$$(\lambda x.x^n)I \multimap_{\mathtt{m}} x^n[x \leftarrow I] \multimap_{\mathtt{e}}^n I^n$$

So n substitution steps $\multimap_{\mathtt{e}}$ after just one multiplicative step $\multimap_{\mathtt{m}}$. It seems even worse than in CBN, while instead, globally, it is a faster mechanism, of a different nature (note that the steps in the sequence are independent, i.e. they are not generated by chains of substitutions occurring one in the other as in CBN). The idea is that the substituted values create or will create new multiplicative redexes, so that if we keep reducing the term we will match the substitution steps in excess (if evaluation terminates, as in the example) and obtain a linear relationship between the two. The point is that in CBV the linear bound holds only for evaluation to normal form, otherwise the gap between $|d|_e$ and $|d|_m$ can be arbitrarily big.

Exponential vs Multiplicative Analysis. We first need some easy invariants.

Lemma 1 (CBV Invariants). *Let t be initial and $d : t \multimap^* u$.*

1. Subterm: *every value in u is a value in t;*
2. Trace: *the number $|u|_{[\,]}$ of explicit substitutions in u is exactly $|d|_m$;*
3. Proper: *every substitution in u contains an answer.*

Proof. Easy inductions on the length of d. Point 1 is used to prove Point 2. □

Let us provide an intuition for the forthcoming proof of the linear bound. An exponential step makes a new copy of a value that will be eventually *consumed* by a multiplicative step, unless the term is divergent. A multiplicative step *consumes* the value in its left subterm. Therefore it is possible to bound the number of exponential steps with the number of consumed values (that is the number of multiplicative steps) plus the number of values in the term, what we call the *value size* of the term.

Definition 1 (Value Size). *The value size $| \cdot |_\lambda$ of a term counts the number of values that are not inside another value. It is defined recursively as follows:* $|x|_\lambda = 0$, $|v|_\lambda = 1$, $|t[x \leftarrow u]|_\lambda = |t|_\lambda + |u|_\lambda$, $|tu|_\lambda = |t|_\lambda + |u|_\lambda$.

In just one surprisingly simple lemma we obtain the main invariant relating $\multimap_{\mathtt{e}}$, $\multimap_{\mathtt{m}}$, and the value size. The corollary uses the previous invariants to instantiate it in the terminating case, obtaining the linear bound.

Lemma 2 (Main Invariant). *Let $d : t \multimap^n u$. Then $|d|_e \leq |d|_m + |u|_\lambda - |t|_\lambda$.*

Proof. By induction over n. Case $n = 0$ is obvious. Otherwise $t \multimap w$ and $e : w \multimap^{n-1} u$ and, by inductive hypothesis, $|e|_e \leq |e|_m + |u|_\lambda - |w|_\lambda$. Cases:

– *the first step is exponential.* Then

$$t = V\langle V'\langle x\rangle[x{\leftarrow}L\langle v\rangle]\rangle \multimap_{\mathtt{e}} V\langle L\langle V'\langle v\rangle[x{\leftarrow}v]\rangle\rangle = w$$

and $|w|_\lambda = |t|_\lambda + 1$. Thus

$$
\begin{aligned}
|d|_e = |e|_e + 1 &\le_{i.h.} |e|_m + |u|_\lambda - |w|_\lambda + 1\\
&= |d|_m + |u|_\lambda - (|t|_\lambda + 1) + 1 = |d|_m + |u|_\lambda - |t|_\lambda
\end{aligned}
$$

– *the first step is multiplicative.* Then

$$t = V\langle L\langle \lambda x.r\rangle L'\langle v\rangle\rangle \multimap_{\mathtt{m}} V\langle L\langle r[x{\leftarrow}L'\langle v\rangle]\rangle\rangle = w$$

and $|w|_\lambda = |t|_\lambda - 1 + |r|_\lambda$. Thus

$$
\begin{aligned}
|d|_e = |e|_e &\le_{i.h.} |e|_m + |u|_\lambda - |w|_\lambda\\
&= |d|_m - 1 + |u|_\lambda - (|t|_\lambda - 1 + |r|_\lambda)\\
&= |d|_m + |u|_\lambda - |t|_\lambda - |r|_\lambda \le |d|_m + |u|_\lambda - |t|_\lambda \quad \square
\end{aligned}
$$

Corollary 1 (Linear Bound for CBV). *Let t be initial and $d : t \multimap^* L\langle v\rangle$. Then $|d|_e \le 2\cdot|d|_m + 1$.*

Proof. By the proper invariant every substitution contains a value plus some substitutions, each one recursively having the same shape, so $|L\langle v\rangle|_\lambda = |L\langle v\rangle|_{[\,]} + 1$, where 1 accounts for the value v. By the trace invariant $|L\langle v\rangle|_{[\,]} = |d|_m$, and so $|L\langle v\rangle|_\lambda \le |d|_m + 1$. Then the main invariant gives: $|d|_e \le |d|_m + |L\langle v\rangle|_\lambda - |t|_\lambda \le |d|_m + |L\langle v\rangle|_\lambda \le |d|_m + |d|_m + 1 = 2 \cdot |d|_m + 1$. $\hfill\square$

Invariance of the CEK machine. Our result on CBV has an implicit by-product. In [19] it is shown that Plotkin's calculus, whose steps can be identified with our $\multimap_{\mathtt{m}}$ steps, is invariant, *i.e.* polynomially related to models like Turing machines or random access machines, see the introduction of [11] for a presentation of the topic. Then, our result implies that the CBV LSC is invariant. In [10] it is shown that the CEK abstract machine [21] is linearly related to the CBV LSC. Therefore, the CEK is invariant. Such a result—albeit expected—is new.

Right-to-Left CBV. In this section we studied left-to-right CBV. The dual right-to-left strategy can be obtained by simply redefining the grammar of evaluation context as

$$V ::= \langle \cdot\rangle \mid VL\langle v\rangle \mid tV \mid V[x{\leftarrow}t]$$

Our proof for the bound with practical values holds unchanged also for the right-to-left strategy. However, it is unclear how right-to-left CBV behaves with theoretical values, as the typical quadratic example for theoretical left-to-right CBV, given by $\delta\delta$, is linear when evaluated with theoretical right-to-left CBV:

$$
\begin{aligned}
\delta\delta &\multimap_{\mathtt{m}} (x_1 x_1)[x_1{\leftarrow}\delta] && \multimap_{\mathtt{e}}\\
&\quad (x_1\delta)[x_1{\leftarrow}\delta] && \multimap_{\mathtt{e}}\\
&\quad (\delta\delta)[x_1{\leftarrow}\delta] && \multimap_{\mathtt{m}}\\
&\quad (x_2 x_2)[x_2{\leftarrow}\delta][x_1{\leftarrow}\delta] && \multimap_{\mathtt{e}} \qquad\qquad (3)\\
&\quad (x_2\delta)[x_2{\leftarrow}\delta][x_1{\leftarrow}\delta] && \multimap_{\mathtt{e}}\\
&\quad (\delta\delta)[x_2{\leftarrow}\delta][x_1{\leftarrow}\delta] && \multimap_{\mathtt{m}}\\
&\quad (x_3 x_3)[x_3{\leftarrow}\delta][x_2{\leftarrow}\delta][x_1{\leftarrow}\delta] && \multimap_{\mathtt{e}} \ldots
\end{aligned}
$$

Note indeed that this is essentially the same evaluation as in (2). We do not know if for theoretical right-to-left CBV $\multimap_m$ and $\multimap_e$ are linearly related. We believe so, but the two proof techniques developed in this paper do not apply.

4   Call-by-Need Analysis

For call-by-need (CBNeed), the analysis is different and technically more involved. At first sight, CBNeed is very similar to call-by-name: the length of substitution sequences is bounded by the number $|d|_m$ of multiplicative steps previously performed, and the bound is easily reached. There is however a fundamental difference. While in CBN *any* substitution sequence can have length $|d|_m$, in CBNeed it is *the concatenation of all chains* that is bound by (twice) $|d|_m$. As for call-by-value, there is a *matching*, or *consumption* phenomenon: firing a substitution chain of length k *consumes* k preceding multiplicative steps, decreasing the bound *for the chains to come* (note that in CBV multiplicative steps consume exponential steps, while here it is the other way around). More precisely, the chains are bound by the number of *unevaluated substitutions* rather than by the number of preceding multiplicative steps, according to the following scheme, that can be seen as a simple form of amortized analysis:

1. every multiplicative step produces an *unevaluated substitution*;
2. the first time an unevaluated substitution substitutes somewhere it changes status and becomes *evaluated*;
3. chains of substitution steps are bound by the number of unevaluated substitutions, that is always $\leq |d|_m$ and only *globally* equal to $|d|_m$.

Our proof will use a calculus enriched with labels on substitutions, to explicitly trace unevaluated substitutions. The labels will have no effect on the dynamics of the calculus, and are only meant as an aid for the proof.

The CBNeed Calculus. For the sake of clarity, we start by introducing the calculus, and then we start over introducing its labeled version. Terms, values, and answers are defined as before. CBNeed evaluation contexts are defined by:

$$N ::= \langle \cdot \rangle \mid Nt \mid N[x \leftarrow t] \mid N'\langle x \rangle[x \leftarrow N]$$

Note that CBNeed evaluation contexts extend the weak head contexts for call-by-name with a clause $(N'\langle x \rangle[x \leftarrow N])$ that turns them into *hereditarily weak head contexts*. This new clause is how sharing will be implemented by the strategy $\multimap := \multimap_m \cup \multimap_e$ defined by:

Rule at Top Level	Contextual closure
$L\langle \lambda x.t \rangle u \mapsto_m L\langle t[x \leftarrow u] \rangle$	$N\langle t \rangle \multimap_m N\langle u \rangle$ iff $t \mapsto_m u$
$N\langle x \rangle[x \leftarrow L\langle v \rangle] \mapsto_e L\langle N\langle v \rangle[x \leftarrow v] \rangle$	$N\langle t \rangle \multimap_e N\langle u \rangle$ iff $t \mapsto_e u$

The multiplicative rule is taken from the CBN calculus. Therefore the definiens of a substitution is not necessarily an answer. The exponential rule come instead from the CBV calculus, and requires arguments to be evaluated to answers before being substituted, reflecting the *by need* content of the strategy.

Now that the calculus is defined, let us evaluate again $\delta\delta$. Using CBNeed and theoretical values it would evaluate exactly in the same way as for CBN. Practical values, instead, give:

$$
\begin{array}{ll}
\delta\delta \multimap_m (x_1 x_1)[x_1 \leftarrow \delta] & \multimap_e \\
(\delta x_1)[x_1 \leftarrow \delta] & \multimap_m \\
(x_2 x_2)[x_2 \leftarrow x_1][x_1 \leftarrow \delta] & \multimap_e \\
(x_2 x_2)[x_2 \leftarrow \delta][x_1 \leftarrow \delta] & \multimap_e \\
(\delta x_2)[x_2 \leftarrow \delta][x_1 \leftarrow \delta] & \multimap_m \\
(x_3 x_3)[x_3 \leftarrow x_2][x_2 \leftarrow \delta][x_1 \leftarrow \delta] & \multimap_e \\
(x_3 x_3)[x_3 \leftarrow \delta][x_2 \leftarrow \delta][x_1 \leftarrow \delta] & \multimap_e \\
(\delta x_3)[x_3 \leftarrow \delta][x_2 \leftarrow \delta][x_1 \leftarrow \delta] & \multimap_m \dots
\end{array}
\tag{4}
$$

Where it is easily seen that for any $d : \delta\delta \multimap^* t$ we have $|d|_e \leq 2 \cdot |d|_m$. We are going to show that—in contrast to CBV—this bound holds for *any* CBNeed derivation, *i.e.* the derivation does not need to end on a normal form.

The labeled CBNeed Calculus. The labeled language is:

$$ t, u, w, r ::= x \mid v \mid tu \mid t[x \leftarrow u]^\circ \mid t[x \leftarrow u]^\bullet; \qquad v ::= \lambda x.t; $$

A *white* substitution $t[x \leftarrow u]^\circ$ represents an unevaluated substitution, that has never substituted its content yet. A *black* substitution $t[x \leftarrow u]^\bullet$ instead is an already evaluated substitution, *i.e.* one that has already acted on some variable occurrence. An invariant of evaluation will be that black substitutions contain values. We use $t[x \leftarrow u]^*$ for $t[x \leftarrow u]^\circ$ or $t[x \leftarrow u]^\bullet$. Of course, we need to redefine also substitution and evaluation contexts, duplicating the cases for substitution:

$$
\begin{array}{l}
L ::= \langle \cdot \rangle \mid L[x \leftarrow t]^\circ \mid L[x \leftarrow t]^\bullet; \\
N, M ::= \langle \cdot \rangle \mid Nt \mid N[x \leftarrow t]^\circ \mid N[x \leftarrow t]^\bullet \mid N\langle x \rangle [x \leftarrow N]^\circ \mid N\langle x \rangle [x \leftarrow N]^\bullet.
\end{array}
$$

According to the informal semantics, the rewriting rules are:

Rule at Top Level	Contextual closure
$L\langle \lambda x.t \rangle u \mapsto_m L\langle t[x \leftarrow u]^\circ \rangle$	$N\langle t \rangle \multimap_m N\langle u \rangle$ iff $t \mapsto_m u$
$N\langle x \rangle [x \leftarrow L\langle v \rangle]^\circ \mapsto_{eo} L\langle N\langle v \rangle [x \leftarrow v]^\bullet \rangle$	$N\langle t \rangle \rightarrow_{eo} N\langle u \rangle$ iff $t \mapsto_{eo} u$
$N\langle x \rangle [x \leftarrow L\langle v \rangle]^\bullet \mapsto_{e\bullet} L\langle N\langle v \rangle [x \leftarrow v]^\bullet \rangle$	$N\langle t \rangle \rightarrow_{e\bullet} N\langle u \rangle$ iff $t \mapsto_{e\bullet} u$

The rewriting relation is $\multimap := \multimap_m \cup \rightarrow_{eo} \cup \rightarrow_{e\bullet}$. Let $\rightarrow_{e*}$ stay for $\rightarrow_{eo}$ or $\rightarrow_{e\bullet}$. A term is *black-proper* if every black substitution contains a value.

Lemma 3 (Invariants). *Let t be a λ-term and $d : t \multimap^* u$.*

1. Subterm: *every value in u is a value in t.*
2. Black-Proper: *u is black-proper.*

Proof. By induction on the length k of $t \multimap^k u$. □

Since the reduction rules only duplicate values, we obtain that every duplicated subterm along a $\multimap$-execution is a subterm of the initial term.

Multiplicative vs Exponential Analysis. Essentially, we prove two facts that refine the abstract properties providing the quadratic bound for CBN. We use $|t|_\circ$ for the number of white substitutions in t and $|d|_{eo}$ for the number of $\to_{eo}$ steps in d.

Lemma 4 (White Trace). *Let t be initial and $d : t \multimap^* u$. Then $|u|_\circ = |d|_m - |d|_{eo}$.*

Proof. By induction on the length k of d.

1. *Base case, i.e. $k = 0$.* Then $|u|_\circ = 0$ because t is a λ-term (it has no explicit substitution) and $|d|_\circ = |d|_{eo} = 0$, so the statement holds.
2. *Inductive case, i.e. $k > 0$.* Then $t \multimap^{k-1} w \multimap u$ and let e be the derivation $t \multimap^{k-1} w$. By *i.h.*, $|w|_\circ = |e|_m - |e|_{eo}$. Cases of $w \multimap u$:

   (a) $w \multimap_m u$. The step creates a new white substitution and does not duplicate/erase any other white substitution, so $|u|_\circ = |w|_\circ + 1$. Since $|d|_m = |e|_m + 1$ and $|d|_{eo} = |e|_{eo}$, the statement holds.
   (b) $w \to_{eo} u$. By the subterm property (Lemma 3.1) the copied value has no substitution, so we have $|u|_\circ = |w|_\circ - 1$. Since $|d|_m = |e|_m$ and $|d|_{eo} = |e|_{eo} + 1$, the statement holds.
   (c) $w \to_{e\bullet} u$. By the subterm property the copied value has no substitution, so $|u|_\circ = |w|_\circ$. Since $|d|_m = |e|_m$ and $|d|_{eo} = |e|_{eo}$, the statement holds. □

By means of an omitted lemma (Lemma 6, page 48, in the appendix) we obtain the following bounds on substitution sequences.

Lemma 5. *Let t be an initial term and $t \multimap^* u$.*

1. *Black Constant Bound: If $u \to_{e*} \to_{e*} w$ then the second step is not black.*
2. *White Syntactic Bound: If $u \to_{eo}^k w$ then $k \leq |u|_\circ$.*

The first point states that sequences of $\to_{e\bullet}$ steps are degenerated, as they have at most length one, and can only appear after multiplicative steps. The second point is a refined version of the syntactic bound for CBN (see Sect. 2).

Proof. The first point is given by the omitted Lemma 6.4. The second point is by induction on k. If $k = 0$ the statement trivially holds. If $u \to_{eo} r \to_{eo}^{k-1} w$ by the subterm property (Lemma 3.1) the substitution step does not duplicate any substitution and turns exactly one white substitution into a black one. So, $|r|_\circ = |u|_\circ - 1$. By *i.h.* we obtain $k - 1 \leq |u|_\circ - 1$ and so $k \leq |u|_\circ$. □

Theorem 2 (Linear Bound for CBNeed). *Let t be initial and $d : t \multimap^* u$. Then $|d|_e \leq 2 \cdot |d|_m$.*

Proof. Given that $\multimap_m$ is evidently terminating, and according to Lemma 5, d writes uniquely as (where $\to_{e\bullet}^{(1)}$ means 0 or 1 steps of $\to_{e\bullet}$):

$$t = t_1 \multimap_m^{a_1} w_1 \to_{e\bullet}^{(1)} u_1 \to_{eo}^{b_1} t_2 \ldots t_k \multimap_m^{a_k} w_k \to_{e\bullet}^{(1)} u_k \to_{eo}^{b_k} u$$

Clearly $|d|_{e\bullet} \leq |d|_m$. Since $|d|_e = |d|_{e\circ} + |d|_{e\bullet}$, we are left to show that $|d|_{e\circ} \leq |d|_m$. Let $d_i : t \multimap^* w_i$ be the prefix of d ending on w_i (including a_j and b_j for $j < i$, plus a_i, but not b_i). Note that defining $b_0 := 0$ we obtain $|d_i|_{e\circ} = \sum_{j=0}^{i-1} b_j$ for $i \in \{1, \ldots, k\}$. Now we can easily estimate the generic term b_i and conclude:

$$b_i \leq_{Lemma\ 5} |u_i|_\circ =_{Lemma\ 4} |d_i|_m - |d_i|_{e\circ} = |d_i|_m - \sum_{j=0}^{i-1} b_j$$

$$|d|_{e\circ} = \sum_{i=0}^{k} b_i = b_k + \sum_{i=0}^{k-1} b_i \leq |d_k|_m - \sum_{j=0}^{k-1} b_j + \sum_{i=0}^{k-1} b_i = |d_k|_m = |d|_m \quad \square$$

On the Need of Labels. In fact, labels are not strictly necessary. It is possible to prove a linear relationship on the original CBNeed calculus, and the proof, along the same lines, is also slightly simpler (the role of white substitutions is played by those substitution whose content is a term of the form $L\langle x \rangle$). The price to pay however is that such an alternative analysis provides only a laxer—despite always linear—bound, as the multiplicative constant is higher (3 instead of 2). We preferred to use labels because the analysis they provide is *tight*, as it is shown by the $\delta\delta$ example, that reaches the bound given by Theorem 2.

Let us conclude with a comment. The call-by-need LSC can be seen as a variant of Chang and Felleisen's calculus [15], that is a λ-calculus without explicit substitutions implementing call-by-need by micro-step evaluation and only one contextual rewriting rule. The result we just obtained shows that a syntax having an explicit constructor for substitutions may provide insights that are not accessible using the traditional syntax of λ-calculus.

Acknowledgements. To Pablo Barenbaum, for discussions and help with some technical details.

References

1. Plotkin, G.D.: Call-by-name, call-by-value and the lambda-calculus. Theor. Comput. Sci. 1(2), 125–159 (1975)
2. Landin, P.J.: The Mechanical Evaluation of Expressions. The Computer Journal 6(4), 308–320 (1964)
3. Crank, E., Felleisen, M.: Parameter-passing and the lambda calculus. In: POPL, pp. 233–244 (1991)
4. Ronchi Della Rocca, S., Paolini, L.: The Parametric λ-Calculus. Springer, Heidelberg (2004)
5. Pierce, B.C.: Types and Programming Languages. MIT Press, Cambridge (2002)
6. Pfenning, F., Simmons, R.J.: Substructural operational semantics as ordered logic programming. In: LICS, pp. 101–110 (2009)
7. Accattoli, B.: An abstract factorization theorem for explicit substitutions. In: RTA, pp. 6–21 (2012)
8. Accattoli, B., Dal Lago, U.: On the invariance of the unitary cost model for head reduction. In: RTA, pp. 22–37 (2012)
9. Accattoli, B., Bonelli, E., Kesner, D., Lombardi, C.: A nonstandard standardization theorem. In: POPL, pp. 659–670 (2014)

10. Accattoli, B., Barenbaum, P., Mazza, D.: Distilling abstract machines. In: ICFP 2014 (accepted, 2014)
11. Accattoli, B., Dal Lago, U.: Beta Reduction is Invariant, Indeed. In: LICS/CSL 2014 (accepted, 2014)
12. Launchbury, J.: A natural semantics for lazy evaluation. In: POPL, pp. 144–154 (1993)
13. Ariola, Z.M., Felleisen, M.: The call-by-need lambda calculus. J. Funct. Program. 7(3), 265–301 (1997)
14. Maraist, J., Odersky, M., Wadler, P.: The call-by-need lambda calculus. J. Funct. Program. 8(3), 275–317 (1998)
15. Chang, S., Felleisen, M.: The call-by-need lambda calculus, revisited. In: Seidl, H. (ed.) Programming Languages and Systems. LNCS, vol. 7211, pp. 128–147. Springer, Heidelberg (2012)
16. Danvy, O., Zerny, I.: A synthetic operational account of call-by-need evaluation. In: PPDP, pp. 97–108 (2013)
17. Girard, J.Y.: Linear logic. Theoretical Computer Science 50, 1–102 (1987)
18. Accattoli, B.: Proof nets and the call-by-value lambda-calculus. In: LSFA, pp. 11–26 (2012)
19. Dal Lago, U., Martini, S.: On constructor rewrite systems and the lambda-calculus. In: Albers, S., Marchetti-Spaccamela, A., Matias, Y., Nikoletseas, S., Thomas, W. (eds.) ICALP 2009, Part II. LNCS, vol. 5556, pp. 163–174. Springer, Heidelberg (2009)
20. Wadsworth, C.P.: Some unusual λ-calculus numeral systems. In: Seldin, J., Hindley, J. (eds.) To H. B. Curry: Essays on Combinatory Logic, Lambda Calculus and Formalism, Academic Press (1980)
21. Felleisen, M., Friedman, D.P.: Control operators, the SECD-machine, and the lambda-calculus. In: 3rd Working Conference on the Formal Description of Programming Concepts (August 1986)

Proofs Appendix

Lemma 6. *Let N be a call-by-need context.*

1. *If $N\langle x\rangle = M\langle y\rangle$ with x not bound by N and y not bound by M, then $N = M$ and $x = y$.*
2. *A term of the form $N\langle x\rangle$, with x not bound by N, is not of the form $M\langle v\rangle$.*
3. *Suppose $N\langle x_0\rangle$ is black-proper for some variable x_0, and $N\langle v\rangle \to_{e*} t$. Then the step is not $\to_{e\bullet}$.*
4. *If $t \to_{e*} \to_{e*} u$ then the second step is not black.*

Proof.

1. By induction on N.
   (a) *Empty context, i.e. $N = \langle\cdot\rangle$:* then M must be $\langle\cdot\rangle$ and $x = y$.
   (b) *Left of an application, i.e. $N = N' t$:* suppose $N'\langle x\rangle\, t$ is of the form $M\langle y\rangle$. Then $N'\langle x\rangle$ must be of the form $M'\langle y\rangle$, with $M = M' t$, and we conclude by *i.h.*.
   (c) *Left of a white or black substitution, i.e. $N = N'[z\leftarrow t]^*$:* suppose the $N\langle x\rangle = N'\langle x\rangle[z\leftarrow t]^*$ is also of the form $M\langle y\rangle$. There are two possibilities:

     i. *The hole of M is on the left of the substitution.* That is, $M = M'[z{\leftarrow}t]^*$ and $N'\langle x\rangle = M'\langle y\rangle$. We conclude by *i.h.*.

     ii. *The hole of M is inside the substitution.* Then it must be that $M = M'\langle z\rangle[z{\leftarrow}M'']^*$. It follows $N'\langle x\rangle = M'\langle z\rangle$, with z not bound by M'. By *i.h.*, we conclude $x = z$, which is absurd since x is not bound by N. Hence this case is impossible.

(d) *Inside a white or black substitution, i.e. $N = N'\langle z\rangle[z{\leftarrow}N'']^*$:* suppose that $N\langle x\rangle = N'\langle z\rangle[z{\leftarrow}N''\langle x\rangle]^*$ is also of the form $M\langle y\rangle$. As in the previous case, there are two possibilities:

     i. *The hole of M is on the left of the substitution.* That is, $M = M'[z{\leftarrow}t]^*$. In particular, we must have $N'\langle z\rangle = M'\langle y\rangle$ with z now not bound by N'. By *i.h.*, we conclude $y = z$ which is absurd. Hence this case is impossible.

     ii. *The hole of M is inside the substitution.* That is, $M=M'\langle z\rangle[z{\leftarrow}M'']^*$. Hence we have that $N'\langle z\rangle[z{\leftarrow}N''\langle x\rangle]^* = M'\langle z\rangle[z{\leftarrow}M''\langle y\rangle]^*$. On one hand, this implies $N'\langle z\rangle = M'\langle z\rangle$ with z now free on both sides, which by *i.h.* gives us $N' = M'$. On the other, we obtain that $N''\langle x\rangle = M''\langle y\rangle$, that lets us conclude by resorting again to the *i.h.*.

2. By induction on N.

(a) *Empty context, i.e. $N = \langle\cdot\rangle$:* a variable cannot be of the form $M\langle v\rangle$.

(b) *Left of an application, i.e. $N = N't$:* suppose $N\langle x\rangle = N'\langle x\rangle t$ is also of the form $M\langle v\rangle$. Then $N'\langle x\rangle$ must be of the form $M'\langle v\rangle$, which is impossible by *i.h.*.

(c) *Left of a white or black substitution, i.e. $N = N'[z{\leftarrow}t]^*$:* suppose that $N\langle x\rangle = N'\langle x\rangle[z{\leftarrow}t]^*$ is also of the form $M\langle v\rangle$. There are two possibilities:

     i. *The hole of M is on the left of the substitution.* That is, $M = M'[z{\leftarrow}t]^*$ and $N'\langle x\rangle = M'\langle v\rangle$. This is impossible by *i.h.*.

     ii. *The hole of M is inside the substitution.* That is, $M=M'\langle z\rangle[z{\leftarrow}M'']^*$. It follows that $N'\langle x\rangle = M'\langle z\rangle$, with z not bound by M'. By point 1 of this lemma, we conclude $x = z$, which is absurd since x is not bound by N. Hence this case is impossible.

(d) *Inside a white or black substitution, i.e. $N = N'\langle z\rangle[z{\leftarrow}N'']^*$:* suppose that $N\langle x\rangle = N'\langle z\rangle[z{\leftarrow}N''\langle x\rangle]^*$ is also of the form $M\langle v\rangle$. There are two possibilities:

     i. *The hole of M is on the left of the substitution.* That is, $M = M'[z{\leftarrow}t]^*$. In particular, we must have $N'\langle z\rangle = M'\langle v\rangle$ with z now not bound by N'. By *i.h.*, this is impossible.

     ii. *The hole of M is inside the substitution.* That is, $M=M'\langle z\rangle[z{\leftarrow}M'']^*$. Hence we have that $N'\langle z\rangle[z{\leftarrow}N''\langle x\rangle]^* = M'\langle z\rangle[z{\leftarrow}M''\langle v\rangle]^*$. From this we obtain that $N''\langle x\rangle = M''\langle v\rangle$ that is impossible by *i.h.*.

3. By induction on N.

(a) *Empty context, i.e. $N = \langle\cdot\rangle$:* trivial, since v is a normal form.

(b) *Left of an application, i.e. $N = N'u$:* any $\rightarrow_{e\bullet}$ redex in $N'\langle v\rangle u$ must be internal to $N'\langle v\rangle$, and we conclude this is impossible by *i.h.*.

(c) *Left of a white or black substitution, i.e. $N = N'[x{\leftarrow}u]^*$:* so $N\langle v\rangle$ is a substitution $N'\langle v\rangle[x{\leftarrow}u]^*$. There are three possibilities for a $\rightarrow_{e\bullet}$ step:

    i. *The $\rightarrow_{e\bullet}$ step takes place on the left of the substitution, i.e.* internal to $N'\langle v\rangle$. This is impossible by *i.h.*.

    ii. *The $\rightarrow_{e\bullet}$ step takes place inside the substitution.* It must then be that $N'\langle v\rangle$ is of the form $M'\langle x\rangle$. By point 2 of this lemma, this is impossible.

    iii. *The $\rightarrow_{e\bullet}$ step is at the root.* In this case the substitution is black. Suppose $N'\langle v\rangle[x\leftarrow u]^{\bullet} \mapsto_{e\bullet} t$. Then $N'\langle v\rangle$ must be of the form $M'\langle x\rangle$, which is impossible by point 2 of this lemma.

(d) *Inside a white or black substitution, i.e.* $N = N'\langle x\rangle[x\leftarrow N'']^*$: so $N\langle v\rangle$ is $N'\langle x\rangle[x\leftarrow N''\langle v\rangle]^*$. There are three possibilities for a $\rightarrow_{e\bullet}$ step:

    i. *The $\rightarrow_{e\bullet}$ step takes place on the left of the substitution. i.e.* internal to $N'\langle x\rangle$. This means $N'\langle x\rangle$ can be written as $M\langle M'\langle y\rangle[y\leftarrow u]^{\bullet}\rangle$. Since the term is black-proper, u must be a value v'. Note also that $M'' := M\langle M'\langle y\rangle[y\leftarrow\langle\cdot\rangle]^{\bullet}\rangle$ is a call-by-need context. Then $N'\langle x\rangle$ can be written as of the form $M''\langle v'\rangle$. This is impossible by point 2 of this lemma.

    ii. *The $\rightarrow_{e\bullet}$ step takes place inside the substitution. i.e.* internal to $N''\langle v\rangle$. This is impossible by *i.h.*.

    iii. *The $\rightarrow_{e\bullet}$ step is at the root.* In this situation the substitution is black. Since $N'\langle x\rangle[x\leftarrow N''\langle v\rangle]^{\bullet}$ is black-proper by hypothesis, we know $N''\langle v\rangle$ must be a value, which implies that $N'' = \langle\cdot\rangle$. By hypothesis we also know that $N\langle x_0\rangle = N'\langle x\rangle[x\leftarrow N''\langle x_0\rangle]^{\bullet} = N'\langle x\rangle[x\leftarrow x_0]^{\bullet}$ is black-proper for some variable x_0. This is absurd, as the term is supposed to be black-proper but the black substitution contains a variable. Hence this case is impossible.

4. Let $t = N'\langle N\langle x\rangle[x\leftarrow L\langle v\rangle]^*\rangle \rightarrow_{e*} N'\langle L\langle N\langle v\rangle[x\leftarrow v]^{\bullet}\rangle\rangle \rightarrow_{e*} u$. By induction on N'. Cases:

(a) *Empty context, i.e.* $N' = \langle\cdot\rangle$ and

$$t = N\langle x\rangle[x\leftarrow L\langle v\rangle]^* \rightarrow_{e*} L\langle N\langle v\rangle[x\leftarrow v]^{\bullet}\rangle \rightarrow_{e*} u$$

It is easily seen that $L\langle N\langle v\rangle[x\leftarrow v]^{\bullet}\rangle \rightarrow_{e*} u$ because there exists w s.t. $N\langle v\rangle[x\leftarrow v]^{\bullet} \rightarrow_{e*} w$ and $L\langle w\rangle = u$ (variables bound by L can only occur in v and evaluation contexts do not go under abstractions). Then we are in the hypotheses of Point 3, that allows to conclude.

(b) *Left of an application, i.e.* $N' = N'' u$: then any $\rightarrow_{e*}$ redex in $N''\langle L\langle N\langle v\rangle[x\leftarrow v]^{\bullet}\rangle\rangle u$ is internal to the left subterm, and we conclude using the *i.h.*.

(c) *Left of a white or black substitution, i.e.* $N' = N''[y\leftarrow w]^*$. Note that the second step cannot be an action of the substitution $[y\leftarrow w]^*$, because its left term is $N''\langle L\langle N\langle v\rangle[x\leftarrow v]^{\bullet}\rangle\rangle$ — i.e. a value in a CBNeed context — and by Point 2 it cannot be of the form $N'''\langle y\rangle$. Then the second step takes place in the left subterm and we conclude by the *i.h.*.

(d) *Inside a white or black substitution, i.e.* $N' = N''\langle x\rangle[x\leftarrow N''']^*$. Note that $[x\leftarrow N''']^*$ is necessarily white, as reduction took place inside it. If the second substitution step takes place inside N''' we conclude by the *i.h.*. Otherwise, the step is an action of $[x\leftarrow N''']^{\circ}$, that is a white step.

Logics with Copy and Remove

Carlos Areces[1], Hans van Ditmarsch[2],
Raul Fervari[1], and François Schwarzentruber[3]

[1] FaMAF, Universidad Nacional de Córdoba & CONICET, Argentina
{areces,fervari}@famaf.unc.edu.ar
[2] LORIA, CNRS - Université de Lorraine, France & IMSc, Chennai, India
hans.van-ditmarsch@loria.fr
[3] ENS Rennes, France
francois.schwarzentruber@ens-rennes.fr

Abstract. We propose a logic with the dynamic modal operators copy
and remove. The copy operator replicates a given model, and the remove
operator removes paths in a given model. We show that the product
update by an action model (with Boolean pre-conditions) in dynamic
epistemic logic decomposes in copy and remove operations. We also show
that copy and remove operators (of path of length 1) can be expressed by
action models. We investigate the complexity of the satisfiability problem
of syntactic fragments of the logic with copy and remove operations.

Keywords: modal logic, dynamic epistemic logic, complexity,
expressivity.

1 Introduction

In modal logic we interpret a modal operator by way of an accessibility relation in a given model. Over the past decades some logics have been proposed
in which the modality is, instead, interpreted by a transformation of the model.
In such logics the modality can be seen as interpreted by a binary relation between pointed Kripke models, where the second argument of the relation is the
transformed model. We could mention sabotage logic here [10], wherein states or
arrows are deleted from a model. Or we could mention dynamic epistemic logics [13] that focus on such model changing operators in view of modeling change
of knowledge or belief (the standard interpretation for the basic modalities in
that setting). In [1,2,6] a new line of contributions to model-transforming logics, motivated by van Benthem's sabotage logic is developed. Our contribution
advances that last line of work, while linking it to dynamic epistemic logics.

Action model logic ($\mathcal{AML}$) [4] is a well-known dynamic epistemic logic to
model information change. Action model logic is an extension of basic epistemic
logic with a dynamic modal operator for the execution of actions. This operator
is parameterized by an action model, a semantic object which typically models
a multi-agent information changing scenario. These actions models are treated
as syntactic objects in modal operators. Action models are complex structures,

U. Kohlenbach et al. (Eds.): WoLLIC 2014, LNCS 8652, pp. 51–65, 2014.
© Springer-Verlag Berlin Heidelberg 2014

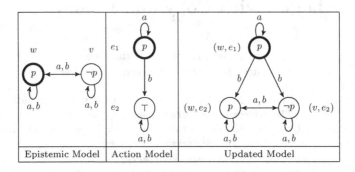

Fig. 1. Agent a privately learns that p

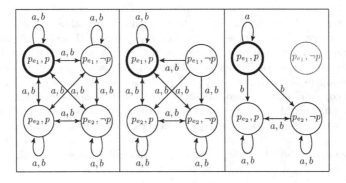

Fig. 2. The result of one copy and two remove operations on the epistemic model of Figure 1, again resulting in the same (bisimilar) updated model

which also leads to high computational complexity (deciding model checking is PSPACE-complete, while deciding satisfiability is NEXPTIME-complete [3]).

In this contribution we propose modal logics with primitive actions called copy and remove. We investigate some of their model theoretic properties and their complexity, and, as an example of what one can do with such logics, we give an embedding of action model logic into our logic: we show that every action model (with propositional pre-conditions) can be simulated by a combination of the copy and remove operators. This is in line with the previously known result that, on the class of finite models, action model execution corresponds to model restriction ('remove') on a bisimilar copy ('copy') of the initial model [11]. The delete we propose is akin to the generalized arrow updates of [9], continuing the work started in [8], that are also known to have equal expressivity as action model logic. But the copy and remove operators we propose are more procedural, whereas these mentioned results are more of a declarative nature.

In Figure 1 we show an epistemic model (a Kripke model), an action model, and the result of executing that action model in that epistemic model. The epistemic model represents that agents a and b are uncertain whether an atomic proposition p is true (and that they have common knowledge of that uncertainty). The actual world, or designated state, of the model is where p is true (shown with a thick circle in the Figure). The action model represents that agent a learns that p is true, whereas agent b (incorrectly) believes that nothing happens—of which a is aware. In short: a privately learns that p. In action models, the valuations of propositional variables are replaced by pre-conditions, in this case p and $\top$ (the formula that is always true). Action models update Kripke models by mean of a restricted modal product, where the domain is limited to the state-action pairs where the pre-conditions of the actions hold. Therefore, there are only three (and not four) such pairs in the updated model: the pair (w, e_1) is missing as the pre-condition of e_1, the formula p, is not true in the state w. The arrows in the product are updated according to the principle that there is a (labeled) arrow between two state-action pairs if there was such an arrow linking both the first arguments and the second arguments. One can now establish that in the resulting model a knows that p (there is only an a-arrow from w to itself), whereas b still believes that a, b are ignorant whether p.

By means of the copy and remove actions of the logics that we propose, we can alternatively describe the effect of this action model. This is depicted in Figure 2. First, we replicate the original epistemic model as many times as there are actions in the action model (twice in this case). We identify each copy with a (fresh) propositional variable corresponding to an action in the action model (e.g., p_{e_1} corresponds to e_1). Thus we obtain the leftmost model in Figure 2. Then, we first remove all the edges (arrows) that point to state-action alternatives wherein the action cannot be executed in the state. Finally, between the remaining state-action pairs we remove all edges that are ruled out according to the accessibility relation in the action model. Thus we obtain the rightmost model in Figure 2.

2 Copy and Remove

In this section we introduce $\mathcal{ML}(\mathsf{cp}, \mathsf{rm})$, a language which can remove edges and create copies of a model.

Definition 1 (Syntax). *Given* PROP, *an infinite and countable set of propositional symbols, and* AGT, *a finite set of agents, let us define the set* FORM *of* $\mathcal{ML}(\mathsf{cp}, \mathsf{rm})$*-formulas, together with a set* PATH *of path expressions.*

$$\text{FORM} ::= \bot \mid p \mid \neg\varphi \mid \varphi \wedge \varphi' \mid \Diamond_a\varphi \mid \mathsf{rm}(\pi)\varphi \mid \mathsf{cp}(\bar{p}, q)\varphi,$$

where $\bar{p} = \langle p_1, \ldots, p_n \rangle$ *is any finite sequence of propositional symbols (all distinct among them) that do not appear in any occurrence of* cp *in* φ, $q \in \bar{p}$, $a \in$ AGT, $\varphi, \varphi' \in$ FORM, *and* $\pi \in$ PATH.

$$\text{PATH} ::= a \mid \pi; \pi' \mid \varphi?,$$

where $a \in$ AGT, $\pi, \pi' \in$ PATH *and* φ *is a Boolean formula.*

We also define the following syntactic fragments: $\mathcal{ML}(\mathsf{cp})$, the fragment with the cp operator but without rm; $\mathcal{ML}(\mathsf{rm})$, the fragment with the rm operator but without cp; $\mathcal{ML}(\mathsf{rm}^-)$, the fragment with rm with path expressions only of the form $\pi = \varphi?; a; \psi?$ but without cp; and $\mathcal{ML}(\mathsf{cp}, \mathsf{rm}^-)$, the fragment with rm with path expressions only of the form $\pi = \varphi?; a; \psi?$ and with cp.

Definition 2 (Models). *A model $\mathcal{M}$ is a triple $\mathcal{M} = \langle W, R, V \rangle$, where W is a non-empty set; $R \subseteq \mathsf{AGT} \times W^2$ is an accessibility relation (we will often write R_a to refer to the set $\{(w, v) \in W^2 \mid (a, w, v) \in R\}$); and $V : \mathsf{PROP} \to \mathcal{P}(W)$ is a valuation. A pair $\mathcal{M}, w$ where w is a state in $\mathcal{M}$ is called a pointed model.*

We represent a path as a sequence $w_0 a_0 w_1 a_1 \ldots w_{n-1} a_{n-1} w_n$ where w_i are states and a_i are agents. Let us now define the set $\mathcal{P}^{\mathcal{M}}(\pi)$ of π-paths in the model $\mathcal{M}$ by induction on π. $\mathcal{P}^{\mathcal{M}}(a)$ contains paths representing a-edges. $\mathcal{P}^{\mathcal{M}}(\pi; \pi')$ contains concatenations of a π-path and a π'-path. In such a concatenation, the last state w of the π-path has to be the first state of the π'-path. $\mathcal{P}^{\mathcal{M}}(\varphi?)$ contains paths of length 0, made of one state, which satisfies φ.

Definition 3 (Paths and Updated Models). *Let $\mathcal{M} = \langle W, R, V \rangle$ a model and $\pi \in \mathsf{PATH}$. We define the set of π-paths $\mathcal{P}^{\mathcal{M}}_{\pi}$ of $\mathcal{M}$ inductively as*

$$
\begin{aligned}
\mathcal{P}^{\mathcal{M}}(a) &= \{wau \mid (w, u) \in R_a\} \\
\mathcal{P}^{\mathcal{M}}(\pi; \pi') &= \{SwS' \mid Sw \in \mathcal{P}^{\mathcal{M}}(\pi) \text{ and } wS' \in \mathcal{P}^{\mathcal{M}}(\pi')\} \\
\mathcal{P}^{\mathcal{M}}(\varphi?) &= \{w \mid \mathcal{M}, w \models \varphi\}.
\end{aligned}
$$

Let $a \in \mathsf{AGT}$, we define $edges_a(P)$ that is the set of a-edges of the path P. Formally, $edges_a(P) = \{(a, w, u) \mid wau \text{ is a subsequence of } P\}$.

Given a model $\mathcal{M} = \langle W, R, V \rangle$, a path expression π, and $\bar{p} = \langle p_1, \ldots, p_n \rangle$, we define the updated models

$\mathcal{M}_{\mathsf{rm}(\pi)} = \langle W, R_{\mathsf{rm}(\pi)}, V \rangle$, *where*
$\quad R_{\mathsf{rm}(\pi)} = R \setminus \bigcup_{a \in \mathsf{AGT}, P \in \mathcal{P}^{\mathcal{M}}(\pi)} edges_a(P)$

$\mathcal{M}_{\mathsf{cp}(\bar{p})} = \langle W_{\mathsf{cp}(\bar{p})}, R_{\mathsf{cp}(\bar{p})}, V_{\mathsf{cp}(\bar{p})} \rangle$, *where*
$\quad W_{\mathsf{cp}(\bar{p})} = \{(w, q) \mid w \in W \text{ and } q \in \bar{p}\}$
$\quad R_{\mathsf{cp}(\bar{p})} = \{(a, (w, q), (w', q')) \mid (a, w, w') \in R\}$
$\quad V_{\mathsf{cp}(\bar{p})}(p) = \{(w, q) \mid w \in V(p)\} \text{ for } p \neq q$
$\quad V_{\mathsf{cp}(\bar{p})}(q) = \{(w, q) \mid w \in W\}$.

Now we can define the semantics of the operators introduced in Definition 1.

Definition 4 (Semantics). *Given a pointed model $\mathcal{M}, w$ and a formula φ we say that $\mathcal{M}, w$ satisfies φ, and write $\mathcal{M}, w \models \varphi$, when*

$$
\begin{aligned}
\mathcal{M}, w &\models p & &\text{iff } w \in V(p) \\
\mathcal{M}, w &\models \neg\varphi & &\text{iff } \mathcal{M}, w \not\models \varphi \\
\mathcal{M}, w &\models \varphi \wedge \psi & &\text{iff } \mathcal{M}, w \models \varphi \text{ and } \mathcal{M}, w \models \psi \\
\mathcal{M}, w &\models \Diamond_a \varphi & &\text{iff for some } v \in W \text{ s.t. } (w, v) \in R_a, \ \mathcal{M}, v \models \varphi \\
\mathcal{M}, w &\models \mathsf{rm}(\pi)\varphi & &\text{iff } \mathcal{M}_{\mathsf{rm}(\pi)}, w \models \varphi \\
\mathcal{M}, w &\models \mathsf{cp}(\bar{p}, q)\varphi & &\text{iff } \mathcal{M}_{\mathsf{cp}(\bar{p})}, (w, q) \models \varphi.
\end{aligned}
$$

φ *is satisfiable if for some pointed model* $\mathcal{M}, w$ *we have* $\mathcal{M}, w \models \varphi$. *When the left side is empty,* $\models \varphi$ *means that* φ *holds in any model. We further define* $\mathsf{cp}(\bar{p})\varphi$ *as an abbreviation for* $\bigwedge_{q\in\bar{p}} \mathsf{cp}(\bar{p}, q)\varphi$.

Bisimulation is a classical notion introduced to investigate the expressive power of modal languages. The conditions required for $\mathcal{ML}(\mathsf{cp}, \mathsf{rm})$ turn out to be very natural: paths deleted via rm traversing a particular state are characterized by the information in successors and predecessors of such point. Hence, it is enough to consider the conditions for the basic temporal logic $\mathcal{ML}(\lozenge^{-1})$ (see [5]):

(Atomic Harmony) for all $p \in \mathsf{PROP}$, $w \in V(p)$ iff $w' \in V'(p)$;
(Zig) if $(w, v) \in R$ then for some v', $(w', v') \in R'$ and vZv';
(Zag) if $(w', v') \in R'$ then for some v, $(w, v) \in R$ and vZv'.
(Zig^{-1}) if $(v, w) \in R_a$ then for some v', $(v', w') \in R_a'$ and vZv';
(Zag^{-1}) if $(v', w') \in R_a'$ then for some v, $(v, w) \in R_a$ and vZv'.

Let $\underline{\leftrightarrow}_{\mathcal{ML}(\mathsf{cp},\mathsf{rm})}$ refer to bisimulations for the language $\mathcal{ML}(\mathsf{cp}, \mathsf{rm})$. We can prove that $\mathcal{ML}(\mathsf{cp}, \mathsf{rm})$-bisimilar models satisfy the same formulas.

Theorem 1 (Invariance under bisimulation.). *For all* $\mathcal{ML}(\mathsf{cp}, \mathsf{rm})$-*formula* φ, *we have* $\mathcal{M}, w \underline{\leftrightarrow}_{\mathcal{ML}(\mathsf{cp},\mathsf{rm})} \mathcal{M}', w'$ *implies* $\mathcal{M}, w \models \varphi$ *iff* $\mathcal{M}', w' \models \varphi$.

From the tree model property for $\mathcal{ML}(\lozenge^{-1})$ it immediately follows:

Corollary 1. *The language* $\mathcal{ML}(\mathsf{cp}, \mathsf{rm})$ *has the tree model property.*

3 Relation between Action Models and $\mathcal{ML}(\mathsf{cp}, \mathsf{rm}^-)$

In this section we start by formally introducing action model logic and then define an embedding into $\mathcal{ML}(\mathsf{cp}, \mathsf{rm}^-)$. We restrict ourselves to the case where preconditions in action models are Boolean.

Definition 5 (Action Models). *Let* $\mathcal{B}$ *be the set of Boolean formulas over certain set* PROP *of propositional symbols. An action model* $\mathcal{E}$ *is a structure* $\mathcal{E} = \langle E, \rightarrow, \mathsf{pre}, \mathsf{post} \rangle$, *where* E *is a non-empty finite set whose elements are called action points; for each* $a \in \mathsf{AGT}$, $\rightarrow (a) \subseteq E \times E$ *is an equivalence relation (we will often write* $\rightarrow_a$ *rather than* $\rightarrow (a)$*);* $\mathsf{pre} : E \rightarrow \mathcal{B}$ *is a pre-condition function; and* $\mathsf{post} : E \rightarrow \mathsf{PROP} \rightarrow \{\top, \bot\}$ *is a post-condition function. Let* e *be an action point in* $\mathcal{E}$, *the pair* $(\mathcal{E}, e)$ *is a pointed action model.*

Action models in action model logic appear as modalities. We will call $\mathcal{AML}$ the fragment where action models have only pre-conditions, i.e., action models of the shape $\langle E, \rightarrow, \mathsf{pre} \rangle$, and use $\mathcal{AML}^+$ for the full language.

Definition 6 (Syntax). *Let* PROP *be a countable, infinite set of propositional symbols and* AGT *a finite set of agent symbols. The set* FORM *of formulas of* $\mathcal{AML}$ *and* $\mathcal{AML}^+$ *over* PROP *and* AGT *is defined as:*

$$\mathsf{FORM} ::= \bot \mid p \mid \neg\varphi \mid \varphi \wedge \psi \mid \lozenge_a\varphi \mid [\alpha]\varphi,$$

where $p \in$ PROP, $a \in$ AGT, $\varphi, \psi \in$ FORM *and* $\alpha \in$ ACT. *The set of actions* ACT *is defined as* ACT $::= \mathcal{E}, e \mid \alpha \cup \beta$, *with* $\mathcal{E}, e$ *an action pointed model and* $\alpha, \beta \in$ ACT. $\langle \alpha \rangle \varphi$ *is a shorthand for* $\neg[\alpha]\neg\varphi$.

Definition 7 (Semantics). *Given an epistemic pointed model* $\mathcal{M}, w$ *with* $\mathcal{M} = \langle W, R, V \rangle$, *an action pointed model* $\mathcal{E}, e$ *with* $\mathcal{E} = \langle E, \rightarrow, \mathsf{pre}, \mathsf{post} \rangle$, *and a formula* φ *we say that* $\mathcal{M}, w \models \varphi$ *when*

$$\mathcal{M}, w \models [\alpha]\varphi \quad \textit{iff for all } \mathcal{M}', w' \textit{ s.t. } \mathcal{M}, w[\![\alpha]\!]\mathcal{M}', w' \textit{ we have } \mathcal{M}', w' \models \varphi$$
$$\mathcal{M}, w[\![\mathcal{E}, e]\!]\mathcal{M}', w' \textit{ iff } \mathcal{M}, w \models \mathsf{pre}(e) \textit{ and } \mathcal{M}', w' = (\mathcal{M} \otimes \mathcal{E}), (w, e)$$
$$[\![\alpha \cup \beta]\!] \qquad = [\![\alpha]\!] \cup [\![\beta]\!].$$

where $(\mathcal{M} \otimes \mathcal{E})$ *is defined as* $\langle W', R', V' \rangle$, *with:*

$$
\begin{aligned}
W' &= \{(v, d) \in W \times E \mid \mathcal{M}, v \models \mathsf{pre}(d)\} \\
((v, d), (u, f)) \in R'_a &\textit{ iff } (v, u) \in R_a \textit{ and } d \rightarrow_a f \\
V'(p) &= \{(v, d) \mid \mathcal{M}, v \models \mathsf{post}(e)(p)\}.
\end{aligned}
$$

If $\mathcal{E}$ *does not have post-conditions then* $V'(p) = \{(v, d) \mid v \in V(p)\}$.

We now show how to embed $\mathcal{AML}$ into $\mathcal{ML}(\mathsf{cp}, \mathsf{rm}^-)$. First, define the short-hand $\mathsf{rm}(\pi_1 \odot \pi_2)\varphi$ for $\mathsf{rm}(\pi_1)\mathsf{rm}(\pi_2)\varphi$. Notice that if π_1 and π_2 are paths of size 1, and given that we are only considering Boolean tests, then $\odot$ is commutative.

Definition 8. *Let* $\mathcal{E} = \langle E, \rightarrow, \mathsf{pre} \rangle$ *be an action model with* $E = \{e_1, \ldots, e_n\}$. *We define the translation* Tr *from* $\mathcal{AML}$-*formulas to* $\mathcal{ML}(\mathsf{cp}, \mathsf{rm}^-)$-*formulas as:*

$$\mathsf{Tr}([\mathcal{E}, e_1]\varphi) = \mathsf{pre}(e_1) \rightarrow \mathsf{cp}(\langle p_{e_1} \ldots p_{e_n} \rangle)\mathsf{rm}(\rho)\mathsf{rm}(\sigma)\mathsf{Tr}(\varphi),$$

where
$$
\begin{aligned}
\rho &\equiv \bigodot_{e \in E, a \in \mathsf{AGT}} \top?; a; (p_{e_i} \wedge \neg\mathsf{pre}(e_i))? \\
\sigma &\equiv \bigodot_{e_i, e_j \in E, a \in \mathsf{AGT}} p_{e_i}?; a; p_{e_j}? \quad \textit{if } e_i \not\rightarrow_a e_j.
\end{aligned}
$$

Tr *commutes with all other formulas.*

Proposition 1. *Let* φ *be an* $\mathcal{AML}$-*formula, then* φ *and* $\mathsf{Tr}(\varphi)$ *are equivalent.*

Proof (Sketch). The antecedent $\mathsf{pre}(e_1)$ is exactly the same clause as for model updates (considering the pointed action model $\mathcal{E}, e_1$ as the desired update). For each action $e_i \in E$, we consider a propositional symbol p_{e_i}. The operation $\mathsf{cp}(\langle p_{e_1} \ldots p_{e_n} \rangle)$ replicates the original model as many times as actions in E (notice that we can always use isomorphic action models to ensure that the propositional symbols used by cp are new). This operation generates the cartesian product $W \times E$. However, the model $\mathcal{M} \otimes \mathcal{E}$ does not consider the whole cartesian product. To cut the unwanted part of the model we introduce $\mathsf{rm}(\rho)$. The path expression ρ characterizes all the edges we introduced by the previous $\mathsf{cp}(\langle p_{e_1} \ldots p_{e_n} \rangle)$ pointing to p_{e_i}-states which do not satisfy the corresponding $\mathsf{pre}(e_i)$. In the same way that it is done in $\mathcal{AML}$ product updates, we remove all arrows pointing to those states. Once we have constructed the domain, it remains to restrict the obtained accessibility relation. This is done by $\mathsf{rm}(\sigma)$. Remember that $((v,d),(u,f)) \in R'_a$ in $\mathcal{M} \otimes \mathcal{E}$ if and only if $(v,u) \in R_a$ and $d \to_a f$. The first part trivially holds in the translation, because cp does not introduce edges between copies of elements that were not related in the original model. Then $\mathsf{rm}(\sigma)$ deletes all the a-edges (w_i, w_j) such that in the action model there is no a-edge from e_i to e_j, for all $a \in \mathsf{AGT}$.

The obtained model is not $\mathcal{M} \otimes \mathcal{E}$, but it is bisimilar according to bisimulation for $\mathcal{ML}$, which is the notion used in $\mathcal{AML}$. As a result, they represent the same information for the agents.   □

In Figure 2 we see the encoding above applied to a concrete update. The first step of the translation replicates as many copies of the original epistemic model, as actions belonging to the domain of the action model, obtaining the leftmost model. This is done via a copy operation. Next, evaluating $\mathsf{rm}(\rho)$ (defined as in Definition 8), we remove all the edges pointing to states where at the same time p_{e_1} holds and $\mathsf{pre}(e_1)$ does not hold, and we get the model in the center of Figure 2. Last, we need to evaluate $\mathsf{rm}(\sigma)$. This removes those edges that have been added by the copy operation, but are not connected in the original action model. Thereby, we remove all the undesirable accessibility edges, obtaining the rightmost model, which is bisimilar to the updated model of Figure 1 (the state labeled by $\{p_{e_1}, \neg p\}$ is not longer accessible).

We show now that copy and remove can be seen as action models in $\mathcal{AML}^+$. This is valuable, as it demonstrates that action models have a certain decomposition: an action model can be described as the composition of simpler action models. This decomposition can be obtained by translating first into $\mathcal{ML}(\mathsf{cp}, \mathsf{rm}^-)$ and then considering copy and remove again as basic action models.

Consider the copy action $\mathsf{cp}(\bar{p})$, and let Q be the set of all propositional symbols occurring in $\bar{p}$. The copy operator can be modeled as an action model $\mathcal{E}(\mathsf{cp}(\bar{p})) = \langle E, \to, \mathsf{pre}, \mathsf{post} \rangle$ such that (for all $q \in E = Q$, $a \in \mathsf{AGT}$):

$$
\begin{aligned}
E &= Q & \mathsf{pre}(q) &= \top \\
\to_a &= E \times E & \mathsf{post}(q)(q) &= \top \\
& & \mathsf{post}(q)(p) &= \bot \quad \text{for } p \in Q \setminus \{q\}.
\end{aligned}
$$

We note that for all $r \in \mathsf{PROP} \setminus Q$ the value is not affected at the execution of this action, as the finite subset of propositional symbols that is assigned a post-condition is the set Q. Consider the translation $': \mathcal{ML}(\mathsf{cp}) \to \mathcal{AML}$ such that $(\mathsf{cp}(\bar{p})\phi)' = [\mathcal{E}(\mathsf{cp}(\bar{p}))]\phi'$ and commutes with all other operators. Then:

Proposition 2. *For all $\varphi \in \mathcal{ML}(\mathsf{cp})$, φ and φ' are equivalent.*

Next, we study the remove action. The action model $\mathcal{E}(\mathsf{rm}(\phi?; a; \psi?)) = \langle E, \to, \mathsf{pre} \rangle$ is defined as

$$
\begin{aligned}
E &= \{00, 10, 01, 11\} & \mathsf{pre}(00) &= \neg\phi \wedge \neg\psi, \\
\to_a &= (E \times E) \setminus \{(10, 01), (10, 11), (11, 01), (11, 11)\} & \mathsf{pre}(10) &= \phi \wedge \neg\psi, \\
\to_b &= (E \times E) \quad \text{for all } b \neq a & \mathsf{pre}(01) &= \neg\phi \wedge \psi, \\
& & \mathsf{pre}(11) &= \phi \wedge \psi.
\end{aligned}
$$

This action model corresponds to the operation of removing all $\phi \xrightarrow{a} \psi$ arrows. Consider the translation $'' : \mathcal{ML}(\mathsf{rm}) \to \mathcal{AML}$ such that $(\mathsf{rm}(\phi?; a; \psi?)\theta)'' = [\mathcal{E}(\mathsf{rm}(\phi?; a; \psi?))]\theta''$ and commutes with all other operators.

Proposition 3. *For all $\varphi \in \mathcal{ML}(\mathsf{rm})$, φ and φ'' are equivalent.*

4   Complexity of Deciding Satisfiability

The following result has been proved already in [6]:

Theorem 2. *Deciding if a formula in $\mathcal{ML}(\mathsf{cp})$ is satisfiable is* PSPACE-*complete.*

We will show that $\mathcal{ML}(\mathsf{rm})$ can be translated into $\mathcal{ML}(\Diamond^{-1})$, the basic modal logic $\mathcal{ML}$ with the past operator $\Diamond^{-1}$. It is easy to see that as tests are Boolean, if two tests are consecutive in a path expression (e.g., $\varphi_1?; \varphi_2?$), we can replace them by a single test (e.g., $(\varphi_1 \wedge \varphi_2)?$). If two agents are consecutive in a path expression (e.g., $a_1; a_2$) we can add a trivial test between them (e.g., $a_1; \top?; a_2$). Thus, without loss of generality we assume that all delete operators have the form

$$\mathsf{rm}(\varphi_1?; a_1; \varphi_2?; a_2; \ldots; a_{n-1}; \varphi_n?)\psi,$$

where $\varphi_i?$ are arbitrary Boolean formulas, and $a_i \in \mathsf{AGT}$. We introduce reduction axioms to get an $\mathcal{ML}(\Diamond^{-1})$-formula, and conclude that any $\mathcal{ML}(\mathsf{rm})$-formula, is equivalent to an $\mathcal{ML}(\Diamond^{-1})$-formula.

First, let us define the abbreviations $\Diamond_{i,j}$, $\Diamond_{i,j}^{-1}$, for a fix path expression $\pi = \varphi_1?; a_1; \ldots; a_{n-1}; \varphi_n?$:

$$\Diamond_{i,j} = \begin{cases} \top & j < i \\ \Diamond_{a_i}\varphi_{i+1} & i = j \\ \Diamond_{a_i}(\varphi_{i+1} \wedge \Diamond_{i+1,j}) & i < j \end{cases} \quad \Diamond_{i,j}^{-1} = \begin{cases} \top & j < i \\ \Diamond_{a_i}^{-1}\varphi_i & i = j \\ \Diamond_{a_j}^{-1}(\Diamond_{i,j-1}^{-1} \wedge \varphi_j) & i < j \end{cases}$$

Now define $rm_i^{\pi} = \Diamond_{1,i-1}^{-1} \wedge \varphi_i \wedge \Diamond_{i,n-1}$. Informally rm_i^{π} means "the current state is at position i in a path that matches $\pi = \varphi_1?; a_1; \varphi_2?; a_2; \ldots; a_{n-1}; \varphi_n?$ which is going to be deleted". For instance, rm_i^{π}, $1 \leq i \leq n$ are defined as:

$$\begin{aligned} rm_1^{\pi} &= \varphi_1 \wedge (\Diamond_{a_1}\varphi_2 \wedge (\Diamond_{a_2}\varphi_3 \ldots \wedge \Diamond_{a_{n-2}}(\varphi_{n-1} \wedge \Diamond_{a_{n-1}}\varphi_n)\ldots)) \\ rm_2^{\pi} &= \Diamond_{a_1}^{-1}\varphi_1 \wedge \varphi_2 \wedge (\Diamond_{a_2}\varphi_3 \ldots \wedge \Diamond_{a_{n-2}}(\varphi_{n-1} \wedge \Diamond_{a_{n-1}}\varphi_n)\ldots) \\ &\quad \ldots \\ rm_{n-1}^{\pi} &= \Diamond_{a_{n-2}}^{-1}(\Diamond_{a_1}^{-1}(\ldots(\Diamond_{a_1}^{-1}\varphi_1 \wedge \varphi_2) \wedge \varphi_3)\ldots) \wedge \varphi_{n-1} \wedge \Diamond_{a_{n-1}}\varphi_n \\ rm_n^{\pi} &= \Diamond_{a_{n-1}}^{-1}(\Diamond_{a_{n-2}}^{-1}(\ldots(\Diamond_{a_1}^{-1}\varphi_1 \wedge \varphi_2) \wedge \varphi_3 \ldots) \wedge \varphi_{n-1}) \wedge \varphi_n. \end{aligned}$$

Lemma 3. *Let $\mathcal{M} = \langle W, R, V \rangle$ be a model, $w \in W$ and $\pi = \varphi_1?; a_1; \varphi_2?; \ldots; \varphi_n?$ a path expression. Let i be such that $0 \leq i \leq n$, then*

$$\mathcal{M}, w \models rm_i^{\pi} \quad \text{iff there is some } P \in \mathcal{P}_{\pi}^{\mathcal{M}} \text{ s.t. } P = w_1 a_1 w_2 \ldots w_n, \ w_i = w$$

and for all $w_j \in P$ we have $\mathcal{M}, w_j \models \varphi_j$.

Definition 9. *Let* $\pi = \varphi_1?; a_1; \varphi_2?; \ldots; \varphi_n?$, $\varphi = \mathsf{rm}(\pi)\theta$ *be an* $\mathcal{ML}(\mathsf{rm}, \Diamond^{-1})$-*formula*[1]. *We define* $\mathsf{Tr}(\varphi)$ *as the* $\mathcal{ML}(\Diamond^{-1})$-*formula resulting of repeatedly applying the following reduction axioms to* φ *(we assume that* $\Diamond_a \psi$ *is written as* $\neg\Box_a \neg\psi$, *and similarly for* $\Diamond^{-1}$).

(1) $\mathsf{rm}(\pi)p \quad\quad\ \leftrightarrow p, \quad p \in \mathsf{PROP}$

(2) $\mathsf{rm}(\pi)\neg\psi \quad\quad \leftrightarrow \neg\mathsf{rm}(\pi)\psi$

(3) $\mathsf{rm}(\pi)(\psi \wedge \psi') \leftrightarrow (\mathsf{rm}(\pi)\psi \wedge \mathsf{rm}(\pi)\psi')$

(4) $\mathsf{rm}(\pi)\Box_a\psi \quad \leftrightarrow \Box_a\mathsf{rm}(\pi)\psi, \quad \text{if } a \notin \pi$

(5) $\mathsf{rm}(\pi)\Box_a^{-1}\psi \quad \leftrightarrow \Box_a^{-1}\mathsf{rm}(\pi)\psi, \quad \text{if } a \notin \pi$

(6) $\mathsf{rm}(\pi)\Box_a\psi \quad \leftrightarrow (\bigwedge_{i\in\{1,\ldots,n-1 \mid a_i = a\}} \neg rm_i^\pi \rightarrow \Box_{a_i}\mathsf{rm}(\pi)\psi)\wedge$
$\quad\quad\quad\quad\quad\quad\quad (\bigwedge_{i\in\{1,\ldots,n-1 \mid a_i = a\}}(rm_i^\pi \rightarrow \Box_{a_i}(rm_{i+1}^\pi \vee \mathsf{rm}(\pi)\psi)))$

(7) $\mathsf{rm}(\pi)\Box_a^{-1}\varphi \quad \leftrightarrow (\bigwedge_{i\in\{1,\ldots,n-1 \mid a_i = a\}} \neg rm_i^\pi \rightarrow \Box_{a_i}^{-1}\mathsf{rm}(\pi)\psi)\wedge$
$\quad\quad\quad\quad\quad\quad\quad (\bigwedge_{i\in\{1,\ldots,n-1 \mid a_i = a\}}(rm_i^\pi \rightarrow \Box_{a_i}^{-1}(rm_{i-1}^\pi \vee \mathsf{rm}(\pi)\psi))).$

Notice that the resulting formula only contains $\Box_a$ and $\Box_a^{-1}$, and does not contain rm. We will prove that the reduction axioms preserves equivalence. The reduction axioms introduced in Definition 9 are justified by the next proposition.

Proposition 4. *Formulas (1) to (7) in Definition 9 are valid.*

The next proposition establishes that we can reduce $\mathcal{ML}(\mathsf{rm}, \Diamond^{-1})$-formulas according to axioms of Definition 9, obtaining an equivalent $\mathcal{ML}(\Diamond^{-1})$-formula. The proof is a direct corollary of Proposition 4.

Proposition 5. *Let* $\mathcal{M} = \langle W, R, V \rangle$ *a model,* $w \in W$ *and* φ *a* $\mathcal{ML}(\mathsf{rm}, \Diamond^{-1})$-*formula. Then* $\mathcal{M}, w \models \varphi$ *iff* $\mathcal{M}, w \models \mathsf{Tr}(\varphi)$.

The next theorem now follows.

Theorem 3. *The satisfiability problem for* $\mathcal{ML}(\mathsf{rm})$ *is decidable.*

The reduction axioms that relate $\Box_a$ and $\mathsf{rm}(\pi)$ produce an exponential blow up in the size of the formula. If we consider only path expressions π of size 1, i.e., we consider the fragment $\mathcal{ML}(\mathsf{rm}^-)$, we can avoid the exponential blow up, and prove that the satisfiability problem is PSPACE-complete.

Proposition 6. *Let* $\mathcal{M} = \langle W, R, V \rangle$ *be a model,* θ, φ *and* ψ *be* $\mathcal{ML}(\mathsf{rm}^-)$-*formulas and* $a \in \mathsf{AGT}$. *Then*

$$\mathcal{M}, w \models \mathsf{rm}(\varphi?; a; \psi?)\Box_a\theta \text{ iff } \mathcal{M}, w \models \Box_a((\psi \wedge \Diamond^{-1}\varphi) \vee \mathsf{rm}(\varphi?; a; \psi?)\theta).$$

We showed that there is a polynomial translation from $\mathcal{ML}(\mathsf{cp}, \mathsf{rm}^-)$ into a dynamic epistemic modal logic with action models with both pre-conditions and post-conditions, that preserves satisfiability. In [3], it is proved that the satisfiability problem for dynamic epistemic modal logic with action models with pre-conditions and *without* post-conditions is in NEXPTIME. We can handle post-conditions in NEXPTIME adapting the tableau method of [3][2].

---

[1] Let $\mathcal{ML}(\mathsf{rm}, \Diamond^{-1})$ be the fragment $\mathcal{ML}(\mathsf{rm})$ extended with the past operator $\Diamond^{-1}$.

[2] A similar result was shown in [12] for public announcement enriched with public assignments which are similar to post-conditions.

Theorem 4. *The satisfiability problem for* $\mathcal{AML}^+$ *is in* NExpTime.

Then we can state:

Corollary 2. *The satisfiability problem for* $\mathcal{ML}(\mathsf{cp}, \mathsf{rm}^-)$ *is in* NExpTime.

As there is a polynomial translation from dynamic epistemic modal logic without post-conditions $\mathcal{AML}$ into $\mathcal{ML}(\mathsf{cp}, \mathsf{rm}^-)$ that preserves satisfiability, the satisfiability problem of a formula in $\mathcal{ML}(\mathsf{cp}, \mathsf{rm}^-)$ is NExpTime-hard.

Theorem 5. *The satisfiability problem for* $\mathcal{ML}(\mathsf{cp}, \mathsf{rm}^-)$ *is* NExpTime-*complete.*

5   Conclusion

We proposed the dynamic modal logic $\mathcal{ML}(\mathsf{cp}, \mathsf{rm})$ which contains copy and remove operators: the copy operator copies an input model, and the remove operator deletes all paths from an input model that are characterized by a given expression. We investigated some model theoretic properties of $\mathcal{ML}(\mathsf{cp}, \mathsf{rm})$ such as bisimulations. In order to give an appropriate notion of bisimulation, we need the same conditions as for the $\Diamond^{-1}$ operator, because we need to differentiate states with respect to the paths that traverse them.

We showed that the action model logic $\mathcal{AML}$, one of the best-known dynamic epistemic logics, can be polynomially embedded in the fragment $\mathcal{ML}(\mathsf{cp}, \mathsf{rm}^-)$ when we consider action models with only Boolean pre-conditions. The restriction to Boolean pre-conditions is certainly a limitation. We consider this to be the first step into a complete understanding of the full language. The embedding simulates every finite action model with a combination of copy and remove operators. As we mentioned, the embedding can be done within $\mathcal{ML}(\mathsf{cp}, \mathsf{rm}^-)$ as it only requires single step removals (i.e., only paths of length one are needed). We showed that the copy and one-step removal themselves correspond to particular action models. As a result we obtain a kind of normal form for action models. By decomposing product updates in sequences of copy and remove operators, it would be possible to characterize large syntactic fragments of $\mathcal{AML}$ with interesting complexities for the satisfiability problem.

We demonstrated that the complexity of the satisfiability of the full language $\mathcal{ML}(\mathsf{cp}, \mathsf{rm})$ is NExpTime-hard. The upper bound of this satisfiability problem is still open, but we conjecture that it is decidable. We proved that satisfiability for the fragment $\mathcal{ML}(\mathsf{rm}^-)$ is decidable, that it is PSpace-complete for $\mathcal{ML}(\mathsf{cp})$, and that it is NExpTime-complete for $\mathcal{ML}(\mathsf{cp}, \mathsf{rm}^-)$.

As future work, we plan to extend the analysis of $\mathcal{AML}$ via its embedding in $\mathcal{ML}(\mathsf{cp}, \mathsf{rm})$. In particular, we will address the general case in which action model pre-conditions can be arbitrary formulas of lower complexity. The main challenge when considering the full language is that when pre-conditions are not Boolean, successive applications of the rm operator are no longer independent of each other, and a more involved mapping into $\mathcal{ML}(\mathsf{cp}, \mathsf{rm})$ is required.

Acknowledgments. We thank the reviewers for their comments, and for their encouragement to address the fully modal version. We acknowledge support from ERC project EPS 313360, from EU 7th Framework Programme under grant agreement no. 295261 (MEALS), and the Laboratoire Internationale Associé "INFINIS".

References

1. Areces, C., Fervari, R., Hoffmann, G.: Moving arrows and four model checking results. In: Ong, L., de Queiroz, R. (eds.) WoLLIC 2012. LNCS, vol. 7456, pp. 142–153. Springer, Heidelberg (2012)
2. Areces, C., Fervari, R., Hoffmann, G.: Swap logic. Logic Journal of the IGPL 22(2), 309–332 (2014)
3. Aucher, G., Schwarzentruber, F.: On the complexity of dynamic epistemic logic. In: Proceedings of TARK 2013, Chennai, India (January 2013)
4. Baltag, A., Moss, L., Solecki, S.: The logic of public announcements, common knowledge and private suspicions. In: Gilboa, I. (ed.) TARK 1998, Evanstin, IL, USA, pp. 43–56. Morgan Kaufmann (July 1998)
5. Blackburn, P., de Rijke, M., Venema, Y.: Modal Logic. Cambridge Tracts in Theoretical Computer Science, vol. 53. Cambridge University Press, Cambridge (2001)
6. Fervari, R.: Relation-Changing Modal Logics. PhD thesis, Facultad de Matemática Astronomía y Física, Universidad Nacional de Córdoba, Córdoba, Argentina (March 2014)
7. Goré, R.: Tableau methods for modal and temporal logics. In: Handbook of Tableau Methods, pp. 297–396 (1999)
8. Kooi, B., Renne, B.: Arrow update logic. Review of Symbolic Logic 4(4), 536–559 (2011)
9. Kooi, B., Renne, B.: Generalized arrow update logic. In: Apt, K. (ed.) TARK, pp. 205–211. ACM (2011)
10. van Benthem, J.: An essay on sabotage and obstruction. In: Hutter, D., Stephan, W. (eds.) Mechanizing Mathematical Reasoning. LNCS (LNAI), vol. 2605, pp. 268–276. Springer, Heidelberg (2005)
11. van Ditmarsch, H., French, T.: Simulation and information: Quantifying over epistemic events. In: Meyer, J.-J.C., Broersen, J. (eds.) KRAMAS 2008. LNCS, vol. 5605, pp. 51–65. Springer, Heidelberg (2009)
12. van Ditmarsch, H., Herzig, A., De Lima, T.: Public announcements, public assignments and the complexity of their logic. Journal of Applied Non-Classical Logics 22(3), 249–273 (2012)
13. van Ditmarsch, H., van der Hoek, W., Kooi, B.: Dynamic Epistemic Logic. Kluwer (2007)

Appendix

Proofs of Section 2

Without loss of generality we assume that all remove operators have the normal form $\text{rm}(\varphi_1?; a_1; \varphi_2?; a_2; \ldots; a_{n-1}; \varphi_n?)\psi$, where $\varphi_i?$ are arbitrary Boolean formulas, and $a_i \in \text{AGT}$ (we can always add $\top?$ and conjunctions to get this normal form). We introduce two lemmas that will be helpful in the proof Theorem 1.

Lemma 1. *Let* $\mathcal{M} = \langle W, R, V \rangle$ *and* $\mathcal{M}' = \langle W', R', V' \rangle$ *be models,* $w \in W$, $w' \in W'$, *be such that* $\mathcal{M}, w \underline{\leftrightarrow}_{\mathcal{ML}(\text{cp},\text{rm})} \mathcal{M}', w'$, *and* $\pi = \varphi_1?; a_1; \varphi_2?; a_2; \ldots; a_{n-1}; \varphi_n?$. *Then, for all* $P \in \mathcal{P}^{\mathcal{M}}(\pi)$ *such that* $P = w_0 a_0 \ldots \mathbf{w} a_i \ldots w_n$, *there is some* $P' \in \mathcal{P}^{\mathcal{M}'}(\pi)$, *with* $P' = w_0' a_0 \ldots \mathbf{w}' a_i \ldots w_n'$ *and for all* $j \in \{1, \ldots, n\}$ *we have* $\mathcal{M}, w_j \underline{\leftrightarrow}_{\mathcal{ML}(\text{cp},\text{rm})} \mathcal{M}', w_j'$.

Proof. Given some $P \in \mathcal{P}_\pi^{\mathcal{M}}$, we need to find some $P' \in \mathcal{P}_\pi^{\mathcal{M}'}$ satisfying the lemma. Let us construct such P'.

Suppose $P = w_0 a_0 \ldots \mathbf{w} a_i \ldots w_n$. Notice that we have the subpath $w a_i w_{i+1}$, which means $(w, w_{i+1}) \in R_{a_i}$. Because $\mathcal{M}, w \underline{\leftrightarrow}_{\mathcal{ML}(\text{cp},\text{rm})} \mathcal{M}', w'$, by (zig) there is some w_{i+1}' such that $(w', w_{i+1}') \in R_{a_i}'$ and $\mathcal{M}, w_{i+1} \underline{\leftrightarrow}_{\mathcal{ML}(\text{cp},\text{rm})} \mathcal{M}', w_{i+1}'$. For this reason, $\mathcal{M}, w_{i+1} \models \psi$ if and only if $\mathcal{M}', w_{i+1}' \models \psi$, for all ψ (in particular φ_{i+1}). Then, w_{i+1} is a good choice in order to construct P'. We can repeat this process to build the subpath $w' a_i w_{i+1}' \ldots w_n'$. In order to choose w_{i-1}, we can proceed in the same way but using (zig^{-1}), and repeating the process until we reach w_1'. Putting all together, we have constructed the right P'.

For the other direction use (zag) and (zag^{-1}). $\square$

Lemma 2. *Let* $\mathcal{M} = \langle W, R, V \rangle$ *and* $\mathcal{M}' = \langle W', R', V' \rangle$ *be two models,* $w \in W$ *and* $w' \in W'$. *Then* $\mathcal{M}, w \underline{\leftrightarrow}_{\mathcal{ML}(\text{cp},\text{rm})} \mathcal{M}', w'$ *implies* $\mathcal{M}_{\text{cp}(\bar{p})}, (w, q) \underline{\leftrightarrow}_{\mathcal{ML}(\text{cp},\text{rm})} \mathcal{M}'_{\text{cp}(\bar{p})}, (w', q)$.

Proof. We have to define a bisimulation $Z \subseteq W_{\text{cp}(\bar{p})} \times W'_{\text{cp}(\bar{p})}$. Because we have $\mathcal{M}, w \underline{\leftrightarrow}_{\mathcal{ML}(\text{cp},\text{rm})} \mathcal{M}', w'$, we define:

$$Z = \{((v, q), (v', q)) \mid (v, q), (v', q) \in W_{\text{cp}(\bar{p})}, \text{ s.t. } \mathcal{M}, v \underline{\leftrightarrow}_{\mathcal{ML}(\text{cp},\text{rm})} \mathcal{M}', v'\}.$$

(atomic harmony) holds because $(v, q)Z(v', q)$ if and only if v and v' satisfy (atomic harmony) in the original models, and (v, q) and (v', q) are both labeled by the symbol q. For (zig), suppose we have $(v, q)Z(v', q)$ and $((v, q), (u, r)) \in (R_{\text{cp}(\bar{p})})_a$. Then we know $(v, u) \in R_a$. Because $\mathcal{M}, v \underline{\leftrightarrow}_{\mathcal{ML}(\text{cp},\text{rm})} \mathcal{M}', v'$, by (zig) there is some u' such that $(v', u') \in R_a'$. Hence, we have $((v', q), (u', r)) \in (R'_{\text{cp}(\bar{p})})_a$. (zag) is straightforward. $\square$

Then we can state:

Theorem 1 (Invariance under bisimulation.). *For all* $\mathcal{ML}(\text{cp}, \text{rm})$-*formula* φ, *we have* $\mathcal{M}, w \underline{\leftrightarrow}_{\mathcal{ML}(\text{cp},\text{rm})} \mathcal{M}', w'$ *implies* $\mathcal{M}, w \models \varphi$ *iff* $\mathcal{M}', w' \models \varphi$.

Proof. The proof is by structural induction. Let $\mathcal{M} = \langle W, R, V \rangle$ and $\mathcal{M}' = \langle W', R', V' \rangle$, such that $\mathcal{M}, w \leftrightarrow_{\mathcal{ML}(\mathsf{cp},\mathsf{rm})} \mathcal{M}', w'$.

We only prove the inductive cases for rm and cp.

$\mathbf{rm}(\pi)\varphi$: Suppose $\mathcal{M}, w \models \mathsf{rm}(\pi)\varphi$, then $\mathcal{M}_{\mathsf{rm}(\pi)}, w \models \varphi$, where $\mathcal{M}_{\mathsf{rm}(\pi)} = \langle W, R_{\mathsf{rm}(\pi)}, V \rangle$, $R_{\mathsf{rm}(\pi)} = R \setminus \bigcup_{P \in \mathcal{P}_\pi^{\mathcal{M}}, a \in \mathsf{AGT}} edges_a(P)$. $\mathcal{M}, w \leftrightarrow_{\mathcal{ML}(\mathsf{cp},\mathsf{rm})} \mathcal{M}', w'$ by hypothesis, then (by Lemma 1) there is $P \in \mathcal{P}_\pi^{\mathcal{M}}$ iff there is $P' \in \mathcal{P}_\pi^{\mathcal{M}'}$. Hence $\mathcal{M}_{\mathsf{rm}(\pi)}, w \leftrightarrow_{\mathcal{ML}(\mathsf{cp},\mathsf{rm})} \mathcal{M}'_{\mathsf{rm}(\pi)}, w'$, and by I.H. $\mathcal{M}'_{\mathsf{rm}(\pi)}, w' \models \varphi$. As a result, $\mathcal{M}', w' \models \mathsf{rm}(\pi)\varphi$.

$\mathbf{cp}(\bar{p}, q)\varphi$: Suppose $\mathcal{M}, w \models \mathsf{cp}(\bar{p}, q)\varphi$. Then we have $\mathcal{M}_{\mathsf{cp}(\bar{p})}, (w, q) \models \varphi$. By $\mathcal{M}, w \leftrightarrow_{\mathcal{ML}(\mathsf{cp},\mathsf{rm})} \mathcal{M}', w'$ and Lemma 2, we have $\mathcal{M}_{\mathsf{cp}(\bar{p})}, (w, q) \leftrightarrow_{\mathcal{ML}(\mathsf{cp},\mathsf{rm})} \mathcal{M}'_{\mathsf{cp}(\bar{p})}, (w', q)$. By I.H. $\mathcal{M}'_{\mathsf{cp}(\bar{p})}, (w', q) \models \varphi$. Therefore, $\mathcal{M}', w' \models \mathsf{cp}(\bar{p}, q)\varphi$.

$\square$

Proofs of Section 4

Theorem 3. *Deciding if a formula in $\mathcal{ML}(\mathsf{cp})$ is satisfiable is* PSPACE-*complete.*

Proof (Sketch). Adapt the classic tableau-based algorithm for the basic modal logic (see [7]) to manage sequences of propositional symbols which represent possible copies of the model. As for the original algorithm, it takes polynomial time. $\square$

Lemma 3. *Let* $\mathcal{M} = \langle W, R, V \rangle$ *be a model,* $w \in W$ *and* $\pi = \varphi_1?; a_1; \varphi_2?; \ldots; \varphi_n?$ *a path expression. Let* i *be such that* $0 \leq i \leq n$, *then*

$$\mathcal{M}, w \models rm_i^\pi \text{ iff there is some } P \in \mathcal{P}_\pi^{\mathcal{M}} \text{ s.t. } P = w_1 a_1 w_2 \ldots w_n, \ w_i = w$$

and for all $w_j \in P$ *we have* $\mathcal{M}, w_j \models \varphi_j$.

Proof. The proof is by induction on the length of π:

$\pi = \varphi_1?$: $\mathcal{M}, w \models rm_1^\pi$ if and only if $\mathcal{M}, w \models \varphi_1$ (by definition of rm_i^π). But $\mathcal{P}_{\varphi_1?}^{\mathcal{M}} = \{v \mid \mathcal{M}, v \models \varphi_1\}$ (all the paths are singletons satisfying φ_1), then $w \in \mathcal{P}_{\varphi_1?}^{\mathcal{M}}$.

$\pi = \varphi_1?; a_1; \varphi_2?; \ldots; \varphi_n?$: Suppose $\mathcal{M}, w \models rm_i^\pi$. By definition of rm^π, we have $\mathcal{M}, w \models \Diamond_{1,i-1}^{-1} \wedge \varphi_i \wedge \Diamond_{i,n-1}$. Now, we know:
1. $\mathcal{M}, w \models \varphi_i$.
2. $\mathcal{M}, w \models \Diamond_{1,i-1}^{-1}$, then by definition of $\Diamond_{i,j}^{-1}$ we have $\mathcal{M}, w \models \Diamond_{a_{i-1}}^{-1}(\Diamond_{1,i-2}^{-1} \wedge \varphi_{i-1})$. By definition of $\models$, there is some $v \in W$ such that $(v, w) \in R_{a_{i-1}}$ and $\mathcal{M}, v \models \Diamond_{1,i-2}^{-1} \wedge \varphi_{i-1}$. Let us define $\pi_1 = \varphi_1?; a_1; \varphi_2?; \ldots; \varphi_{i-1}?$. Then, by definition of rm_i^π, we have $\mathcal{M}, v \models rm_{i-1}^{\pi_1}$, and by I.H., there is a path $P_1 \in \mathcal{P}_{\pi_1}^{\mathcal{M}}$ such that $P_1 = w_1 a_1 \ldots w_{i-1}$, with $w_{i-1} = v$ and for all $w_j \in P_1$, $\mathcal{M}, w_j \models \varphi_j$ ($0 \leq j \leq i-1$).
3. $\mathcal{M}, w \models \Diamond_{i,n-1}$, then by definition of $\Diamond_{i,j}$ we have $\mathcal{M}, w \models \Diamond_{a_i}(\varphi_{i+1} \wedge \Diamond_{i+1,n-1})$. By definition of $\models$, there is some $t \in W$ such that $(w, t) \in R_{a_i}$

and $\mathcal{M}, t \models \varphi_{i+1} \wedge \Diamond_{i+1,n-1}$. Let us define $\pi_2 = \varphi_{i+1}?; a_{i+1}; \ldots; \varphi_n?$. Then, by definition of rm^π, we have $\mathcal{M}, t \models rm_{i+1}^{\pi_2}$, and by I.H., there is a path $P_2 \in \mathcal{P}_{\pi_2}^{\mathcal{M}}$ such that $P_2 = w_{i+1} a_{i+1} \ldots w_n$, with $w_{i+1} = t$ and for all $w_j \in P_2$, $\mathcal{M}, w_j \models \varphi_j$ $(i+1 \leq j \leq n)$.

Notice that $\pi = \pi_1; a_{i-1}; \varphi_i?; a_i; \pi_2$. It remains to choose $P = P_1 a_{i-1} w_i a_i P_2$ and we have what we wanted.

$\square$

Proposition 4. *Formulas (1) to (7) in Definition 9 are valid.*

Proof. We prove each of them separately:

1. Suppose $\mathcal{M}, w \models rm(\pi)p$. By definition of $\models$, we have $\mathcal{M}_{rm(\pi)}, w \models p$. Because $rm(\pi)$ keeps the same valuation in the updated model, $w \in V(p)$. Then (by $\models$), $\mathcal{M}, w \models p$.

2. Follows from the self-duality of rm, which is trivial given that it is a global operator.

3. Suppose $\mathcal{M}, w \models rm(\pi)(\psi \wedge \psi')$. Then, by definition of $\models$, $\mathcal{M}_{rm(\pi)}, w \models (\psi \wedge \psi')$, which means $\mathcal{M}_{rm(\pi)}, w \models \psi$ and $\mathcal{M}_{rm(\pi)}, w \models \psi'$. Applying again definition of $\models$, we have $\mathcal{M}, w \models rm(\pi)\psi$ and $\mathcal{M}, w \models rm(\pi)\psi'$, iff $\mathcal{M}, w \models rm(\pi)\psi \wedge rm(\pi)\psi'$.

4. (5 is straightforward). Suppose $\mathcal{M}, w \models rm(\pi)\square_{a_i}\psi$. Applying definition of $\models$ twice, we have that for all v such that $(w, v) \in (R_{rm(\pi)})_{a_i}$, $\mathcal{M}_{rm(\pi)}, v \models \psi$. We assume $a_i \notin \pi$, then $(w, v) \in (R_{rm(\pi)})_{a_i}$ iff $(w, v) \in R_{a_i}$, then we have for all v such that $(w, v) \in R_{a_i}$, $\mathcal{M}_{rm(\pi)}, v \models \psi$, iff for all v such that $(w, v) \in R_{a_i}$, $\mathcal{M}, v \models rm(\pi)\psi$. Hence by $\models$, $\mathcal{M}, w \models \square_{a_i} rm(\pi)\psi$.

6. (7 is straightforward). Let $\mathcal{M} = \langle W, R, V \rangle$ be a model, $w \in W$, and let $rm(\pi)\square_{a_i}\psi$ be an $\mathcal{ML}(rm, \Diamond^{-1})$-formula with $\pi = \varphi_1?; a_1; \varphi_2?; \ldots; \varphi_n?$, such that $a_i \in \pi$. We want to prove

$$\mathcal{M}, w \models rm(\pi)\square_{a_i}\psi \text{ iff } \mathcal{M}, w \models \delta \wedge \delta'$$

where
$$\delta = \bigwedge_{k \in \{1, \ldots, n-1 \mid a_k = a_i\}} \neg rm_k^\pi \rightarrow \square_{a_k} rm(\pi)\psi$$
$$\delta' = \bigwedge_{k \in \{1, \ldots, n-1 \mid a_k = a_i\}} (rm_k^\pi \rightarrow \square_{a_k}(rm_{k+1}^\pi \vee rm(\pi)\psi)).$$

Let us suppose that $\mathcal{M}, w \models rm(\pi)\square_{a_i}\psi$. Then, by definition of $\models$, we have that for all $v \in W$ such that $(w, v) \in (R_{rm(\pi)})_{a_i}$, $\mathcal{M}_{rm(\pi)}, v \models \psi$. We will check the two conjuncts δ and δ' separately (for the other direction of the iff, we can assume the two conjuncts together and use the same steps):

1. Suppose $\mathcal{M}, w \models \bigwedge_{k \in \{1, \ldots, n-1 \mid a_k = a_i\}} \neg rm_k^\pi$. By definition of $\models$, we have $\mathcal{M}, w \not\models \bigvee_{k \in \{1, \ldots, n-1 \mid a_k = a_i\}} rm_k^\pi$. It means that there is no $P \in \mathcal{P}_\pi^\mathcal{M}$ satisfying Lemma 3, such that $w \in P$, hence no deletions have been done traversing w. Then for all $v \in W$, $(w, v) \in R_{a_i}$ iff $(w, v) \in (R_{rm(\pi)})_{a_i}$. Because we have for all $v \in W$ such that $(w, v) \in (R_{rm(\pi)})_{a_i}$, $\mathcal{M}_{rm(\pi)}, v \models \psi$, then for all $v \in W$ such that $(w, v) \in R_{a_i}$, $\mathcal{M}_{rm(\pi)}, v \models \psi$. Therefore, we have for all $v \in W$ such that

$(w, v) \in R_{a_i}$, $\mathcal{M}, v \models \mathrm{rm}(\pi)\psi$, then (by $\models$) $\mathcal{M}, w \models \Box_{a_i}\mathrm{rm}(\pi)\psi$.

2. Suppose now for some arbitrary k, $\mathcal{M}, w \models rm_k^\pi$, where $k \in \{1, \ldots, n-1 \mid a_k = a_i\}$. By Lemma 3 it means that there is a path traversing w that has been deleted. We also know $\mathcal{M}_{\mathrm{rm}(\pi)}, w \models \Box_{a_k}\psi$ by assumption and $k = i$, then for all $v \in W$ such that $(w, v) \in (R_{\mathrm{rm}(\pi)})_{a_k}$, $\mathcal{M}_{\mathrm{rm}(\pi)}, v \models \psi$. Then, for all $u \in W$ such that $(w, u) \in R_{a_k}$, either $\mathcal{M}_{\mathrm{rm}(\pi)}, u \models \psi$ or $u \in P$, with $P \in \mathcal{P}_\pi^\mathcal{M}$, and u is at position $k + 1$ (because w is at position $k = i$), i.e., $\mathcal{M}, u \models rm_{k+1}^\pi$ (by Lemma 3). Therefore, $\mathcal{M}, w \models \Box_{a_k}(rm_{k+1}^\pi \lor \mathrm{rm}(\pi)\psi)$. $\qquad\square$

Proposition 6. *Let $\mathcal{M} = \langle W, R, V \rangle$ be a model, θ be a $\mathcal{ML}(\mathrm{rm}^-)$-formula, φ and ψ be Boolean formulas and $a \in \mathsf{AGT}$. Then*

$$\mathcal{M}, w \models \mathrm{rm}(\varphi?; a; \psi?)\Box_a\theta \text{ iff } \mathcal{M}, w \models \Box_a((\psi \land \Diamond^{-1}\varphi) \lor \mathrm{rm}(\varphi?; a; \psi?)\theta).$$

Proof. Let us suppose that $\mathcal{M}, w \models \mathrm{rm}(\varphi?; a; \psi?)\Box_a\theta$. Then, we have that for all $v \in W$ s.t. $(w, v) \in (R_{\mathrm{rm}(\varphi?;a;\psi?)})_a$, $\mathcal{M}_{\mathrm{rm}(\varphi?;a;\psi?)}, w \models \theta$ $\otimes$. Let u be s.t. $(w, u) \in R_a$, and let suppose $\mathcal{M}, u \models \neg(\psi \land \Diamond^{-1}\varphi)$. This means that $(w, u) \in R$ iff $(w, u) \in (R_{\mathrm{rm}(\varphi?;a;\psi?)})_a$. Then (by $\otimes$) $\mathcal{M}_{\mathrm{rm}(\varphi?;a;\psi?)}, u \models \theta$ iff (by $\models$) $\mathcal{M}, u \models \mathrm{rm}(\varphi?; a; \psi?)\theta$, iff $\mathcal{M}, w \models \Box_a(\neg(\psi \land \Diamond_a^{-1}\varphi) \to \mathrm{rm}(\varphi?; a; \psi?)\theta)$. $\qquad\square$

Theorem 4. *The satisfiability problem for $\mathcal{AML}^+$ (i.e., action models with postconditions) is in* NExpTime.

Proof (Sketch). In the following σ denotes a symbol for worlds. Σ', Σ'', etc. denote sequences of pointed action models. The symbol $\checkmark$ means that the world survives a sequence of pointed action models.

- $\frac{(\sigma\ \Sigma'\ p)}{(\sigma\ \epsilon\ p)}$ is replaced by $\frac{(\sigma\ \Sigma'\ p)}{(\sigma\ \Sigma''\ p)}$ and $\frac{(\sigma\ \Sigma'\ \neg p)}{(\sigma\ \epsilon\ \neg p)}$ is replaced by $\frac{(\sigma\ \Sigma'\ \neg p)}{(\sigma\ \Sigma''\ \neg p)}$ where $\Sigma' = \Sigma''; \Sigma'''$ such that Σ''' is the longest sequence of pointed action models where p is not modified in the preconditions of current actions;
- Add the rules: $\frac{(\sigma\ \Sigma'\ \checkmark)}{(\sigma\ \Sigma'\ p)}$ if the post-condition in the initial action of the last pointed action model in Σ' makes p true and $\frac{(\sigma\ \Sigma'\ \checkmark)}{(\sigma\ \Sigma'\ \neg p)}$ if it makes p false.

The resulting tableau method can still be turned into a non-deterministic algorithm running in exponential time. $\qquad\square$

Unranked Second-Order Anti-Unification

Alexander Baumgartner and Temur Kutsia

Research Institute for Symbolic Computation
Johannes Kepler University Linz, Austria
{abaumgar,kutsia}@risc.jku.at

Abstract. In this work we study anti-unification for unranked hedges, permitting context and hedge variables. Hedges are sequences of unranked terms. The anti-unification problem of two hedges $\tilde{s}$ and $\tilde{q}$ is concerned with finding their generalization, a hedge $\tilde{g}$ such that both $\tilde{s}$ and $\tilde{q}$ are substitution instances of $\tilde{g}$. Second-order power is gained by using context variables to generalize vertical differences at the input hedges. Hedge variables are used to generalize horizontal differences. An anti-unification algorithm is presented, which computes a generalization of input hedges and records all the differences. The computed generalizations are least general among a certain class of generalizations.

1 Introduction

The anti-unification problem for two terms t_1 and t_2 requires finding their generalization: a term such that both t_1 and t_2 are instances of it under some substitutions. The interesting generalizations are least general ones (lggs). Anti-unification algorithms are supposed to compute lggs.

In 1970, Plotkin [20] and Reynolds [21] independently came up with essentially the same anti-unification algorithm. It was designed for first-order ranked terms (i.e., where function symbols have a fixed arity) in the syntactic case. Since then, a number of algorithms and their modifications have been developed, addressing the problem in various theories (e.g., [1, 2, 5, 7, 10, 13, 19]) and from the point of view of different applications (e.g., [4, 9, 11, 15, 17, 18, 22]).

In this paper, we consider anti-unification for hedges, which are finite sequences of unranked terms. Such terms are constructed from function symbols that do not have a fixed arity. We permit two kinds of variables: first-order, for hedges, and second-order, for contexts. Contexts that we consider here are hedges with a single occurrence of the distinguished symbol "hole". They are functions which can apply to another context or to a hedge, which are then "plugged" in the place of the hole.

Some applications of anti-unification indeed require higher-order features. For instance, reuse of proofs in program verification needs anti-unification with higher-order variables [18]. A restricted use of higher-order variables in generalizations turned out to be helpful for analogy making with Heuristic-Driven Theory Projection [15]. Anti-unification with combinator terms plays a role in replaying program derivations [12].

First-order anti-unification for ranked terms has been used to detect software code clones [9, 17]. It helps to achieve high-precision for clones obtained, essentially, by renaming and reformatting, but ranked anti-unification is not strong enough to detect

U. Kohlenbach et al. (Eds.): WoLLIC 2014, LNCS 8652, pp. 66–80, 2014.
© Springer-Verlag Berlin Heidelberg 2014

clones obtained by omitting/inserting pieces of statements in the code. Unranked anti-unification can detect similarities not only between renamed parts of a hedge, but also between parts which differ from each other by inserting or omitting subparts, as was indicated in [16]. These features can be useful also for comparison of XML documents, which can be abstracted by unranked trees.

However, one important restriction of existing hedge anti-unification algorithms, such as, e.g., [8, 16, 23], is that the languages used in these algorithms do not permit higher-order variables. This imposes a natural restriction on solutions: The computed lggs do not reflect similarities between input hedges, if those similar pieces are located under distinct heads or at different depths. For instance, $f(a, b)$ and $g(h(a, b))$ are generalized by a single variable, although both terms contain a and b and a more natural generalization could be, e.g. $X(a, b)$, where X is a higher-order variable. In applications, it is often desirable to detect these similarities.

This is the problem we address here, permitting the use of context variables to abstract vertical differences between trees, and hedge variables used to abstract horizontal differences. The algorithm described in this paper first constructs a "skeleton" of a generalization of the input hedges, which corresponds to a hedge embedded into each of the input hedges. Next, it inserts context and/or hedge variables into the skeleton, which are supposed to uniformly generalize (vertical and horizontal) differences between input hedges, to obtain an lgg (with respect to the given skeleton). The skeleton computation function is the parameter of the algorithm: One can compute an lgg which contains, for instance, a constrained longest common subforest [3], or an agreement subhedge/subtree [14] of the input hedges.

In this paper we focus on the step of computing an lgg of two hedges, when the skeleton is already constructed. We assume that the latter is given in the form of an admissible alignment, which is a certain sequence of symbols occurring in both hedges, together with the positions these symbols occur in. We need to restrict variable occurrences in the generalization to guarantee that for each admissible alignment a unique lgg is computed, leading to the notion of rigid generalizations. We develop an algorithm which takes two hedges and an admissible alignment and computes a rigid lgg of the hedges with respect to that alignment. The computed lgg is unique modulo variable renaming. Moreover, we can return not only the generalization, but also the differences between the input hedges, which tells us how one can obtain the original hedges from the generalization. The algorithm runs in quadratic time and requires linear space with respect to the size of the input. This result means that, for instance, if the skeleton is a constrained longest common subhedge of the input hedges in the sense of [24], then both skeleton and generalization computation can be done in quadratic time, because the time complexity of computing a constrained lcs is quadratic.

In some cases, the skeleton can be constructed in multiple ways, giving rise to several admissible alignments. It requires that the generalizations computed for each alignment should be compared to each other, to make the obtained set minimal. This problem requires matching with context and hedge variables in the minimization step and goes beyond the scope of this paper.

A prototype implementation of the algorithm is available from http://www.risc.jku.at/projects/stout/software/urauc.php.

Example 1. The hedge $(X(a), f(X(g(a, x), c), x))$ is a generalization of two hedges $(h(a), f(h(g(a, b, b), c), b, b))$ and $(a, f(g(a, d), c, d))$. Dotted and dashed nodes indicate differences, while the solid ones form the admissible alignment. The first hedge can be obtained from the generalization by replacing the context variable X with the context $h(\circ)$ and the hedge variable x with the hedge (a, b). To obtain the second hedge, we need to replace X with the hole (i.e., to eliminate X) and to replace x by d.

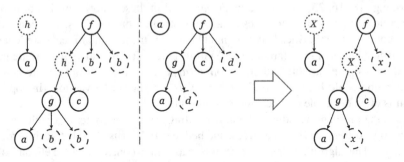

2 Preliminaries

Given pairwise disjoint countable sets of unranked function symbols $\mathcal{F}$ (symbols without fixed arity), hedge variables $\mathcal{V}_H$, unranked context variables $\mathcal{V}_C$, and a special symbol $\circ$ (the hole), we define *terms*, *hedges*, and *contexts* by the following grammar:

$$
\begin{aligned}
t &:= x \mid f(\tilde{s}) \mid X(\tilde{s}) && \text{(terms)} \\
\tilde{s} &:= t_1, \dots, t_n && \text{(hedges)} \\
\tilde{c} &:= \tilde{s}_1, \circ, \tilde{s}_2 \mid \tilde{s}_1, f(\tilde{c}), \tilde{s}_2 \mid \tilde{s}_1, X(\tilde{c}), \tilde{s}_2 && \text{(contexts)}
\end{aligned}
$$

where $x \in \mathcal{V}_H$, $f \in \mathcal{F}$, $X \in \mathcal{V}_C$, and $n \geq 0$.

Hedges are finite sequences of terms, constructed over $\mathcal{F}$ and $\mathcal{V}_H \cup \mathcal{V}_C$. A term can be seen as a singleton hedge. A context can be seen as a hedge over $\mathcal{F} \cup \{\circ\}$ and $\mathcal{V}_H \cup \mathcal{V}_C$, where the hole occurs exactly once. A singleton context is then a term over $\mathcal{F} \cup \{\circ\}$ and $\mathcal{V}_H \cup \mathcal{V}_C$ with a single hole in it. To improve readability, we put non-singleton hedges and contexts between parenthesis.

We use the letters x, y, z for hedge variables and X, Y, Z for context variables. By f, g, h, a, b, c, d, e we denote function symbols, by $\tilde{s}, \tilde{q}, \tilde{r}, \tilde{g}, \tilde{h}$ hedges, by $\tilde{c}, \tilde{d}$ arbitrary contexts and by $\dot{c}, \dot{d}$ singleton contexts. We use ϕ, ψ for a context variable or a function symbol. The empty hedge is denoted by ϵ. Terms of the form $a(\epsilon)$ are written as just a. Examples of a term, a hedge, and a context are, respectively, $f(f(a), b)$, $(x, X(a, x), f(f(a), b))$, and $(x, X(a, x), f(f(\circ), b))$.

A context $\tilde{c}$ can apply to a hedge $\tilde{s}$, denoted by $\tilde{c}[\tilde{s}]$, obtaining a hedge by replacing the hole in $\tilde{c}$ with $\tilde{s}$. For example, $(x, X(a, x), f(f(\circ), b))[a, X(a)] = (x, X(a, x), f(f(a, X(a)), b))$. Application of a context to a context is defined similarly.

The length of a hedge $\tilde{s}$, denoted $|\tilde{s}|$, is the number of elements in it. We denote by $\tilde{s}|_i$ the ith element of $\tilde{s}$ and by $\tilde{s}|_i^j$ the subhedge $(\tilde{s}|_i, \dots, \tilde{s}|_j)$. If $i > j$ then $\tilde{s}|_i^j$ is the empty hedge. The set of all function symbols which appear in a hedge $\tilde{s}$ (resp., in a context $\tilde{c}$) is denoted by $\mathcal{F}(\tilde{s})$ (resp., by $\mathcal{F}(\tilde{c})$). We overload the notation $\mathcal{F}(A)$ for the set of all function symbols which appear in a set of hedges and contexts A.

A *substitution* is a mapping from hedge variables to hedges and from context variables to contexts, which is identity almost everywhere. When substituting a context variable X by a context, the context will be applied to the argument hedge of X. The symbols σ, ϑ are used to denote a substitution. Substitutions can be applied to hedges and contexts in the usual way. We use postfix notation for application, writing, e.g., $\tilde{s}\sigma$ for the application of σ to $\tilde{s}$. For example, if $\sigma = \{x \mapsto \epsilon, y \mapsto (a, x), X \mapsto g(\circ)\}$ is a substitution, then $(X(x), y, f(X(y), c))\sigma = (g, a, x, f(g(a, x), c))$. The notion *range* of a substitution σ is standard and denoted by $\mathrm{Ran}(\sigma)$.

A hedge $\tilde{s}$ is the *instance* of a hedge $\tilde{q}$ if there exists a substitution σ with $\tilde{q}\sigma = \tilde{s}$. We say that $\tilde{q}$ is more general than $\tilde{s}$ if $\tilde{s}$ is an instance of $\tilde{q}$ and denote this by $\tilde{q} \preceq \tilde{s}$. If $\tilde{q} \preceq \tilde{s}$ and $\tilde{s} \preceq \tilde{q}$, then we write $\tilde{q} \simeq \tilde{s}$. If $\tilde{q} \preceq \tilde{s}$ and $\tilde{s} \not\preceq \tilde{q}$, then we say that $\tilde{q}$ is strictly more general than $\tilde{s}$ and write $\tilde{q} \prec \tilde{s}$. A hedge $\tilde{g}$ is a *generalization* of the hedges $\tilde{s}$ and $\tilde{q}$ if $\tilde{s}$ and $\tilde{q}$ are instances of $\tilde{g}$.

The word representation of a hedge is defined by the concatenation of the depth-first pre-order traversal of the constituent terms. For instance, $afgagbbc$ is the word representation of $(a, f(g(a, g(b, b)), c))$. Generalizations contain a common subsequence of the word representation of the input hedges. We will use this property in the formulation of our anti-unification algorithm. Observe, e.g., the hedges from example 1:

$$
\begin{array}{l}
(\ h(\mathbf{a}),\ \mathbf{f}(\ h(\mathbf{g}(\mathbf{a},b,b),\mathbf{c}),b,b)) \\
\hline
(\quad\ \mathbf{a}\ ,\mathbf{f}(\quad\ \mathbf{g}(\mathbf{a},d\ \ ),\mathbf{c}\ ,d\ \ )) \\
\hline
(X(\mathbf{a}),\ \mathbf{f}(X(\mathbf{g}(\mathbf{a},x\ \ ),\mathbf{c}),x\ \ ))
\end{array}
$$

The set of *positions* of a hedge $\tilde{s} = (t_1, \ldots t_n)$, denoted $pos(\tilde{s})$, is a set of strings of positive integers. It is defined as $pos(\tilde{s}) := \bigcup_{i=1}^{n}\{i \cdot p \mid p \in pos_T(t_i)\}$, where $\cdot$ stands for concatenation. $pos_T(t)$ is defined as $pos_T(x) := \{\lambda\}$ and $pos_T(\phi(\tilde{q})) := \{\lambda\} \cup pos(\tilde{q})$, where λ is the empty string. For example, $pos(f(a, g(b, c))) = \{1, 1\cdot1, 1\cdot2, 1\cdot2\cdot1, 1\cdot2\cdot2\}$ and $pos(a, f(b, g(c)), d) = \{1, 2, 2\cdot1, 2\cdot2, 2\cdot2\cdot1, 3\}$. In the latter hedge, the symbol g stands at the position $2\cdot2$ and c occurs at the position $2\cdot2\cdot1$.

- Two symbols $\mathsf{s}_1, \mathsf{s}_2 \in \mathcal{F} \cup \mathcal{V}_H \cup \mathcal{V}_C$ of a hedge are *horizontal consecutive* if the corresponding positions $I_{\mathsf{s}_1} \cdot i_{\mathsf{s}_1}$ and $I_{\mathsf{s}_2} \cdot i_{\mathsf{s}_2}$ are in the relation $I_{\mathsf{s}_1} = I_{\mathsf{s}_1}$ and $i_{\mathsf{s}_1} + 1 = i_{\mathsf{s}_2}$.
- Two symbols $\mathsf{s}_1, \mathsf{s}_2 \in \mathcal{F} \cup \mathcal{V}_H \cup \mathcal{V}_C$ of a hedge $\tilde{s}$ are in a *vertical chain* if their positions I_{s_1} and I_{s_2} are in the relation $I_{\mathsf{s}_1} \cdot 1 = I_{\mathsf{s}_2}$ and $I_{\mathsf{s}_1} \cdot 2 \notin pos(\tilde{s})$.

For example, in $(a, f(X(a, b)))$, the occurrence of a at position 1 and the occurrence of f at 2 are horizontal consecutive, as well as a at $2\cdot1\cdot1$ and b at $2\cdot1\cdot2$. The occurrence of f at 2 and the occurrence of X at $2\cdot1$ are in vertical chain.

With $<$ we denote the (strict) *lexicographic ordering* and with $\sqsubset$ the (strict) *ancestor relation* on positions, e.g., $1\cdot2\cdot1 < 1\cdot2\cdot2$, $1\cdot2\cdot1 < 1\cdot2\cdot1\cdot2$, and $1\cdot2\cdot1 \sqsubset 1\cdot2\cdot1\cdot2$. The relation $\sqsubseteq$ is defined as $\sqsubset \cup =$.

Given three positions I_1, I_2 and I_3, the ternary relation $\bowtie$ is defined as

$$I_1 \bowtie_{I_3} I_2 :\Longleftrightarrow \text{ there is } I_4 \neq \lambda \text{ such that } I_4 \sqsubset I_1 \text{ and } I_4 \sqsubset I_2 \text{ and } I_4 \not\sqsubset I_3 \text{ and}$$
$$I_1, I_2, I_3 \text{ are pairwise not in } \sqsubseteq.$$

This relation tests whether I_1 and I_2 have a common ancestor which is not an ancestor of I_3. None of these positions should be an ancestor of another. For instance, $1{\cdot}1 \bowtie_2 1{\cdot}2$, but *not* $1 \bowtie_3 2$, $1{\cdot}1 \bowtie_2 1{\cdot}1{\cdot}2$, $1{\cdot}1 \bowtie_2 1{\cdot}1$, and $1{\cdot}1 \bowtie_{1{\cdot}3} 1{\cdot}2$. A real world example of this relation would be two sisters and one of their uncles.

3 The Skeletons: Admissible Alignments

In this section we introduce the concept of admissible alignments, which are used later as skeletons to compute corresponding generalizations. For simplicity, we formulate all the notions and the algorithm for two hedges. The extension to more hedges is straightforward. Hedges to be generalized are assumed to be variable disjoint.

Given two hedges $\tilde{s}$ and $\tilde{q}$, an *alignment* is a sequence of the form $a_1 \langle I_1, J_1 \rangle \ldots a_m \langle I_m, J_m \rangle$ such that $I_1 < \cdots < I_m$, $J_1 < \cdots < J_m$, and a_k is the symbol at position I_k in $\tilde{s}$ and at position J_k in $\tilde{q}$ for all $1 \leqslant k \leqslant m$.

An alignment represents common function symbols inside of two hedges with the corresponding positions, respecting the ordering $<$. It is a common subsequence of the word representation of those hedges with some additional information about the positions. The empty alignment is denoted by $\mathfrak{e}$.

Collisions in an alignment $\mathfrak{a}$ of two hedges are defined as follows:

- A collision appears at two elements $a_k \langle I_k, J_k \rangle$, $a_l \langle I_l, J_l \rangle$ of $\mathfrak{a}$ if either $(I_k \sqsubset I_l$ and $J_k \not\sqsubseteq J_l)$ or $(I_k \not\sqsubseteq I_l$ and $J_k \sqsubset J_l)$.
- A collision appears at three elements $a_k \langle I_k, J_k \rangle$, $a_l \langle I_l, J_l \rangle$, $a_n \langle I_n, J_n \rangle$ of $\mathfrak{a}$ if $I_k \bowtie_{I_n} I_l$ and $J_l \bowtie_{J_k} J_n$.

For instance, the alignment $f \langle 2, 1 \rangle b \langle 2{\cdot}1, 1{\cdot}2 \rangle c \langle 2{\cdot}2, 2 \rangle$ of the hedges $(a, f(b, c))$ and $(f(a, b), c)$ contains a collision at the two elements $f \langle 2, 1 \rangle$ and $c \langle 2{\cdot}2, 2 \rangle$. The alignment $a \langle 1, 1{\cdot}1 \rangle b \langle 2{\cdot}1, 1{\cdot}2 \rangle c \langle 2{\cdot}2, 2 \rangle$ of the same hedges has a collision at its three elements.

An alignment $\mathfrak{a}$ is called *admissible* if there are no collisions in it. Note that for any two elements $a_k \langle I_k, J_k \rangle$, $a_l \langle I_l, J_l \rangle$ of an admissible alignment $\mathfrak{a}$, $I_k < I_l$ iff $J_k < J_l$ and $I_k \sqsubset I_l$ iff $J_k \sqsubset J_l$.

Admissible alignments are related to generalization by the following theorem:

Theorem 1. *Let $\mathfrak{a} = a_1 \langle I_1, J_1 \rangle \ldots a_m \langle I_m, J_m \rangle$ be an alignment of $\tilde{s}$ and $\tilde{q}$ such that for all $1 \leqslant k \leqslant m$ the function symbol a_k is unique in $\tilde{s}$ and unique in $\tilde{q}$. $\mathfrak{a}$ is admissible iff there exists a generalization $\tilde{g}$ of $\tilde{s}$ and $\tilde{q}$ with $\mathcal{F}(\tilde{g}) = \{a_1, \ldots, a_m\}$.*

Notice that requiring uniqueness of the function symbols $a_1, \ldots, a_m$ in theorem 1 does not impose a loss of generality. One can simply rename those symbols with fresh ones and restore the original function symbols afterwards.

From this theorem, we get that for any admissible alignment $\mathfrak{a}$ of two hedges there exists a generalization $\tilde{g}$ of those hedges which contains all the corresponding function symbols. The other direction is also true: For any generalization $\tilde{g}$ of two hedges there exists their admissible alignment containing all the function symbols which appear in $\tilde{g}$.

We call such a $\tilde{g}$ a *supporting generalization* of $\tilde{s}$ and $\tilde{q}$ with respect to $\mathfrak{a}$.

Example 2. Let $\tilde{s} = (a, a(b, b))$ and $\tilde{q} = (a(a(b(b))), b, b)$.
- $b\langle 2 \cdot 2, 1 \cdot 1 \cdot 1 \rangle a\langle 1, 1 \rangle$ is not an alignment of $\tilde{s}, \tilde{q}$.
- $a\langle 1, 1 \rangle a\langle 2, 1 \cdot 1 \rangle b\langle 2 \cdot 1, 1 \cdot 1 \cdot 1 \rangle b\langle 2 \cdot 2, 3 \rangle$ is a non-admissible alignment of $\tilde{s}, \tilde{q}$.
- $a\langle 1, 1 \cdot 1 \rangle b\langle 2 \cdot 1, 2 \rangle b\langle 2 \cdot 2, 3 \rangle$ is an admissible alignment of $\tilde{s}, \tilde{q}$, with $(X(a(x)), Y(b, b))$ being a corresponding supporting generalization.
- $(x, a(y, Y(b)), z)$ is a supporting generalization of $\tilde{s}, \tilde{q}$, with respect to $a\langle 2, 1 \rangle b\langle 2 \cdot 2, 1 \cdot 1 \cdot 1 \rangle$.

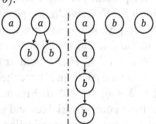

4 Computing Least General Rigid Generalizations

We aim at solving the following problem: Given two hedges $\tilde{s}$ and $\tilde{q}$ and their admissible alignment a, compute a least general supporting generalization $\tilde{g}$ of $\tilde{s}$ and $\tilde{q}$ with respect to a. However, least general supporting generalizations might not be unique. For instance, for (a, b, a) and (b, c) with the admissible alignment $b\langle 2, 1 \rangle$, we have two supporting lggs (x, b, x, y) and (x, b, y, x).

Therefore, we are interested in a special class of supporting generalizations, which we call rigid generalizations. Given two hedges $\tilde{s}, \tilde{q}$ and their admissible alignment a, a hedge $\tilde{g}$ is called a *rigid generalization* of $\tilde{s}$ and $\tilde{q}$ with respect to a, if $\tilde{g}$ is a supporting generalization of $\tilde{s}$ and $\tilde{q}$ with respect to a such that the following conditions hold:

- There exist substitutions σ, ϑ with $\tilde{g}\sigma = \tilde{s}$ and $\tilde{g}\vartheta = \tilde{q}$ such that all the contexts in σ and ϑ are singleton contexts.
- No context variable in $\tilde{g}$ applies to the empty hedge.
- $\tilde{g}$ doesn't contain horizontal consecutive hedge variables.
- $\tilde{g}$ doesn't contain vertical chains of variables.
- $\tilde{g}$ doesn't contain context variables with a hedge variable as the first or the last argument (i.e., no subterms of the form $X(x, \dots)$ and $X(\dots, x)$).

This definition puts some restrictions on the usage of the variables. Especially, our very general concept of context variables demands for some restrictions. For instance, $X(a, b)$ is a rigid generalization of $f(g(a, b, c))$ and (a, b) with respect to $a\langle 1 \cdot 1 \cdot 1, 1 \rangle b\langle 1 \cdot 1 \cdot 2, 2 \rangle$, while $X(a, b, x)$ and $X(Y(a, b))$ are not rigid generalizations.

A rigid generalization $\tilde{g}$ of $\tilde{s}$ and $\tilde{q}$ with respect to a is called a *rigid lgg* of $\tilde{s}$ and $\tilde{q}$ with respect to a, if there is no rigid generalization $\tilde{h}$ of $\tilde{s}$ and $\tilde{q}$ with respect to a which satisfies $\tilde{g} \prec \tilde{h}$.

Note that two hedges might have a supporting generalization which is less general than their rigid lgg with respect to the same admissible alignment. For instance, $X(a) \prec X(X(a))$ and both of them are generalizations of $f(f(a))$ and $g(g(g(g(a))))$ with respect to $a\langle 1 \cdot 1 \cdot 1, 1 \cdot 1 \cdot 1 \cdot 1 \cdot 1 \rangle$, but only $X(a)$ is a rigid generalization.

From now on, we concentrate on computing least general rigid generalizations of two variable-disjoint hedges with respect to an admissible alignment.

An *anti-unification problem* (AUP) is a triple of the form $x \colon \tilde{s} \triangleq \tilde{q}; X \colon \tilde{c} \triangleq \tilde{d}; a$, where

- x is a hedge variable and $\tilde{s}, \tilde{q}$ are hedges,
- X is a context variable and $\tilde{c}, \tilde{d}$ are contexts,
- a is an admissible alignment of $\tilde{s}$ and $\tilde{q}$.

We present our anti-unification algorithm as a rule-based algorithm that works on triples P; S; σ, where the problem set P is a set of AUPs, the store S is a set of AUPs with empty alignments, σ is a substitution which keeps track of the generalization computed so far, and for all pairs of AUPs $\{x\colon \tilde{s}_1 \triangleq \tilde{q}_1;\; X\colon \tilde{c}_1 \triangleq \tilde{d}_1;\; \mathfrak{a}_1,\; y\colon \tilde{s}_2 \triangleq \tilde{q}_2;\; Y\colon \tilde{c}_2 \triangleq \tilde{d}_2;\; \mathfrak{a}_2\} \subseteq P \cup S$ holds $x \neq y$ and $X \neq Y$.

As all the AUPs in S have the empty alignment, we write $x\colon \tilde{s} \triangleq \tilde{q};\; X\colon \tilde{c} \triangleq \tilde{d}$ instead of $x\colon \tilde{s} \triangleq \tilde{q};\; X\colon \tilde{c} \triangleq \tilde{d};\; \mathfrak{e}$ for an AUP of S. In the rules below, we use the symbols Y, Z for fresh context variables and y, z for fresh hedge variables. The brackets $[\,]$, as before, are used for context application. The symbol $\uplus$ stands for disjoint union. Furthermore, $i^{\cdot\cdot}$ denotes $i - 1$ and i^{++} denotes $i + 1$.

Spl-H: Split Hedge

$\{x\colon \tilde{s} \triangleq \tilde{q};\; X\colon \tilde{c} \triangleq \tilde{d};\; a_1\langle i_1 \cdot I_1, j_1 \cdot J_1\rangle \ldots a_k\langle i_k \cdot I_k, j_k \cdot J_k\rangle$
$\quad a_{k+1}\langle i_{k+1} \cdot I_{k+1}, j_{k+1} \cdot J_{k+1}\rangle \ldots a_m\langle i_m \cdot I_m, j_m \cdot J_m\rangle\} \uplus P;\; S;\; \sigma \implies$
$\{y\colon \tilde{s}|_{i_1}^{i_k} \triangleq \tilde{q}|_{j_1}^{j_k};\; Y\colon \circ \triangleq \circ;\; a_1\langle (i_1 - i_1^{\cdot\cdot}) \cdot I_1, (j_1 - j_1^{\cdot\cdot}) \cdot J_1\rangle \ldots$
$\quad a_k\langle (i_k - i_1^{\cdot\cdot}) \cdot I_k, (j_k - j_1^{\cdot\cdot}) \cdot J_k\rangle\} \cup$
$\{z\colon \tilde{s}|_{i_k^{++}}^{i_m} \triangleq \tilde{q}|_{j_k^{++}}^{j_m};\; Z\colon \circ \triangleq \circ;\; a_{k+1}\langle (i_{k+1} - i_k) \cdot I_{k+1}, (j_{k+1} - j_k) \cdot J_{k+1}\rangle \ldots$
$\quad a_m\langle (i_m - i_k) \cdot I_m, (j_m - j_k) \cdot J_m\rangle\} \cup P;$
$\{x\colon \epsilon \triangleq \epsilon;\; X\colon \tilde{c}[\tilde{s}|_1^{i_1^{\cdot\cdot}}, \circ, \tilde{s}|_{i_m^{++}}^{|\tilde{s}|}] \triangleq \tilde{d}[\tilde{q}|_1^{j_1^{\cdot\cdot}}, \circ, \tilde{q}|_{j_m^{++}}^{|\tilde{q}|}]\} \cup S;\; \sigma\{x \mapsto (Y(y), Z(z))\},$

If $i_1 \neq i_{k+1}$ and $j_1 \neq j_{k+1}$, and, moreover, $i_1 = i_k$ or $j_1 = j_k$, for $1 \leqslant k < m$.

Abs-L: Abstract Left Context

$\{x\colon (\tilde{s}_l, \phi(\tilde{s}), \tilde{s}_r) \triangleq \tilde{q};\; X\colon \tilde{c} \triangleq \tilde{d};\; a_1\langle i \cdot I_1, J_1\rangle \ldots a_m\langle i \cdot I_m, J_m\rangle\} \uplus P;\; S;\; \sigma \implies$
$\{x\colon \tilde{s} \triangleq \tilde{q};\; X\colon \tilde{c}[\tilde{s}_l, \phi(\circ), \tilde{s}_r] \triangleq \tilde{d};\; a_1\langle I_1, J_1\rangle \ldots a_m\langle I_m, J_m\rangle\} \uplus P;\; S;\; \sigma,$

where $I_1 \neq \lambda$, $\phi(\tilde{s})$ is the term at position i in $(\tilde{s}_l, \phi(\tilde{s}), \tilde{s}_r)$, and $\tilde{s}_l, \tilde{s}_r$ are hedges.

Abs-R: Abstract Right Context

$\{x\colon \tilde{s} \triangleq (\tilde{q}_l, \phi(\tilde{q}), \tilde{q}_r);\; X\colon \tilde{c} \triangleq \tilde{d};\; a_1\langle I_1, j \cdot J_1\rangle \ldots a_m\langle I_m, j \cdot J_m\rangle\} \uplus P;\; S;\; \sigma \implies$
$\{x\colon \tilde{s} \triangleq \tilde{q};\; X\colon \tilde{c} \triangleq \tilde{d}[\tilde{q}_l, \phi(\circ), \tilde{q}_r];\; a_1\langle I_1, J_1\rangle \ldots a_m\langle I_m, J_m\rangle\} \uplus P;\; S;\; \sigma,$

where $J_1 \neq \lambda$, $\phi(\tilde{q})$ is the term at position j in $(\tilde{q}_l, \phi(\tilde{q}), \tilde{q}_r)$, and $\tilde{q}_l, \tilde{q}_r$ are hedges.

App-A: Apply Alignment

$\{x\colon (\tilde{s}_l, a_1(\tilde{s}), \tilde{s}_r) \triangleq (\tilde{q}_l, a_1(\tilde{q}), \tilde{q}_r);\; X\colon \tilde{c} \triangleq \tilde{d};$
$\quad a_1\langle i, j\rangle a_2\langle i \cdot I_2, j \cdot J_2\rangle \ldots a_m\langle i \cdot I_m, j \cdot J_m\rangle\} \uplus P;\; S;\; \sigma \implies$
$\{y\colon \tilde{s} \triangleq \tilde{q};\; Y\colon \circ \triangleq \circ;\; a_2\langle I_2, J_2\rangle \ldots a_m\langle I_m, J_m\rangle\} \cup P;$
$\{x\colon \epsilon \triangleq \epsilon;\; X\colon \tilde{c}[\tilde{s}_l, \circ, \tilde{s}_r] \triangleq \tilde{d}[\tilde{q}_l, \circ, \tilde{q}_r]\} \cup S;\; \sigma\{x \mapsto a_1(Y(y))\},$

where $a_1(\tilde{s}), a_1(\tilde{q})$ are the terms at the positions i, j and $\tilde{s}_l, \tilde{s}_r, \tilde{q}_l, \tilde{q}_r$ are hedges.

Sol-H: Solve Hedge

$\{x\colon \tilde{s} \triangleq \tilde{q};\; X\colon \circ \triangleq \circ;\; \mathfrak{e}\} \uplus P;\; S;\; \sigma \implies P;\; \{x\colon \tilde{s} \triangleq \tilde{q};\; X\colon \circ \triangleq \circ\} \cup S;\; \sigma\{X \mapsto \circ\}.$

Res-C: **Restore Context**

$P; \{x\colon \epsilon \triangleq \epsilon;\ X\colon (\tilde{s}_l, \dot{c}, \tilde{s}_r) \triangleq (\tilde{q}_l, \dot{d}, \tilde{q}_r)\} \cup S;\ \sigma \Longrightarrow$
$P; \{x\colon \epsilon \triangleq \epsilon;\ X\colon \dot{c} \triangleq \dot{d},\ y\colon \tilde{s}_l \triangleq \tilde{q}_l;\ Y\colon \circ \triangleq \circ,\ z\colon \tilde{s}_r \triangleq \tilde{q}_r;\ Z\colon \circ \triangleq \circ\} \cup S;$
$\quad \sigma\{X \mapsto (y, X(\circ), z)\},$

if not $\epsilon = \tilde{s}_l = \tilde{s}_r = \tilde{q}_l = \tilde{q}_r$. $\dot{c}, \dot{d}$ are singleton contexts.

Mer-S: **Merge Store**

$P; \{x_1\colon \tilde{s} \triangleq \tilde{q};\ X_1\colon \tilde{c} \triangleq \tilde{d},\ x_2\colon \tilde{s} \triangleq \tilde{q};\ X_2\colon \tilde{c} \triangleq \tilde{d}\} \cup S;\ \sigma \Longrightarrow$
$P; \{x_1\colon \tilde{s} \triangleq \tilde{q};\ X_1\colon \tilde{c} \triangleq \tilde{d}\} \cup S;\ \sigma\{x_2 \mapsto x_1, X_2 \mapsto X_1\}.$

Clr-S: **Clear Store**

$P; \{x\colon \epsilon \triangleq \epsilon;\ X\colon \circ \triangleq \circ\} \cup S;\ \sigma \ \Longrightarrow\ P;\ S;\ \sigma\{x \mapsto \epsilon, X \mapsto \circ\}.$

The idea of the store is to keep track of already solved AUPs in order to generalize the same AUPs in the same way, as it is illustrated in the Mer-S rule.

To compute generalizations of $\tilde{s}$ and $\tilde{q}$ with respect to an admissible alignment $\mathfrak{a}$, the procedure starts with $\{x\colon \tilde{s} \triangleq \tilde{q};\ X\colon \circ \triangleq \circ;\ \mathfrak{a}\};\ \varnothing;\ \varepsilon$, where x and X are fresh variables, and applies the rules exhaustively. We denote this procedure by $\mathfrak{G}$. The intuition is that at i's step of such a derivation, $X(x)\sigma_i$ is supposed to be a generalization of $\tilde{s}$ and $\tilde{q}$, with the idea that when the process stops with σ in the last step, then $X(x)\sigma$ is a rigid lgg of $\tilde{s}$ and $\tilde{q}$ with respect to $\mathfrak{a}$.

Before discussing the properties of $\mathfrak{G}$, we briefly explain informally what the rules do. At each step, each AUP $x\colon \tilde{s} \triangleq \tilde{q};\ X\colon \tilde{c} \triangleq \tilde{d};\ \mathfrak{a}$ in P represents the hedges $\tilde{c}[\tilde{s}]$ and $\tilde{d}[\tilde{q}]$ which are to be generalized, such that the generalization contains the function symbols from $\mathfrak{a}$. They are split according to the occurrences of alignment elements: All symbols from $\mathfrak{a}$ are in $\tilde{s}$ and $\tilde{q}$. None of them appear in $\tilde{c}$ and $\tilde{d}$.

Such an AUP can be transformed by one of the first four rules: Spl-H, Abs-L, Abs-R, or App-A. The eventual goal of these transformations is to reach the occurrences of the first alignment element in $\tilde{s}$ and $\tilde{q}$. In the course of the transformation, $\tilde{c}$ and $\tilde{d}$ are getting extended with contexts above those occurrences.

When the symbols in $\mathfrak{a}$ are distributed in more than one term both in $\tilde{s}$ and in $\tilde{q}$, then we use the Spl-H rule to select subhedges of $\tilde{s}$ and $\tilde{q}$ which contain all the alignment elements. (The other parts of $\tilde{s}$ and $\tilde{q}$ are moved to the store, since they will not contribute a symbol to the generalization.) Furthermore, by this rule, each of these subhedges are split into two smaller subhedges: From the $\tilde{s}$ side these are $\tilde{s}|_{i_1}^{i_k}$ and $\tilde{s}|_{i_k^{++}}^{i_m}$, and from the $\tilde{q}$ side they are $\tilde{q}|_{j_1}^{j_k}$ and $\tilde{q}|_{j_k^{++}}^{j_m}$. The split point k is decided by the following criteria:

- $\tilde{s}|_{i_1}^{i_k}$ and $\tilde{q}|_{j_1}^{j_k}$ contain the first $k > 0$ elements of $\mathfrak{a}$.
- $\tilde{s}|_{i_k^{++}}^{i_m}$ and $\tilde{q}|_{j_k^{++}}^{j_m}$ contain the elements of $\mathfrak{a}$ starting from $k + 1$. There exists at least one such element.
- $\tilde{s}|_{i_1}^{i_k}$ or $\tilde{q}|_{j_1}^{j_k}$ is a term (a singleton hedge), and the $k + 1$'st element of $\mathfrak{a}$ does not belong to it.

The process will continue by generalizing $\tilde{s}|_{i_1}^{i_k}$ and $\tilde{q}|_{j_1}^{j_k}$ with respect to the first k-element prefix of $\mathfrak{a}$, and generalizing $\tilde{s}|_{i_k^{++}}^{i_m}$ and $\tilde{q}|_{j_k^{++}}^{j_m}$ with respect to the elements of $\mathfrak{a}$

starting from $k + 1$. Note that in the next step Spl-H is not applicable to the AUP with $\tilde{s}|_{i_1}^{i_k}$ and $\tilde{q}|_{j_1}^{j_k}$. This is because at least one of them is a single term which completely contains the alignment elements. Therefore either Abs-L, Abs-R, or App-A applies.

Consider the hedges $(g(a), f(a, g(b)), c, g(b), e)$ and $(e, e, h(a, e), f(b), a, c, d, b)$ and the admissible alignment $a\langle 2{\cdot}1, 3{\cdot}1\rangle b\langle 2{\cdot}2{\cdot}1, 4{\cdot}1\rangle c\langle 3, 6\rangle b\langle 4{\cdot}1, 8\rangle$ of them.

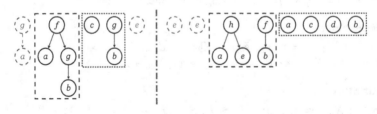

The dashed nodes in the figure above denote the parts which are moved into the store. The dashed rectangle denotes $\tilde{s}|_{i_1}^{i_k}$ and $\tilde{q}|_{j_1}^{j_k}$ and the dotted one denotes $\tilde{s}|_{i_k^{++}}^{i_m}$ and $\tilde{q}|_{j_k^{++}}^{j_m}$.

When all symbols in a belong to one term in $\tilde{s}$ or in $\tilde{q}$ (or maybe both), but the root of that term is *not* the symbol a_1 from the first element of a, then an attempt is made to get deeper to that term, to reach the subterm whose top symbol is the a_1 from a. This descent is carried out by Abs-L or Abs-R, depending whether we are searching for the subterm with a_1 in the top in $\tilde{s}$ or in $\tilde{q}$.

When all symbols in a belong to one term in $\tilde{s}$ and one term in $\tilde{q}$, and these terms have the same root symbol which is exactly the a_1 from the first element of a, then a_1 is moved to the generalization. This is what the App-A rule does. The process will continue with generalizing the hedges under the occurrences of a_1 in $\tilde{s}$ and $\tilde{q}$.

When the alignment is empty in x: $\tilde{s} \triangleq \tilde{q}$; X: $\tilde{c} \triangleq \tilde{d}$; ϵ in P, then the hedge there will not contribute a symbol in the generalization. Moreover, both $\tilde{c}$ and $\tilde{d}$ are holes, because only App-A can make the alignment empty, and it makes the contexts in the obtained AUP the hole. Such AUPs are considered solved, as their generalization is just x they contain. They should be put in the store, which keeps information about the differences between the hedges to be generalized. At the same time, the context variable X can be deleted, as it just stand for the hole. This is what the Sol-H rule does.

The other three rules work on the store. Clr-S removes the empty AUP from the store and eliminates the corresponding variables form the generalization. Mer-S guarantees that the same AUPs are generalized with the same variables, making sure that the same differences in the input hedges are generalized uniformly. Finally, the Res-C rule guarantees that each context variable in the generalization generalizes singleton contexts in the input hedges: A property required for rigid generalizations.

We define two substitutions obtained by a set S of AUPs:

$$\sigma_{\text{L}}(S) ::= \{x \mapsto \tilde{s}, X \mapsto \tilde{c} \mid x \colon \tilde{s} \triangleq \tilde{q}; X \colon \tilde{c} \triangleq \tilde{d}; a \in S\}$$

$$\sigma_{\text{R}}(S) ::= \{x \mapsto \tilde{q}, X \mapsto \tilde{d} \mid x \colon \tilde{s} \triangleq \tilde{q}; X \colon \tilde{c} \triangleq \tilde{d}; a \in S\}$$

Example 3. Let $\tilde{s} = f(a, f(b, b))$ and $\tilde{q} = (b, f(a, b), b)$ be the input hedges with the admissible alignment $a = f\langle 1, 2\rangle a\langle 1{\cdot}1, 2{\cdot}1\rangle b\langle 1{\cdot}2{\cdot}1, 2{\cdot}2\rangle$. Then the algorithm $\mathfrak{G}$ starts with the initial system

$$\{x\colon f(a, f(b, b)) \triangleq (b, f(a, b), b);\ X\colon \circ \triangleq \circ;\ f\langle 1, 2\rangle a\langle 1{\cdot}1, 2{\cdot}1\rangle b\langle 1{\cdot}2{\cdot}1, 2{\cdot}2\rangle\};\ \varnothing;\ \varepsilon$$

and computes $\varnothing;\ S;\ \sigma$, where

$$S = \{y\colon \epsilon \triangleq \epsilon;\ Y\colon f(\circ, b) \triangleq \circ,\ z\colon \epsilon \triangleq b;\ Z\colon \circ \triangleq \circ\},$$
$$\sigma = \{x \mapsto (z, f(a, Y(b)), z), X \mapsto \circ, \ldots\}.$$

$X(x)\sigma = (z, f(a, Y(b)), z)$ generalizes $\tilde{s}$ and $\tilde{q}$ with respect to $\mathfrak{a}$. From the store S we can read $\sigma_{L}(S) = \{z \mapsto \epsilon, Y \mapsto f(\circ, b), \ldots\}$ and $\sigma_{R}(S) = \{z \mapsto b, Y \mapsto \circ, \ldots\}$. Then we have $X(x)\sigma\sigma_{L}(S) = \tilde{s}$ and $X(x)\sigma\sigma_{R}(S) = \tilde{q}$.

Now we turn to discussing the properties of $\mathfrak{G}$. Termination is the first of them:

Theorem 2 (Termination of $\mathfrak{G}$). *The system $\mathfrak{G}$ terminates on any input.*

The substitutions $\sigma_{L}(S)$ and $\sigma_{R}(S)$ are used to characterize the invariant of $\mathfrak{G}$:

Lemma 1 (Generalization Invariant). *Let P_0; S_0; σ_0 such that for all x_0: $\tilde{s}_0 \triangleq \tilde{q}_0$; X_0: $\tilde{c}_0 \triangleq \tilde{d}_0$; $\mathfrak{a}_0 \in P_0$ the variables x_0, X_0 only appear together as term $X_0(x_0)$ in σ_0. If P_0; S_0; $\sigma_0 \Longrightarrow^* P_n$; S_n; σ_n is a derivation in $\mathfrak{G}$ then for all x_0: $\tilde{s}_0 \triangleq \tilde{q}_0$; X_0: $\tilde{c}_0 \triangleq \tilde{d}_0$; $\mathfrak{a}_0 \in P_0 \cup S_0$ holds*

- $X_0(x_0)\sigma_0\sigma_{L}(P_0 \cup S_0) = X_0(x_0)\sigma_n\sigma_{L}(P_n \cup S_n)$,
- $X_0(x_0)\sigma_0\sigma_{R}(P_0 \cup S_0) = X_0(x_0)\sigma_n\sigma_{R}(P_n \cup S_n)$.

This lemma has a corollary which states that for the invariant, the initial substitution is irrelevant:

Corollary 1. *If P_0; S_0; $\vartheta_0 \Longrightarrow^* P_n$; S_n; $\vartheta_0\vartheta_1 \ldots \vartheta_n$ is a derivation in $\mathfrak{G}$ then for all x_0: $\tilde{s}_0 \triangleq \tilde{q}_0$; X_0: $\tilde{c}_0 \triangleq \tilde{d}_0$; $\mathfrak{a}_0 \in P_0 \cup S_0$ holds*

- $X_0(x_0)\sigma_{L}(P_0 \cup S_0) = X_0(x_0)\vartheta_1 \ldots \vartheta_n\sigma_{L}(P_n \cup S_n)$,
- $X_0(x_0)\sigma_{R}(P_0 \cup S_0) = X_0(x_0)\vartheta_1 \ldots \vartheta_n\sigma_{R}(P_n \cup S_n)$.

The soundness theorem shows that $\mathfrak{G}$ indeed computes rigid generalizations. Besides, the store keeps the information which indicates how to obtain the initial hedges from the generalization:

Theorem 3 (Soundness of $\mathfrak{G}$). *Let P be a set of AUPs of the form $\{x\colon \tilde{s} \triangleq \tilde{q};\ X\colon \circ \triangleq \circ;\ \mathfrak{a}\}$. Every exhaustive rule application in $\mathfrak{G}$ yields a derivation $P;\ \varnothing;\ \varepsilon \Longrightarrow^+ \varnothing;\ S;\ \sigma$ where $\tilde{g} = X(x)\sigma$ is a rigid generalization of $\tilde{s}$ and $\tilde{q}$ with respect to $\mathfrak{a}$ and the store S records all the differences such that $\tilde{g}\sigma_{L}(S) = \tilde{s}$ and $\tilde{g}\sigma_{R}(S) = \tilde{q}$.*

The next theorem is the Completeness Theorem. It, essentially, says that for the given alignment, a rigid generalization $\mathfrak{G}$ computes is least general among all rigid generalizations of the input hedges.

Theorem 4 (Completeness of $\mathfrak{G}$). *Let $\tilde{g}$ be a rigid generalization of $\tilde{s}$ and $\tilde{q}$ with respect to $\mathfrak{a}$. Then there exists a derivation $\{x \colon \tilde{s} \triangleq \tilde{q};\ X \colon \circ \triangleq \circ;\ \mathfrak{a}\};\ \varnothing;\ \varepsilon \Longrightarrow^+ \varnothing;\ S;$ σ obtained by $\mathfrak{G}$ such that $\tilde{g} \leq X(x)\sigma$.*

There is a nondeterminism in the algorithm. The Uniqueness Theorem says that different transformations compute generalizations which are equivalent modulo $\simeq$, i.e., differ from each other only by variable renaming:

Theorem 5 (Uniqueness modulo $\simeq$). *Let $\mathfrak{a}$ be an admissible alignment of $\tilde{s}$ and $\tilde{q}$. If $\{x_1 \colon \tilde{s} \triangleq \tilde{q};\ X_1 \colon \circ \triangleq \circ;\ \mathfrak{a}\};\ \varnothing;\ \varepsilon \Longrightarrow^+ \varnothing;\ S_1;\ \sigma_1$ and $\{x_2 \colon \tilde{s} \triangleq \tilde{q};\ X_2 \colon \circ \triangleq \circ;$ $\mathfrak{a}\};\ \varnothing;\ \varepsilon \Longrightarrow^+ \varnothing;\ S_2;\ \sigma_2$ are two exhaustive derivations in $\mathfrak{G}$, then $X_1(x_1)\sigma_1 \simeq X_2(x_2)\sigma_2$.*

Finally, the complexity analysis reveals that the algorithm runs in quadratic time and requires linear space:

Theorem 6 (Complexity of $\mathfrak{G}$). *The anti-unification algorithm $\mathfrak{G}$ has $O(n^2)$ time complexity and $O(n)$ space complexity, where n is the number of symbols in the input.*

Acknowledgments. This research has been supported by the Austrian Science Fund (FWF) under the project SToUT (P 24087-N18).

References

1. Alpuente, M., Escobar, S., Meseguer, J., Ojeda, P.: A modular equational generalization algorithm. In: Hanus, M. (ed.) LOPSTR 2008. LNCS, vol. 5438, pp. 24–39. Springer, Heidelberg (2009)
2. Alpuente, M., Escobar, S., Meseguer, J., Ojeda, P.: Order-sorted generalization. Electr. Notes Theor. Comput. Sci. 246, 27–38 (2009)
3. Amir, A., Hartman, T., Kapah, O., Shalom, B.R., Tsur, D.: Generalized LCS. Theor. Comput. Sci. 409(3), 438–449 (2008)
4. Armengol, E., Plaza, E.: Bottom-up induction of feature terms. Machine Learning 41(3), 259–294 (2000)
5. Baader, F.: Unification, weak unification, upper bound, lower bound, and generalization problems. In: Book, R.V. (ed.) RTA 1991. LNCS, vol. 488, pp. 86–97. Springer, Heidelberg (1991)
6. Baumgartner, A., Kutsia, T.: Unranked Second-Order Anti-Unification. Technical report no. 14-05, RISC, JKU Linz (March 2014), http://www.risc.jku.at/publications/download/risc_4966/Baumgartner_Kutsia_2014.pdf
7. Baumgartner, A., Kutsia, T., Levy, J., Villaret, M.: A variant of higher-order anti-unification. In: van Raamsdonk, F. (ed.) RTA. LIPIcs, vol. 21, pp. 113–127. Schloss Dagstuhl - Leibniz-Zentrum fuer Informatik (2013)
8. Boley, H.: Finite domains and exclusions as first-class citizens. In: Dyckhoff, R. (ed.) ELP 1993. LNCS, vol. 798, pp. 37–61. Springer, Heidelberg (1994)
9. Bulychev, P.E., Kostylev, E.V., Zakharov, V.A.: Anti-unification algorithms and their applications in program analysis. In: Pnueli, A., Virbitskaite, I., Voronkov, A. (eds.) PSI 2009. LNCS, vol. 5947, pp. 413–423. Springer, Heidelberg (2010)
10. Burghardt, J.: E-generalization using grammars. Artif. Intell. 165(1), 1–35 (2005)

11. Delcher, A.L., Kasif, S.: Efficient parallel term matching and anti-unification. J. Autom. Reasoning 9(3), 391–406 (1992)
12. Hasker, R.W.: The Replay of Program Derivations. PhD thesis, University of Illionois at Urbana-Champaign (1995)
13. Huet, G.: Résolution d'équations dans des langages d'ordre 1, 2, . . . , ω. PhD thesis, Université Paris VII (September 1976)
14. Kao, M.-Y., Lam, T.W., Sung, W.-K., Ting, H.-F.: An even faster and more unifying algorithm for comparing trees via unbalanced bipartite matchings. J. Algorithms 40(2), 212–233 (2001)
15. Krumnack, U., Schwering, A., Gust, H., Kühnberger, K.-U.: Restricted higher-order anti-unification for analogy making. In: Orgun, M.A., Thornton, J. (eds.) AI 2007. LNCS (LNAI), vol. 4830, pp. 273–282. Springer, Heidelberg (2007)
16. Kutsia, T., Levy, J., Villaret, M.: Anti-unification for unranked terms and hedges. J. Autom. Reasoning 52(2), 155–190 (2014)
17. Li, H., Thompson, S.: Similar code detection and elimination for Erlang programs. In: Carro, M., Peña, R. (eds.) PADL 2010. LNCS, vol. 5937, pp. 104–118. Springer, Heidelberg (2010)
18. Lu, J., Mylopoulos, J., Harao, M., Hagiya, M.: Higher order generalization and its application in program verification. Ann. Math. Artif. Intell. 28(1-4), 107–126 (2000)
19. Pfenning, F.: Unification and anti-unification in the calculus of constructions. In: LICS, pp. 74–85. IEEE Computer Society (1991)
20. Plotkin, G.D.: A note on inductive generalization. Machine Intel. 5(1), 153–163 (1970)
21. Reynolds, J.C.: Transformational systems and the algebraic structure of atomic formulas. Machine Intel. 5(1), 135–151 (1970)
22. Schmid, U.: Inductive Synthesis of Functional Programs. LNCS (LNAI), vol. 2654. Springer, Heidelberg (2003)
23. Yamamoto, A., Ito, K., Ishino, A., Arimura, H.: Modelling semi-structured documents with hedges for deduction and induction. In: Rouveirol, C., Sebag, M. (eds.) ILP 2001. LNCS (LNAI), vol. 2157, pp. 240–247. Springer, Heidelberg (2001)
24. Zhang, K.: Algorithms for the constrained editing problem between ordered labeled trees and related problems. Pattern Recognition 28, 463–474 (1995)

A Appendix

We first illustrate, step by step, how the algorithm $\mathfrak{G}$ computes the rigid lgg for the anti-unification problem in Example 3:

$$\{x: f(a, f(b, b)) \triangleq (b, f(a, b), b); \ X: \circ \triangleq \circ; \ f\langle 1, 2\rangle a\langle 1{\cdot}1, 2{\cdot}1\rangle b\langle 1{\cdot}2{\cdot}1, 2{\cdot}2\rangle\}; \ \varnothing; \ \varepsilon$$

$\Rightarrow_{\text{App-A}}$ $\{y_1: (a, f(b, b)) \triangleq (a, b); \ Y_1: \circ \triangleq \circ; \ a\langle 1, 1\rangle b\langle 2{\cdot}1, 2\rangle\};$
$\qquad \{x: \epsilon \triangleq \epsilon; \ X: \circ \triangleq (b, \circ, b)\}; \ \{x \mapsto f(Y_1(y_1))\}$

$\Rightarrow_{\text{Res-C}}$ $\{y_1: (a, f(b, b)) \triangleq (a, b); \ Y_1: \circ \triangleq \circ; \ a\langle 1, 1\rangle b\langle 2{\cdot}1, 2\rangle\};$
$\qquad \{z_1: \epsilon \triangleq b; \ Z_1: \circ \triangleq \circ, z_2: \epsilon \triangleq b; \ Z_2: \circ \triangleq \circ\}; \ \{x \mapsto (z_1, f(Y_1(y_1)), z_2), X \mapsto \circ\}$

$\Rightarrow_{\text{Mer-S}}$ $\{y_1: (a, f(b, b)) \triangleq (a, b); \ Y_1: \circ \triangleq \circ; \ a\langle 1, 1\rangle b\langle 2{\cdot}1, 2\rangle\};$
$\qquad \{z_1: \epsilon \triangleq b; \ Z_1: \circ \triangleq \circ\}; \ \{x \mapsto (z_1, f(Y_1(y_1)), z_1), X \mapsto \circ\}$

$\Rightarrow_{\substack{\text{Spl-H} \\ \text{Clr-S}}}$ $\{y_2: a \triangleq a; \ Y_2: \circ \triangleq \circ; \ a\langle 1, 1\rangle, \ y_3: (f(b, b)) \triangleq b; \ Y_3: \circ \triangleq \circ; \ b\langle 1{\cdot}1, 1\rangle\};$
$\qquad \{z_1: \epsilon \triangleq b; \ Z_1: \circ \triangleq \circ\}; \ \{x \mapsto (z_1, f(Y_2(y_2), Y_3(y_3)), z_1), X \mapsto \circ\}$

$\Rightarrow_{\substack{\text{App-A} \\ \text{Clr-S}}}$ $\{y_4: \epsilon \triangleq \epsilon; \ Y_4: \circ \triangleq \circ; \ \epsilon, \ y_3: (f(b, b)) \triangleq (b); \ Y_3: \circ \triangleq \circ; \ b\langle 1{\cdot}1, 1\rangle\};$
$\qquad \{z_1: \epsilon \triangleq b; \ Z_1: \circ \triangleq \circ\}; \ \{x \mapsto (z_1, f(a(Y_4(y_4)), Y_3(y_3)), z_1), X \mapsto \circ\}$

$\Rightarrow_{\substack{\text{Sol-H} \\ \text{Clr-S}}}$ $\{y_3: f(b, b) \triangleq b; \ Y_3: \circ \triangleq \circ; \ b\langle 1{\cdot}1, 1\rangle\};$
$\qquad \{z_1: \epsilon \triangleq b; \ Z_1: \circ \triangleq \circ\}; \ \{x \mapsto (z_1, f(a, Y_3(y_3)), z_1), X \mapsto \circ\}$

$\Rightarrow_{\text{Abs-L}}$ $\{y_3: (b, b) \triangleq b; \ Y_3: f(\circ) \triangleq \circ; \ b\langle 1, 1\rangle\};$
$\qquad \{z_1: \epsilon \triangleq b; \ Z_1: \circ \triangleq \circ\}; \ \{x \mapsto (z_1, f(a, Y_3(y_3)), z_1), X \mapsto \circ\}$

$\Rightarrow_{\text{App-A}}$ $\{y_5: \epsilon \triangleq \epsilon; \ Y_5: \circ \triangleq \circ; \ \epsilon\}; \ \{y_3: \epsilon \triangleq \epsilon; \ Y_3: f(\circ, b) \triangleq \circ, z_1: \epsilon \triangleq b; \ Z_1: \circ \triangleq \circ\};$
$\qquad \{x \mapsto (z_1, f(a, Y_3(b(Y_5(y_5)))), z_1), X \mapsto \circ\}$

$\Rightarrow_{\substack{\text{Sol-H} \\ \text{Clr-S}}}$ $\varnothing; \{y_3: \epsilon \triangleq \epsilon; \ Y_3: f(\circ, b) \triangleq \circ, z_1: \epsilon \triangleq b; \ Z_1: \circ \triangleq \circ\};$
$\qquad \{x \mapsto (z_1, f(a, Y_3(b)), z_1), X \mapsto \circ\}.$

The proofs of the properties (e.g. termination, soundness, completeness and uniqueness) of $\mathfrak{G}$ can be found in the technical report [6]. Here we give only the proof of theorem 1 and the complexity result.

Theorem 1. *Let* $\mathfrak{a} = a_1\langle I_1, J_1\rangle \ldots a_m\langle I_m, J_m\rangle$ *be an alignment of* $\tilde{s}$ *and* $\tilde{q}$ *such that for all* $1 \leqslant k \leqslant m$ *the function symbol* a_k *is unique in* $\tilde{s}$ *and unique in* $\tilde{q}$. *$\mathfrak{a}$ is admissible iff there exists a generalization* $\tilde{g}$ *of* $\tilde{s}$ *and* $\tilde{q}$ *with* $\mathcal{F}(\tilde{g}) = \{a_1, \ldots, a_m\}$.

Proof. Let $\mathfrak{a} = a_1\langle I_1, J_1\rangle \ldots a_m\langle I_m, J_m\rangle$ be an alignment of $\tilde{s}$ and $\tilde{q}$ such that for all $1 \leqslant k \leqslant m$ the function symbol a_k is unique in $\tilde{s}$ and unique in $\tilde{q}$.

($\Leftarrow$) Assume $\tilde{g}$ is a generalization of $\tilde{s}$ and $\tilde{q}$ with $\mathcal{F}(\tilde{g}) = \{a_1, \ldots, a_m\}$. We will prove by contradiction that there are no collisions in $\mathfrak{a}$ (see definition of admissible alignment). Furthermore we assume that there are at least two elements in $\mathfrak{a}$ because the other cases are trivial by definition.

Case 1: Assume there is a collision at two elements of $\mathfrak{a}$. Then there exists $a_i, a_j \in \{a_1, \ldots, a_m\}$ such that a_i is an ancestor of a_j in $\tilde{s}$ while it is not an ancestor of a_j in $\tilde{q}$. We know that $\tilde{g}$ contains both symbols a_i and a_j.

Case 1.1: a_i is an ancestor of a_j in $\tilde{g}$. Then we have $a_i(\tilde{r}_1, t, \tilde{r}_2)$ being a subterm of $\tilde{g}$ where t is the term which contains a_j and $\tilde{r}_1, \tilde{r}_2$ are arbitrary hedges. By assumption there exists a substitution σ with $a_i, a_j \notin \mathcal{F}(\text{Ran}(\sigma))$ such that a_i is not an ancestor of a_j in $\tilde{g}\sigma$ but by the rule of substitution application $a_i(\tilde{r}_1, t, \tilde{r}_2)\sigma = a_i(\tilde{r}_1\sigma, t\sigma, \tilde{r}_2\sigma)$ the ancestor-descendant relation is preserved which is a contradiction.

Case 1.2: a_i is not an ancestor of a_j in $\tilde{g}$. Then we have $(\tilde{r}_1, t_1, \tilde{r}_2, t_2, \tilde{r}_3)$ being a subhedge of $\tilde{g}$ where t_1 is the term which contains a_i, t_2 is the term which contains a_j and $\tilde{r}_1, \tilde{r}_2, \tilde{r}_3$ are arbitrary hedges. By assumption there exists a substitution σ with $a_i, a_j \notin \mathcal{F}(\text{Ran}(\sigma))$ such that a_i is an ancestor of a_j in $\tilde{g}\sigma$ but this contradicts the rule of substitution application $(\tilde{r}_1, t_1, \tilde{r}_2, t_2, \tilde{r}_3)\sigma = (\tilde{r}_1\sigma, t_1\sigma, \tilde{r}_2\sigma, t_2\sigma, \tilde{r}_3\sigma)$ again.

Case 2: A collision appears at three elements. Let a_i, a_j, a_k be those elements. Without loss of generality, assume that a_i, a_j have a common ancestor ϕ which is not an ancestor of a_k in $\tilde{s}$ and let a_j, a_k have a common ancestor ψ which is not an ancestor of a_i in $\tilde{q}$. By assumption, $\tilde{g}$ contains all three symbols exactly once. It follows that there are substitutions σ_1, σ_2 with $a_i, a_j, a_k \notin \mathcal{F}(\text{Ran}(\sigma_1) \cup \text{Ran}(\sigma_2))$ where $\tilde{g}\sigma_1 = \tilde{s}$ and $\tilde{g}\sigma_2 = \tilde{q}$. By assumption, we know that $\tilde{g}\sigma_1$ contains a subhedge $(t_{ij}, \tilde{s}_k)$, with t_{ij} being the term which contains the symbols ϕ, a_i, a_j, and $\tilde{s}_k$ being a hedge which contains the symbol a_k. This implies that $\tilde{g}$ contains either ϕ or a context variable which can be instantiated to introduce ϕ. It follows that $\tilde{g}$ also contains a subhedge $(t'_{ij}, \tilde{s}'_k)$, with t'_{ij} being the term which contains the symbols a_i, a_j and $\tilde{s}'_k$ being a hedge which contains the symbol a_k. Similarly $\tilde{g}\sigma_2$ contains a subhedge $(\tilde{q}_i, t_{jk})$, with $\tilde{q}_i$ being a hedge which contains the symbol a_i and t_{jk} being the term which contains the symbols ψ, a_j, a_k. Further on $\tilde{g}$ either contains ψ or a context variable, say X, which can be instantiated to introduce ψ. Let us call this metavariable χ. As ψ is an ancestor of both, a_j and a_k in $\tilde{q}$, χ has to be above t'_{ij}. This is a contradiction to the assumption that ψ is not an ancestor of a_i in $\tilde{q}$.

($\Rightarrow$) Proof by construction of an algorithm which computes such a generalization for a given admissible alignment of two hedges. In section 4 we described this algorithm and proved its properties. □

Theorem 6 (Complexity of $\mathfrak{G}$). *The anti-unification algorithm $\mathfrak{G}$ has $O(n^2)$ time complexity and $O(n)$ space complexity, where n is the number of symbols in the input.*

Proof. Let P_0; S_0; $\sigma_0 = \{x \colon \tilde{s} \triangleq \tilde{q}; X \colon \circ \triangleq \circ; \mathfrak{a}\}$; $\varnothing$; ε be the initial state of $\mathfrak{G}$ and P_{i-1}; S_{i-1}; $\sigma_{i-1} \Longrightarrow P_i$; S_i; σ_i an arbitrary rule application. By theorem 5 we can arrange the rule applications as we like to obtain a maximal derivation. First the rules Spl-H, Abs-L/R, App-A and Sol-H are applied exhaustively. This are the only rules that operate on P_{i-1} and furthermore they do not have conditions on S_{i-1} or σ_{i-1} such that P_0; S_0; $\sigma_0 \Longrightarrow^+ \varnothing$; S_j; σ_j, for some j. Afterwards they are not applicable again and Res-C is applied exhaustively $\varnothing$; S_j; $\sigma_j \Longrightarrow^*_{\text{Res-C}} \varnothing$; S_k; σ_k. It transforms all the contexts in the store to terms. The rules Clr-S and Mer-S operate on S_k but they only remove AUPs from there, such that Res-C will not be applicable again. Finally we postpone the application of Mer-S to the very end, leading to a partial derivation $\varnothing$; S_k; $\sigma_k \Longrightarrow^*_{\text{Clr-S}} \varnothing$; S_l; $\sigma_l \Longrightarrow^*_{\text{Mer-S}} \varnothing$; S_n; σ_n where no more rule is applicable because Mer-S does not introduce any AUPs, to which another rule could apply.

Now we analyze the first phase P_0; S_0; $\sigma_0 \Longrightarrow^+ \varnothing$; S_j; σ_j. The rule Spl-H splits an AUP into two AUPs and moves some parts into the store. The space overhead for one application is constant because the two new AUPs in P_i and the one in S_i together exactly cover the original one from P_{i-1}, and four new variables are introduced. It can be applied $O(n)$ many times because both of the new AUPs are nonempty. It needs linear time (by the length of the alignment) to check for applicability and find the position for splitting the AUP. Also the context application needs linear time. The rules Abs-L/R are also applicable $O(n)$ many times. They strictly reduce the size of a hedge in P_i. The space overhead is zero. The test for applicability, the context application as well as the operations on the alignment need linear time. App-A is applicable $O(n)$ many times as well and one application needs linear time and constant space. It strictly reduces the size of a hedge in P_i and one application needs linear time, for the same reasons as the above rules. As Spl-H is applicable at most $O(n)$ many times and doubles the elements of P_i at each application and all the other rules do not increase the length of P_i, Sol-H is applicable $O(n)$ many times too. It follows that the number of introduced variables is $O(n)$ and the size of S_j is also bound by $O(n)$.

We compose the substitution σ_i immediately, but we only keep the mappings for x and X in σ_i, such that $\sigma_i = \{x \mapsto \tilde{r}_i, X \mapsto \tilde{c}_i\}$, for some $\tilde{r}_i, \tilde{c}_i$. As all the introduced variables in Spl-H and App-A are fresh, they only appear once in $\tilde{r}_i$ or $\tilde{c}_i$. This invariant of the first phase leads to $O(n)$ size of σ_i as well as $O(n)$ time for the substitution composition in Spl-H, App-A and Sol-H. All together we get $O(n^2)$ time complexity and $O(n)$ space complexity for the first phase.

The second phase is $\varnothing; S_j; \sigma_j \Longrightarrow^*_{\text{Res-C}} \varnothing; S_k; \sigma_k$. The rule Res-C is applicable only once per AUP leading to $O(n)$ many applications. The space overhead is constant at each application, introducing four fresh variables. It needs linear time at each application. We again compose σ_i immediately and for similar reasons as above, the substitution composition in Res-C only needs $O(n)$ time, leading to an overall time complexity of $O(n^2)$ and space complexity $O(n)$.

From the $O(n)$ size of the store, it follows that also the store cleaning rule is applicable $O(n)$ many times and the overall time complexity of this phase is $O(n^2)$, as we compose substitutions immediately like before. The space overhead for Clr-S is zero.

It remains to show that $\varnothing; S_l; \sigma_l \Longrightarrow^*_{\text{Mer-S}} \varnothing; S_n; \sigma_n$ only needs $O(n^2)$ time. Therefore we postpone the substitution composition. Comparing $O(n) * O(n)$ AUPs in the store needs $O(n^2)$ time and removing an AUP from the store needs constant time using a linked list. As the size of the store is bound by $O(n)$ and Mer-S removes one AUP at each application, there are $O(n)$ postponed substitution compositions. Each of them of constant size as they all are just variable renamings. This leads to linear space overhead and we have to compose $O(n)$ substitutions where each composition needs $O(n)$ time. This concludes our complexity analysis where we showed that the algorithm runs in $O(n^2)$ time using $O(n)$ space.   □

Reinstatement and the Requirement of Maximal Specificity in Argument Systems*

Gustavo A. Bodanza[1] and Claudio Andrés Alessio[2]

[1] Universidad Nacional del Sur and CONICET, Argentina
ccbodanz@criba.edu.ar
[2] Universidad Católica de Cuyo and CONICET, Argentina
claudioalessio@uccuyo.edu.ar

Abstract. An argument is reinstated when all its defeaters are in turn ultimately defeated. This is a kind of principle governing most argument systems in AI. Nevertheless, some criticisms to this principle have been offered in the literature. Assuming that reinstatement is prima facie acceptable, we analyze some counterexamples in order to identify common causes. As a result, we found that the problem arises when arguments in a chain of attacks are related by specificity. We argue that the reason is that non-maximally specific arguments can be reinstated originating fallacious justifications. Following old intuitions by Carl Hempel about inductive explanations, we propose a requirement of maximal specificity on defeasible arguments and introduce "undermining defeaters" which, in essence, facilitate the rejection of those arguments which do not satisfy the requirement. This ideas are formally defined using the DeLP system for defeasible logic programming.

1 Introduction: Problems with Argument Reinstatement

Argument reinstatement is at the core of most argument systems, especially those which can be treated as instances of Dung's argumentation frameworks ([4]). The intuition is that an argument should be reinstated when all its possible defeaters are in turn defeated outright (cf. [1]). The example below, introduced by Dung as a motivation for his admissibility semantics, illustrates the rationale of reinstating argument A given argument C.

Example 1.
A: (Agent 1:) My government cannot negotiate with your government because your gorvernment doesn't even recognize my government.
B: (Agent 2:) Your government doesn't recognize my government either.
C: (Agent 1:) But your government is a terrorist government.
Then accepting that A and B are mutually attacking arguments and that C attacks B (but not the converse), the reinstatement of A by C makes sense.

On the other hand, other examples suggest that reinstatement cannot be taken as a general principle:

---

* Partially supported by SeCyT of the Universidad Nacional del Sur, and Universidad Católica de Cuyo, Argentina.

U. Kohlenbach et al. (Eds.): WoLLIC 2014, LNCS 8652, pp. 81–93, 2014.
© Springer-Verlag Berlin Heidelberg 2014

Example 2.
A: Tweety flies because it is a bird, and birds tend to fly.
B: Tweety does not fly because it is a penguin, and penguins tend not to fly.
C: Tweety flies because it is a magic penguin, and magic penguins tend to fly.

Horty ([11]) has argued against reinstatement using a similar example. If *A* and *C* are jointly admitted, then a sound conclusion (Tweety flies) could be justified on basis of a weak reason (that flies because it is a bird). Clearly, a stronger reason is that Tweety has a skill that specifically magic penguins have. The acceptance of *A* would be unsound if the model is intended to offer the best explanation for the conclusion it yields. *A* would be acceptable only if all subclasses of birds (including penguins and magic penguins) are equally plausible to fly; but that is not the case here. The fact that *C* reinstates *A*'s conclusion (which is also *C*'s own conclusion) cannot be a reason for *C* to reinstate the whole argument *A*, because *A* does not meet that criterion.

A worse situation arises when the conclusion of the reinstated argument is stronger than that of the reinstating argument, as in the following case (also introduced by Horty):

Example 3.
A: Beth is millionaire because he is a Microsoft employee, and they tend to be million-aire.
B: Beth has less than half a million because he is a new Microsoft employee, and they tend to have less than half a million.
C: Beth has at least half a million because he is a new Microsoft employee in department *X*, and they have at least half a million.

Here *A*'s conclusion is stronger than *C*'s conclusion, in the sense that the last one is logically implied by the first one but not vice versa; that looks counterintuitive. Argument *A* would be reasonably accepted just in case that being millionaire be equally plausible for any subclass of the class of Microsoft employees.

Curiously, all the counterexamples to reinstatement that we found in the literature involve arguments that can be compared by specificity. That motivated the present study, which tries to show that the problem is the way in which specificity is used to establish defeat rather than a problem of the reinstatement principle.

The specificity criterion has been widely discussed in Philosophy of Science. Hempel ([9]) defended a *requirement of maximal specificity* as a condition for the acceptance of probabilistic/inductive-statistical explanations. Early applications of specificity in non-monotonic reasoning in AI were also aware of the intuition that only maximally specific explanations should be accepted, so from the argumentative point of view ([16]) as from the defeasible inheritance networks point of view ([6], [7], [10], [13], [18]). On the other hand, a specificity-based preference criterion among arguments combined with a reinstatement-based warrant procedure was introduced in [19].

Prakken ([17]) argued that reinstatement cannot be applied when statistical reasoning is at stake because more general arguments (like *A* in the above example) just cannot be constructed in a right representation, since the pertinent defaults must be blocked, and so only the most specific arguments remain; hence –Prakken concludes– the problem here is not about reinstatement but one of representation. In our opinion, while finding general principles of representation could be a hard enterprise, the problem can instead be solved by finding general conditions under which the arguments can compete and defeat among them. Accordingly, we will argue that maximally specific arguments "undermine" less specific arguments when their conclusions are not plausible given the total evidence.

The paper is organized as follows. In section 2 a requirement of maximal specificity is formally introduced in terms of the defeasible logic programming language DeLP ([8]). Sections 3 and 4 introduce "undermining" defeaters and their role in a skeptical warrant procedure. Section 5 discuss the view of other authors through more examples, and our conclusions are offered in section 6.

2 The Requirement of Maximal Secificity in Rule-Based Argumentation Systems: The Case of DeLP

We introduce here the requirement of maximal specificity as a demarcation criterion for the acceptance of arguments. As such, it should be used to filter the arguments which are not maximally specific w.r.t. their conclusions as they can leave room for irrelevant explanations. We will formally define our criterion in the context of the particular rule-based argument system DeLP ([8]), where specificity is formally defined as a criterion for argument comparison.

DeLP is based on a first-order language $\mathcal{L}$ that is partitioned in three disjoint sets: a set of facts, a set of strict rules and a set of defeasible rules. *Facts* are literals, i.e. ground atoms (L) or negated ground atoms ($\sim L$, where '$\sim$' represents the classical negation); facts represent particular knowledge. Both *strict* and *defeasible rules* are program rules. Syntactically, strict rules are sequents of the form $L \leftarrow L_1, \ldots, L_n$ and defeasible rules are sequents of the form $L \prec L_1, \ldots, L_n$, where $L, L_1, \ldots, L_n$ are literals. Strict rules represent general, non-defeasible knowledge while defeasible rules represent tentative, defeasible knowledge. A *defeasible logic program* (de.l.p.) $\mathcal{P}$ is a pair (Π, Δ) where Π is a set partitioned in two subsets Π_F, containing only facts, and Π_G, containing only strict rules, and Δ is a set of defeasible rules. Given a de.l.p. $\mathcal{P} = (\Pi, \Delta)$ we say that a literal L is a *defeasible derivation* from Γ in $\mathcal{P}$, in symbols, $\Gamma \mathrel{\vdash\mkern-7mu\sim}_{\mathcal{P}} L$ iff $\Gamma \subseteq \Pi \cup \Delta$ and there exists a sequence of ground (instantiated) literals $L_1, \ldots, L_n$ such that $L_n = L$ and for each L_i, $1 \leq i \leq n$, either $L_i \in \Gamma$ or there exists either a strict rule $(L \leftarrow L_1, \ldots, L_k)$ or a defeasible rule $(L \prec L_1, \ldots, L_k)$ in Γ such that $\{L_1, \ldots, L_k\} \subseteq \{L_1, \ldots, L_{i-1}\}$. If all the rules used in the derivation of A are strict then we say that L is a *strict derivation* from Γ, in symbols, $\Gamma \vdash_{\mathcal{P}} L$. (From now on, we will write '$\mathrel{\vdash\mkern-7mu\sim}$' and '$\vdash$' instead of '$\mathrel{\vdash\mkern-7mu\sim}_{\mathcal{P}}$' and '$\vdash_{\mathcal{P}}$', respectively, when the referenced de.l.p. is obvious.)

Definition 1. *(Argument structure ([8])) Given a de.l.p.* $\mathcal{P} = (\Pi, \Delta)$, *an* argument structure *(in* $\mathcal{P}$*) is a pair* $\langle T, h \rangle$, *where* $T \subseteq \Delta$ *and* h *is a literal (the argument's* conclusion*), and*

1. $\Pi \cup T \mathrel{\vdash\mkern-7mu\sim} h$,
2. $\Pi \cup T \mathrel{\not\vdash\mkern-7mu\sim} \bot$,
3. $\not\exists T'$ $(T' \subset T \land \Pi \cup T' \mathrel{\vdash\mkern-7mu\sim} h)$.

Definition 2. *(Subargument ([8])) An argument structure $\langle T, h \rangle$ is a subargument structure of an argument structure $\langle T', h' \rangle$ if $T \subseteq T'$.*

Definition 3. *(Strictly more specific ([8])) Let $\mathcal{P} = (\Pi, \Delta)$ be a de.l.p. and let F be the set of all literals that have a defeasible derivation from $\mathcal{P}$. Let $\langle T_1, h_1 \rangle$ and $\langle T_2, h_2 \rangle$ be two argument structures obtained from $\mathcal{P}$. $\langle T_1, h_1 \rangle$ is strictly more specific than $\langle T_2, h_2 \rangle$, in symbols, $\langle T_1, h_1 \rangle \succ_{\text{spec}} \langle T_2, h_2 \rangle$ iff*
1. for all $H \subseteq F$, if $H \cup \Pi_G \cup T_1 \vdash\!\!\!\sim h_1$ and $H \cup \Pi_G \not\vdash\!\!\!\sim h_1$ then $H \cup \Pi_G \cup T_2 \vdash\!\!\!\sim h_2$, (every H that "activates" h_1 also 'activates' h_2), and
2. there exists $H \subseteq F$ such that $H \cup \Pi_G \cup T_2 \vdash\!\!\!\sim h_2$, $H \cup \Pi_G \not\vdash\!\!\!\sim h_2$ and $H \cup \Pi_G \cup T_1 \not\vdash\!\!\!\sim h_1$ (some H "activates" h_2 but not h_1).

Using this same specificity criterion, Poole [16] proposes to choose the most specific explanations, i.e. those arguments which are maximal with respect to $\succ_{\text{spec}}$. In this way, Poole leaves no room for reinstatement among arguments compared by specificity. This criterion is near to what we will propose here, but so stated it can have the effect of precluding acceptable arguments even when less specific arguments are not in conflict with the maximally specific ones.

Example 4. Let $\mathcal{P} = (\Pi, \Delta)$ be a de.l.p. representing the knowledge that all lapwings are birds, birds tend to fly, lapwings tend to nest on the ground and Pedro is a lapwing:
$\Pi = \{\ bird(x) \leftarrow lapwing(x),\ lapwing(pedro)\ \}$
$\Delta = \{\ flies(x) \prec bird(x),\ nests\_on\_the\_ground(x) \prec lapwing(x)\ \}$
Then we have the argument structures:
$A = \langle \{flies(pedro) \prec bird(pedro)\},\ flies(pedro) \rangle$,
$B = \langle \{nests\_on\_the\_ground(pedro) \prec lapwing(pedro)\},$
$\qquad nests\_on\_the\_ground(pedro)\rangle$
Since $B \succ_{\text{spec}} A$, choosing only the maximal elements of $\succ_{\text{spec}}$ precludes the acceptable argument A and its conclusion $flies(pedro)$.

Indeed, selecting just the maximal elements of $\succ_{\text{spec}}$ does not seem to be a good approach to the requirement of maximal specificity as proposed in Philosophy of Science for inductive-probabilistic explanations. The intuition in [9] is that what is inferred in a maximally specific explanation about a class G taking into account the total evidence must also be inferred about any subclass H of G with the same probability. Though extrapolating this criterion to defeasible argumentation is difficult since inferences are not obtained with probability measures, we propose that a maximally specific defeasible argument about a class G should at least not be contradictory with the defeasible conclusions obtained about any subclass H of G, considering the total evidence, i.e. the information represented in Π. In terms of DeLP, this means that maximally specific arguments should not have "proper defeaters" as these are indicative of "undermining" evidence.

Definition 4. *(Proper defeater ([8])) An argument structure $\langle S, j \rangle$ is a proper defeater of an argument structure $\langle T, h \rangle$ if for some sub-argument $\langle T', h' \rangle$ of $\langle T, h \rangle$, $\langle S, j \rangle \succ_{\text{spec}} \langle T', h' \rangle$ and $\Pi \cup \{j, h'\} \vdash \bot$. Given a set of argument structures S we also define $def_{prop}(S) =_{df} \{(A, B) : A, B \in S \text{ and } A \text{ is a proper defeater of } B\}$.*

Definition 5. *(Undermining evidence) Given a de.l.p.* $\mathcal{P} = (\Pi, \Delta)$, *a subset* F *of* Π_F *is* undermining evidence *of an argument structure* $\langle T, h \rangle$ *if* $F \cup T' \cup \Pi_G \hspace{2pt}\mid\hspace{-6pt}\sim h'$ *for some proper defeater* $\langle T', h' \rangle$ *of* $\langle T, h \rangle$ *(i.e. F "activates" some proper defeater of the argument).*

Note that having a proper defeater is a sufficient condition for having undermining evidence, though other conditions could be also found (more on this in section 5). Now we can formally state the property of maximal specificity as follows:

Definition 6. (Maximal Specificity (MS)) *Given a de.l.p* $\mathcal{P}$, *we say that an argument structure* $\langle T, h \rangle$ *is* maximally specific (w.r.t. its conclusion h) in $\mathcal{P}$ iff there exists no undermining evidence of $\langle T, h \rangle$ in $\mathcal{P}$.

Requiring MS as a condition for argument warrant implies to reject any argument structure which has some proper defeater. Note that it does not matter whether proper defeaters are defeated or not to reject a non-maximally specific argument; that is why we prefer to highlight the undermining evidence and to use proper defeaters just as a way for detecting it.

Example 5. (Example 4 revisited) Both A and B satisfy MS, A w.r.t. $flies(pedro)$ and B w.r.t. $nests\_on\_the\_ground(pedro)$.

Example 6. (De.l.p. for representing Example 3) Let $\mathcal{P} = (\Pi, \Delta)$ a de.l.p. such that

$\Pi = \{ has\_at\_least\_half\_a\_million(x) \leftarrow millionaire(x),$
$\quad\quad ms\_employee(x) \leftarrow new\_ms\_employee(x),$
$\quad\quad new\_ms\_employee(x) \leftarrow new\_ms\_employee\_dept\_x(x),$
$\quad\quad new\_ms\_employee\_dept\_x(beth)\}$

$\Delta = \{ millionaire(x) \prec ms\_employee(x),$
$\quad\quad \sim has\_at\_least\_half\_a\_million(x) \prec new\_ms\_employee(x),$
$\quad\quad has\_at\_least\_half\_a\_million(x) \prec new\_ms\_employee\_deptX(x)\}$

Then we have the argument structures:

$A = \langle \{millionaire(beth) \prec ms\_employee(beth)\}, millionaire(beth) \rangle,$
$B = \langle \{\sim has\_at\_least\_half\_a\_million(beth) \prec new\_ms\_employee(beth)\},$
$\quad\quad \sim has\_at\_least\_half\_a\_million(beth) \rangle,$
$C = \langle \{has\_at\_least\_half\_a\_million(beth) \prec new\_ms\_employee\_deptX(beth)\},$
$\quad\quad has\_at\_least\_half\_a\_million(beth) \rangle.$

Then C satisfies MS w.r.t. $has\_at\_least\_half\_a\_million(beth)$, and neither A nor B satisfy MS because $\{new\_ms\_employee\_dept\_x(beth)\}$ is undermining evidence for them.

3   Undermining Defeaters

Systems in which arguments interact *only* through proper defeaters can lead to the acceptance of non-maximally specific arguments if the warrant procedure satisfies the reinstatement principle. But this does not necessarily imply that reinstatement is invalid. Those argument systems in which different kinds of defeat are used —including proper defeaters— can be amended to sanction only maximally specific arguments. Our proposal is simple and consists in the introduction of "undermining defeaters", which are based on the main result derived from the notion of 'undermining evidence':

Lemma 1. *Let $\langle T, h \rangle$ and $\langle T', h' \rangle$ be two argument structures such that $\langle T, h \rangle$ is a proper defeater of $\langle T', h' \rangle$. If H is undermining evidence for $\langle T, h \rangle$ then H is undermining evidence for $\langle T', h' \rangle$.*

Proof. Let $\langle T, h \rangle$ be a proper defeater of $\langle T', h' \rangle$ and let H be undermining evidence for $\langle T, h \rangle$. Then H activates some proper defeater $\langle S, j \rangle$ of $\langle T, h \rangle$. Since $\langle S, j \rangle$ is more specific than $\langle T, h \rangle$, H activates $\langle T, h \rangle$. And since $\langle T, h \rangle$ is more specific than $\langle T', h' \rangle$, H also activates $\langle T', h' \rangle$. Then, by Definition 5, H is undermining evidence for $\langle T', h' \rangle$.
□

Definition 7. *(Undermining defeater) Given two arguments structures $\langle T, h \rangle$ and $\langle T', h' \rangle$, we say that $\langle T, h \rangle$ is an* undermining *defeater of $\langle T', h' \rangle$ iff for any subset of facts $F \subseteq \Pi_F$, if $F \cup T \cup \Pi_G \hspace{1pt}\vdash\hspace{-6pt}\sim\hspace{1pt} h$ then $F \cup S \cup \Pi_G \hspace{1pt}\vdash\hspace{-6pt}\sim\hspace{1pt} j$ for some proper defeater $\langle S, j \rangle$ of $\langle T', h' \rangle$ (i.e. if F activates $\langle T, h \rangle$ then F also activates some proper defeater of $\langle T', h' \rangle$). We also define $def_{und}(S) =_{df} \{(A, B) : A, B \in S$ and A is an undermining defeater of $B\}$.*

Undermining defeaters can be viewed as a kind of undercutting defeaters, at least indirectly, since their acceptance implies the use of total evidence which gives the reason that makes the conclusion of the defeated argument not inferable. They are clearly not rebutting defeaters since it cuould be the case that the joint acceptance of both an argument an its undermining defeater does not yield contradiction (for instance, in Example 5 argument C is an undermining defeater of argument A, but it is not a rebutting defeater. See, e.g., [15] for more on the distinction undercutting/rebutting defeater).

Clearly, from Lemma 1 and Definition 7 we have that undermining evidence 'propagate' through a chain of proper defeaters.

Lemma 2. *Let $\langle T, h \rangle$ be a proper defeater of $\langle T', h' \rangle$. Then for every proper defeater $\langle S, j \rangle$ of $\langle T, h \rangle$, $\langle S, j \rangle$ is an undermining defeater of $\langle T', h' \rangle$.*

Proof. Immediate from Definition 7 and Lemma 1. □

As a consequence, in cycles of proper defeaters the ensuing undermining defeaters are indicative of undermining evidence for all the arguments involved in the cycle, including themselves (Fig. 1). Finally, the previous lemmata lead immediately to the following equation.

Theorem 1. *Let S be any set of argument structures, and $def_{prop}(S)^{tr}$ be the transitive closure[1] of $def_{prop}(S)$. Then $def_{und}(S) = def_{prop}(S)^{tr}$.*

Proof. Immediate from Lemma 1 and Lemma 2. □

In the next section, the above result will enable us to think of different ways of representing the rejection of non-maximally specific arguments.

---

[1] The *transitive closure* of a binary relation R is the minimal (w.r.t. $\subseteq$) transitive relation R' such that $R \subseteq R'$.

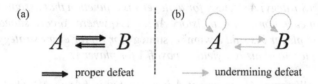

(a) $A \rightleftarrows B$ (b) $A \rightleftarrows B$

$\Longrightarrow$ proper defeat $\longrightarrow$ undermining defeat

Fig. 1. Argumentation framework (a) with proper defeaters only and (b) with undermining defeaters

4 Undermining Defeaters and a Warrant Procedure Satisfying Reinstatement

Non-maximally specific arguments can be rejected through a warrant procedure satisfying reinstatement, which means that reinstatement can be saved from the before mentioned criticisms as a principle of defeasible argumentation. In DeLP, warrant can be determined through a dialectical analysis represented by a two-party game, where a proponent tries to defend an argument and an opponent tries to refute it (we define this game as in [2]). Given the set *Args* of all the argument structures that can be constructed in a de.l.p. $\mathcal{P}$, and once all the defeat relations over *Args* are established, argument warrant can be analyzed through a Dung's style argumentation framework ([4]).

Definition 8. *(Argumentation framework associated with a de.l.p.) Given a de.l.p.* $\mathcal{P}$, *the* argumentation framework associated with $\mathcal{P}$ *is the pair* $\langle Args, attacks \rangle$ *where Args is the set of all the argument structures obtained from* $\mathcal{P}$ *and attacks =* $\bigcup DEF(Args)$, *where* $DEF(Args) = \{def_1, \ldots, def_k\}$, *is the set containing every defeat criterion* $def_i \subseteq Args \times Args$ $(1 \leq i \leq n)$ *defined on Args. (We will assume that* $def_{prop}(Args) \in DEF(Args)$.)

Definition 9. *(Argumentation game) An* argumentation game *on an argumentation framework* $\langle Args, attacks \rangle$ *is a zero-sum extensive game in which:*
1. There are two players, i and $-i$, who play the roles of $\boldsymbol{P}$ *and* $\boldsymbol{O}$, *respectively.*
2. A history in the game is any sequence $A_0, A_1, A_2, \ldots, A_{2k}, A_{2k+1}, \ldots$ *of choices of arguments in Args made by the players in the game.* A_{2k} *corresponds to* $\boldsymbol{P}$ *and* A_{2k+1} *to* $\boldsymbol{O}$, *for* $k = 0, 1, \ldots$. *At any history,* A_0 *is the argument that player* $\boldsymbol{P}$ *intends to defend.*
3. In a history, the choices by a player i at a level $k > 0$ are $C_i(k) = \{A \in Args : (A, A_{k-1}) \in attacks\}$.
4. A history of finite length K, $A_0, \ldots, A_K$, *is terminal if* A_K *corresponds to player j ($j = i$ or $j = -i$) and* $C_{-j}(K+1) = \emptyset$.
5. Payoffs are determined at terminal histories: at $A_0, \ldots, A_K$, $\boldsymbol{P}$'s *payoff is 1 (representing winning) if K is even (i.e.,* $\boldsymbol{O}$ *cannot reply to* $\boldsymbol{P}$'s *last argument), and* -1 *(representing loosing) otherwise. In turn,* $\boldsymbol{O}$'s *payoff at* $A_0, \ldots, A_K$ *is 1 if K is odd and* -1 *otherwise.*

Definition 10. *(Strategy) A strategy for a player i is a function that assigns an element $A_{l+1} \in C_i(l)$ at each non-terminal history $A_0, \ldots, A_l$ where A_l corresponds to player $-i$. A strategy of player i is said a* winning strategy *for i if for every strategy chosen by $-i$, the ensuing terminal history yields a payoff 1 for player i.*

Definition 11. *(Warrant) An argument A is* warranted *in $(Args, attacks)$ iff P has a winning strategy to defend A in the game associated to $(Args, attacks)$.*

Furthermore, different game protocols can be defined to obtain different behaviors. Since we are interested here in the refutation of non-maximally specific arguments, let us see how to do that in systems that incorporate undermining defeaters and in systems based only in proper defeaters. For the first approach we propose the following protocol:

(1) The game ends if, at any level k, a player i advances an argument A such that the argument B moved at level $k - 1$ by player $-i$ is such that A is an undermining defeater of B (i wins).

(2) P is not allowed to advance an argument that was already advanced by either player in the same history.

Rule (1) says that once an undermining defeater is played the game ends (the player who moved the non-maximally specific argument loses). The purpose of this rule is to obligate the players to use only maximally specific arguments. Rule (2), in time, ensures finite, skeptical games. Let us call this protocol PU.

On the other hand, an obvious way to obtain the same behavior in argumentation frameworks where only proper defeaters are defined is by replacing the first rule as follows:

(1') The game ends if, at any level k, a player i advances an argument A such that the argument B moved at level $k - 1$ by player $-i$ is such that there exists a sequence $A_1, \ldots, A_n$ where $A_1 = A$, $A_n = B$ and $(A_h, A_{h+1}) \in def_{prop}$ for every h, $1 \le h < n$ (i wins).

Let us call this protocol PP. Then Theorem 1 clearly ensures the same behavior under protocol PU as under protocol PP.

Example 7. Given an argumentation framework $(Args, attacks)$ where $Args = \{A, B, C, D\}$ and $attacks = \{(C, D)\} \cup def_{prop}(Args)$, where $def_{prop}(Args) = \{(A, B),$

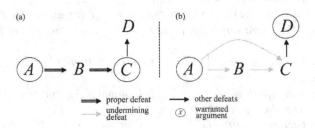

Fig. 2. Argumentation framework (a) without and (b) with undermining defeaters

$(B, C)\}$. Then $def_{und}(Args) = \{(A, B), (B, C), (A, C)\}$. Note that while **P** has winning strategies for defending both A and C in the game associated to $(Args, attacks)$ in the plain game (i.e. not having any added protocol) (Fig. 2, (a)), **P** has winning strategies for defending both A and D so in the game associated to $(Args, attacks)$ under protocol PP as in the game associated to $(Args, attacks \cup def_{und})$ under protocol PU (Fig. 2, (b)).

5 Discussion

The solution we have proposed here works well, in particular, for the simplest version of DeLP [19] where the attack relation is defined only in terms of blocking and proper defeaters. Given a de.l.p. $\mathcal{P} = (\Pi, \Delta)$, $\langle T, h \rangle$ is a *blocking defeater* of $\langle T', h' \rangle$ iff there exists some sub-argument $\langle T'', h'' \rangle$ of $\langle T', h' \rangle$ such that $\Pi \cup \{h, h''\} \vdash \bot$, and $\langle T, h \rangle$ and $\langle T'', h'' \rangle$ are not related by specificity. Define $def_{block} =_{df} \{(A, B) : A$ is a blocking defeater of $B\}$. Then def_{block} is clearly symmetric. Let us now analyze Example 7 in terms of this system. As the attack from C to D is not a case of proper defeater, it must be a case of blocking defeater. Then, by the symmetry of def_{block}, D also attacks C. But note that this does not change the resulting warrant of A and D under protocol PP. On the other hand, some dubious cases that can arise under other specifications of the attack relation are avoided in this system. For example, assume that A is a (non-proper) defeater of B and B is a proper defeater of C. Then it could seem reasonable the reinstatement of C by A, even when C is not maximally specific. Nevertheless, that could not happen in DeLP because the assumption that A is a non-proper defeater of B implies that A and B are blocking defeaters one of each other. Hence, it is easy to see that the reinstatement of C by A is impossible under protocol PP as **P** lacks of a winning strategy for A (**O** can repeat B to refute A, leaving **P** out of moves).

Similar examples would suggest that non-maximally specific arguments should be reinstated anyway. For instance, consider again Example 2 but where argument C is now: "It cannot be concluded that Tweety is a penguin since it was observed under deficient sight conditions during a blizzard". Now C could be seen as an undercutting defeater of B and B as a proper defeater of A, what would lead to the reinstatement of the non-maximally specific argument A. But note that the acceptance of C implies the treatment of 'Tweety is a penguin' not as evidence but as a questionable presumption, hence B should not be treated as a proper defeater of A strictly. Therefore, A is still a maximally specific argument and its reinstatement seems right.

For other cases where C is a (unidirectional, non-proper) defeater of B and B is a proper defeater of A, it is not clear whether A should be reinstated or not. Indeed, it is difficult for us to conceive such an example.

The last example was introduced by Prakken ([17]) to show that, unlike *direct reinstatement*, *indirect reinstatement* is valid. Direct reinstatement is when all three arguments are in conflict on their final conclusions (e.g. Example 2). Indirect reinstatement, on the other hand, is when the reinstating argument C defeats the 'middle' argument B on one of its intermediary conclusions (e.g. Example 3). But this distinction is not related to our solution as it does not focus on the kind of defeaters which are involved and the role of undermining evidence on them, which is the key in the MS property.

The following example was also introduced by Prakken to argue that reinstatement depends "on the nature of the domain, the kind of knowledge involved and the context in which this knowledge is used" ([17]: 93):

Example 8.
A: John will be imprisoned up to 6 years because for theft imprisonment up to 6 years is acceptable, and John has been found guilty of theft.
B: John will be imprisoned for no more than 3 years because for theft out of poverty imprisonment of more than 3 years is not acceptable, and evidence shows that John stole motivated by poverty.
C: John will be imprisoned for more than 4 years because he stole during riots, and for theft during riots, even when poverty is proved, only imprisonment of more than 4 years is acceptable.

Prakken argues that the reinstatement of *A* by *C* is valid here and leads to accept an imprisonment between 4 and 6 years. We disagree at this point since anyway, in our opinion, *C* is a proper defeater of *B* and *B* is a proper defeater of *A*, hence *C* is an undermining defeater of *A*. The total evidence considered in *C* about a more serious crime than theft out of poverty leads to put a minimum of 4 years of imprisonment, leaving the upper limit not established. Indeed, we can imagine even more serious crimes (e.g. murder) which occurrence together with theft would rise the top above 6 years. Hence we think that *C* is the only warranted argument and *A* should not be reinstated.

Nevertheless, there are still open problems to deal with. Our notion of undermining defeat is not completely characterized as it lies on a concept of undermining evidence for which we state sufficient but not necessary conditions. While having a proper defeater is a clear sign of an argument's undermining evidence, in other cases the total evidence should prevent some conclusion without sanctioning the contrary. Horty ([12]) analyses the following example. Assume that a population of ruffed finches, a kind of birds, is distributed among a couple of islands. Their nests are mostly but not entirely confined to Green Island, but there is a particular subspecies known as least ruffed finches whose nests are distributed almost evenly between Green Island and Sand Island. Now, consider a particular individual, Frank, who happens to be a least ruffed finch. What should we conclude about the location of Frank's nest? Though this situation cannot be represented in DeLP because disjunctions cannot occur in the head of a rule, we can adjust the information to the formalism by considering that Green Island and Sand Island conform a group of islands, call it 'Two Islands', so we can get the following representation:

Example 9. Let $\mathcal{P} = (\Pi, \Delta)$ be a de.l.p. representing the knowledge that all least ruffed finches are ruffed finches, ruffed finches tend to nest on Green Island, least ruffed finches tend to nest on Two Islands, nesting on Green Island implies nesting in Two Islands, and Frank is a least ruffed finch:

$$\Pi = \{ \, ruffed\_finch(x) \leftarrow least\_ruffed\_finch(x),$$
$$nests\_on\_TwoIslands(x) \leftarrow nests\_on\_GreenIsland(x),$$
$$least\_ruffed\_finch(frank) \, \}$$

$\Delta = \{\ nests\_on\_GreenIsland(x) -\!\!\prec ruffed\_finch(x),$
$\qquad nests\_on\_TwoIslands(x) -\!\!\prec least\_ruffed\_finch(x)\ \}$

Then we have, among others, the argument structures:
$A = \langle \{nests\_on\_GreenIsland(frank) -\!\!\prec ruffed\_finch(frank)\},$
$\qquad nests\_on\_GreenIsland(frank)\rangle$
$B = \langle \{nests\_on\_TwoIslands(frank) -\!\!\prec least\_ruffed\_finch(frank)\},$
$\qquad nests\_on\_TwoIslands(frank)\rangle$

Though B is more specific than A it is not a proper defeater, hence the conclusion that Frank nests on Green Island is obtained. The formalisms incurs in the fallacy of exclusion, since the information that Frank is a least ruffed finch is obviated, treating Frank just as a ruffed finch. To solve the problem, Horty proposes to add a (meta-level) default expressing that cases of least ruffed finches exclude the application of the default that connects ruffed finches with nesting on Green Island (a kind of undercutting defeater). But this solution requires more representation, while we are inclined to less representation dependence. In our opinion, this must be solved by defining a new kind of undermining defeater which makes appropriate use of the total evidence, so that argument B (or the evidence on which B is built) undercuts, in some specified way, argument A.

More in the line of Horty's solution, the work by Modgil on hierarchical argumentation ([14]) offers another interesting turn to the problem of reinstatement introducing arguments for (meta-level) preference criteria. The model develops a form of meta-argumentation where, for example, if $A\ attacks\ B$ is established on basis of a preference criterion P1, and $B\ attacks\ A$ is established on basis of a preference criterion P2, an argument C supporting the preference of P1 over P2 poses an attack on $B\ attacks\ A$, A resulting reinstated. Note that, under this view, C is not attacking B but *the attack* of B over A. The example of Tweety observed during a blizzard can be interpreted in this terms assuming that the preference criterion is based on an ordering $>$ on the evidence, such that $bird(tweety) > penguin(tweety)$. Then, while B is a proper defeater of A, C expresses a preference of A over B based on $>$, so that C defends A. Examples like Example 9, on the other hand, cannot be solved unless, again, a special kind of undercutting defeater is defined.

This gives rise to the question of what kind of defeaters are undermining defeaters. We have argued that they qualify as undercutting defeaters. As undermining defeaters are based on a total-evidence requirement they can be considered a kind of –in Pollock's terms– subproperty defeaters, just the same as specificity (i.e. proper) defeaters. And subproperty defeaters are all undercutting defeaters. Pollock's words seem to confirm our opinion:

> To the best of my knowledge, there has never been an intuitive example of specificity defeat presented anywhere in the literature that is not an example of the operation of the total-evidence requirement in one of these special varieties of defeasible inference [statistical syllogism, direct inference, various kinds of legal and deontic reasoning], and the latter are all instances of undercutting defeat. Accordingly, I will assume that undercutting defeaters and rebutting defeaters are the only possible kinds of defeaters. ([15]: 236)

Finally, several principles have been introduced in order to validate the argumentation inference of rule-based argumentation systems, mainly consistency and closure ([3], [5]). A formal analysis of the relationship between maximal specificity and these principles is planned as future work.

6 Conclusion

The issue of reinstatement as a principle for argument systems was the subject of a serious criticism ([11]) while its defense (mainly that of [17]) has not been entirely satisfactory in our opinion. The criticism focuses only cases in which specificity is the comparison criterion among arguments. We argued here that the problem is that specificity based argument systems do not incorporate a precise way of defeating all non-maximally specific arguments. We proposed a formal criterion of maximal specificity which, in accordance with early researches about inductive explanations ([9]), is based on the total evidence represented in the knowledge base. Moreover, we introduced *undermining defeaters* and showed how they enable the warrant of only maximally specific arguments in the context of the DeLP system ([8]) defining particular argumentation game protocols.

Acknowledgements. We thank three anonymous referees who made helpful criticisms that improved this work.

References

1. Baroni, P., Giacomin, M.: Semantics of Abstract Argument Systems. In: Rahwan, I., Simari, G.R. (eds.) Argumentation in artificial intelligence, pp. 25–44. Springer Publishing Company (2009)
2. Bodanza, G., Thomé, F., Simari, G.: Argumentation Games for Admissibility and Cogency Criteria. In: Verheij, B., Szeider, S., Woltran, S. (eds.) Computational Models of Argument. Proceedings of COMMA 2012, pp. 153–164. IOS Press (2012)
3. Caminada, M., Amgoud, L.: On the Evaluation of Argumentation Formalisms. Artificial Intelligence 171(5-6), 286–310 (2007)
4. Dung, P.M.: On the Acceptability of Arguments and its Fundamental Role in Nonmonotonic Reasoning, Logic Programming and n-Person Games. Artificial Intelligence 77, 321–358 (1995)
5. Dung, P.M., Thang, P.M.: Closure and Consistency In Logic-Associated Argumentation. Journal of Artificial Intelligence Research 49, 79–109 (2014)
6. Etherington, D.: Formalizing Nonmonotonic Reasoning Systems. Artificial Intelligence 31, 41–85 (1987)
7. Etherington, D., Reiter, R.: On Inheritance Hierarchies with Exceptions. In: Proceedings of the Third National Conference on Artificial Intelligence (AAAI 1983), pp. 104–108 (1983)
8. García, A., Simari, G.: Defeasible Logic Programming: An Argumentative Approach. Theory and Practice of Logic Programming 4(1), 95–138 (2004)
9. Hempel, C.: Maximal Specificity and Lawlikeness in Probabilistic Explanation. Philosophy of Science 35(2), 116–133 (1968)

10. Horty, J.: Some Direct Theories of Nonmonotonic Inheritance. In: Gabbay, D., Hobber, C., Robinson, J. (eds.) Handbook of Logic in Artificial Intelligence and Logic Programming. Nonmonotonic Reasoning and Uncertain Reasoning, vol. 3, pp. 111–187. Oxford University Press (1994)
11. Horty, J.: Argument Construction and Reinstatement in Logics for Defeasible Reasoning. Artificial Intelligence and Law 9, 1–28 (2001)
12. Horty, J.: Reasons as Defaults. Oxford University Press (2012)
13. Horty, J., Thomason, R., Touretzky, D.: A Skeptical Theory of Inheritance in Nonmonotonic Semantic Networks. Artificial Intelligence 42, 311–348 (1990)
14. Modgil, S.: Value Based Argumentation in Hierarchical Argumentation Frameworks. In: Dunne, P., Bench-Capon, T. (eds.) Proc. of Computational Models of Argument, COMMA 2006, Liverpool, UK, September 11-12. Frontiers in Artificial Intelligence and Applications, vol. 144, pp. 297–308. IOS Press (2006)
15. Pollock, J.: Defeasible Reasoning with Variable Degrees of Justification. Artificial Intelligence 133, 233–282 (2001)
16. Poole, D.: On the Comparison of Theories: Preferring the Most Specific Explanation. In: Proc. of the Ninth IJCAI, Los Altos, pp. 144–147 (1985)
17. Prakken, H.: Intuitions and the Modelling of Defeasible Reasoning: Some Case Studies. In: Benferhat, S., Giunchiglia, E. (eds.) Proceedings of the 9th International Workshop on Non-Monotonic Reasoning (NMR 2002), Tolouse, France, April 19-21, vol. 2, pp. 91–102 (2002)
18. Reiter, R., Criscuolo, G.: On Interacting Defaults. In: Proceedings of the Seventh International Joint Conference on Artificial Intelligence (IJCAI 1981), pp. 270–276 (1981)
19. Simari, G., Loui, R.: A Mathematical Treatment of Defeasible Reasoning. Artificial Intelligence 53, 125–157 (1992)

Sequentialization for N-Graphs
via Sub-N-Graphs

Ruan Carvalho, Laís Andrade, Anjolina de Oliveira, and Ruy de Queiroz

Depto. de Estatística e Informática, Universidade Federal Rural de Pernambuco,
52171-900 Recife, Pernambuco, Brazil
Centro de Informática, Universidade Federal de Pernambuco,
50740-560 Recife, Pernambuco, Brazil
rvbc@deinfo.ufrpe.br, {lsa,ago,ruy}@cin.ufpe.br

Abstract. Since proof-nets for MLL$^-$ were introduced by Girard (1987), several studies have been made on its soundness proof. Bellin & Van de Wiele (1995) produced an elegant proof based on properties of subnets (empires and kingdoms) and Robinson (2003) proposed a straightforward generalization of this presentation for proof-nets from sequent calculus for classical logic. This paper extends these studies to obtain a proof of sequentialization theorem for N-Graphs, which is a symmetric natural deduction calculus for classical propositional logic that adopts Danos–Regnier's criteria and has defocussing switchable links, via sub-N-Graphs.

Keywords: N-Graphs, natural deduction, sequent calculus, MLL$^-$, subnets.

1 Preface

Since proof-nets for MLL$^-$ were introduced by Girard [11], several studies have been made on its soundness proof. The first correctness criterion defined for proof-nets was given with the definition of the *no shorttrip condition*: Girard used trips to define empires and proved that if all terminal formulas in a proof-net R are conclusions of times links, then there is at least one terminal which splits R. After Danos–Regnier's work [9] it has become possible to define empires using their newly defined DR graphs and, with this new notion of empires, Girard proved sequentialization for proof-nets with quantifiers [12]. Another important advance was achieved by the introduction of a new type of subnets, namely kingdoms. Once the notion of kingdoms was introduced, Bellin & Van de Wiele produced an elegant proof of the sequentialization theorem using simple properties of subnets [5].

A straightforward generalization of this proof was obtained by Robinson [19]. He pointed out that Danos–Regnier's technique relies only on the format of the rules and does not depend on the logic involved. So he devised a proof system based on the classical sequent calculus and applied the characterization of subnets and the proof of sequentialization for MLL$^-$ to his proof-nets for classical logic. His proof followed the model defined by Bellin & Van de Wiele [5].

U. Kohlenbach et al. (Eds.): WoLLIC 2014, LNCS 8652, pp. 94–108, 2014.
© Springer-Verlag Berlin Heidelberg 2014

However, this generalization does not cover the existence of so called switchable links with one premise and more than one conclusion, and also the absence of axiom links. In such systems subnets are not necessarily closed under hereditary premises. So, if a subnet contains a formula occurrence A and B is above[1] A in the proof-net, then B may not be in this subnet. Other works in linear logic related to these issues are Lafont's interaction nets (which do not have axiom links) [15] and the system of Blute *et al*, which contains such switchable links [6] and inspired the proof-nets for classical logic proposed by Führman & Pym [10]. Hughes also proposed a graphical proof system for classical logic where proofs are combinatorial rather than syntactic: a proof of A is a homomorphism between a coloured graph and a graph associated with A [13]. McKinley, on the other hand, proposed the expansion nets, a system that focus on canonical representation of cut-free proofs [16].

Here we present a new approach to perform the sequentialization for N-Graphs, a multiple conclusion calculi inspired by the proof-nets for the propositional classical logic developed by de Oliveira [17,18], but with a switchable defocussing link and without axiom links. One of the main results of this paper, besides giving a new soundness proof for N-Graphs, is the definition of a generalized method to make surgical cuts in proofs for classical logic. This comes with the fact that the presence of the split node in an N-Graph can occur essentially anywhere in the proof, unlike proof-nets where the split node is always a terminal formula.

The need to identify the split node is at the heart of our proof of the sequentialization. In order to achive that we define the north, the south and the whole empires of a formula occurrence A. The first one corresponds to the empires notion of Girard's and Robinson's proof-nets. The second one is the largest sub-N-Graph which has A as a premise (defined due to the presence of elimination rules in N-Graphs). The last one is the union of the previously defined and it induces a strict ordering over the graph nodes, which will be fundamental to find the split node.

2   N-Graphs

Proposed by de Oliveira [17,18], N-Graphs is a symmetric natural deduction (ND) calculus with the presence of structural rules, similar to the sequent calculus. It is a multiple conclusion proof system for classical logic where proofs are built in the form of directed graphs (*"digraphs"*). Several studies have been developed on N-Graphs since its first publication in 2001 [17], like Alves' development on the geometric perspective and cycle treatment towards the normalization of the system [3] and Cruz's definition of intuitionistic N-Graphs [8]. A normalization algorithm was presented for classical N-Graphs [1], along with the subformula and separation properties [2]. More recently a linear time proof checking algorithm was proposed [4].

---

[1] We say that B is above A when B is a hereditary premise of A.

The system is defined somewhat like the proof-nets. There is the concept of *proof-graphs*, which are all graphs constructed with the valid links where each node is the premise and conclusion of *at most* one link, and the concept of *N-Graphs*, which are the correct proof-graphs, i.e. the proof-graphs that represent valid proofs. These constructions are analogous to the definition of proof-structure and proof-net, respectively.

The links represent atomic steps in a derivation. *Focussing links* are the ones with two premises and one conclusion, as illustrated by Fig. 1($\wedge - I$, $\perp - link$, $\top - focussing\ weak$ and contraction). The *defocussing links* are the ones with one premise and two conclusions, as shown in Fig. 1 ($\vee - E$, $\top - link$, $\perp - defocussing\ weak$ and expansion). All other links are called *simple links* and have only one premise and one conclusion (Fig. 2). The indegree (outdegree) of a vertex v is the number of edges oriented towards (away from) it. The set of vertices with indegree (outdegree) equal to zero is the set of premises (conclusions) of the proof-graph G, and is represented by $PREMIS(G)$ ($CONC(G)$).

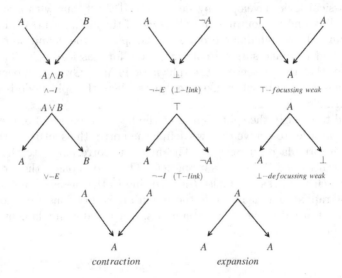

Fig. 1. Focussing and defocussing links

A *logical link* represents a derivation in ND ($\top - link$ acts as the law of the excluded middle). A *structural link* expresses the application of a structural rule as it is done in sequent calculus: it enables weakening a proof ($\top - focussing\ weak$, $\perp - defocussing\ weak$, $\top - simple\ weak$ and $\perp - simple\ weak$), duplicating premises (expansion link) and grouping conclusions in equivalence classes (contraction link). There is no link to emulate the interchange rule because in a proof-graph the order of the premises is not important for the application of derivation rules.

The axioms are represented by proof-graphs with one vertex and no edges. Then, a single node labeled by A is already a valid derivation: it represents an axiom in sequent calculus ($A \vdash A$). So here it makes no sense to talk about the smallest subgraph having A as a conclusion: it would be trivially the vertex v

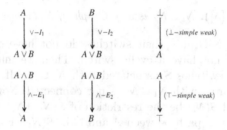

Fig. 2. Simple links

labeled by A. Therefore the notion of kingdoms, as defined and used by Bellin & Van de Wiele [5] for their sequentialization, is useless for N-Graphs.

Similar to Danos-Regnier criterion [9], we define the following subgraphs associated to a proof-graph.

Definition 1 (Switching). *Given a proof-graph G, a switching graph $S(G)$ associated with G is a spanning subgraph[2] of G in which the following edges are removed: one of the two edges of every expansion link and one of the two edges of every contraction link.*

Definition 2 (N-Graph derivation). *A proof-graph G is a N-Graph derivation (or N-Graph for short) iff every switching graph associated with G is acyclic and connected.*

The focussing and defocussing links may also be classified according to their semantics. The links $\wedge - I$, $\bot - link$, $\top - focussing$ *weak* and *expansion* are called *conjunctive*. The *disjunctive* links are: $\vee - E$, $\top - link$, $\bot - defocussing$ *weak* and *contraction*. Here the switchable links (the ones that have one of its edge removed in every switching) draws attention. The links where its geometry contradicts their semantic are switchable. Although focussing, the contraction has a disjunctive semantic; and the expansion is a conjunctive link, even though defocussing. This means the formula occurences this links connect in a proof-graph must be already connected some other way in order to the proof to be sound.

Soundness and completeness of the system were proved through a mapping between N-Graphs and LK (sequent calculus for classical logic) [17,18] and in Section 6 we give a new proof of sequentialization.

3   Sub-N-Graphs

Definition 3 (sub-N-Graph). *We say that H is a subproof-graph of a proof-graph G if H is a subgraph of G and H is a proof-graph. If a vertex v labeled by a formula occurrence A is such that $v \in PREMIS(H)$ ($v \in CONC(H)$), then A is an upper (lower) door of H. If H is also a N-Graph, then it is a sub-N-Graph.*

Let N_1 and N_2 be sub-N-Graphs of a N-Graph N.

---

[2] A spanning subgraph is a subgraph G_1 of G containing all the vertices of G.

Lemma 1 (Union [5]). $N_1 \cup N_2$ *is a* N-Graph *iff* $N_1 \cap N_2 \neq \emptyset$.

Proof. Once N is a N-Graph, their switches do not have cycles, and so any subgraph of N may not have a cyclic switch. Then we must prove only the connectedness of all switches S associated with $N_1 \cup N_2$. If $N_1 \cap N_2 = \emptyset$, then any switching associated with $N_1 \cup N_2$ is not connected. Now let $A \in N_1$, $B \in N_2$, $C \in N_1 \cap N_2$ and $S(N_i)$ be the restriction of $S(N_1 \cup N_2)$ to $S(N_i)$. Since N_1 is a N-Graph, there is a path between A and C in $S(N_1)$. For the same reason, there is a path between B and C in $S(N_2)$. Thus there is a path between A and B in $S(N_1 \cap N_2)$ once $C \in N_1 \cap N_2$. $\square$

Lemma 2 (Intersection [5]). *If* $N_1 \cap N_2 \neq \emptyset$, *then* $N_1 \cap N_2$ *is a* N-Graph.

Proof. As in the previous lemma, it is sufficient to prove the connectivity of $N_1 \cap N_2$. Since $N_1 \cap N_2 \neq \emptyset$, let $A \in N_1 \cap N_2$. If A is the only vertex present in $N_1 \cap N_2$, then it is connected and so is a N-Graph (axiom). Otherwise, let B be any other vertex in $N_1 \cap N_2$, S be a switching of $N_1 \cap N_2$ and S_i be an extension of S for N_i. Once N_1 and N_2 are sub-N-Graphs, there are a path π_1 between A and B in $S(N_1)$ and a path π_2 in $S(N_2)$. If $\pi_1 \neq \pi_2$, then $S_{12} = S_1 \cup S_2$ is a switching for $N_1 \cup N_2$ and $S_{12}(N_1 \cup N_2)$ has a cycle. So $\pi_1 = \pi_2$ and A and B are connected. $\square$

Definition 4 (North, south and whole empires). *Let* A *be a formula occurrence in a* N-Graph N. *The* north *(south) empire of* A, *represented by* $eA^\wedge$ *($eA_\vee$) is the largest sub-N-Graph of* N *having* A *as a* lower *(upper) door. The* whole empire *of* A *(wA) is the union of* $eA^\wedge$ *and* $eA_\vee$.

If we prove that $eA^\wedge$ and $eA_\vee$ exist, then it is immediate the existence of wA by lemma 1. In the following section we give two equivalent constructions of empires and prove some properties.

4   North and South Empires

4.1   Constructions and Existence

Definition 5 ($S^\wedge(N, A)$ and $S_\vee(N, A)$). *Let* A *be a formula occurrence in a* N-Graph N *and* S *an associated switch of* N. *If* A *is a premise of a link with a conclusion* A' *and the edge* (A, A') *belongs to* $S(N)$, *then remove this edge and* $S^\wedge(N, A)$ *is the component that contains* A *and* $\overline{S^\wedge(N, A)}$ *is the other one (if* A *is premise of a disjunctive defocussing link, then* $\overline{S^\wedge(N, A)}$ *has two components). If* A *is not premise of any link in* $S(N)$, *then* $S^\wedge(N, A)$ *is* $S(N)$ *($\overline{S^\wedge(N, A)}$ is empty). $S_\vee(N, A)$ is defined analogously: if* A *is a conclusion of a link with a premise* A'' *and the edge* (A'', A) *belongs to* $S(N)$, *then remove it and* $S_\vee(N, A)$ *is the component which has* A *and* $\overline{S_\vee(N, A)}$ *is the other one (if* A *is conclusion of a conjunctive focussing link, then* $\overline{S_\vee(N, A)}$ *has two components). If* A *is not conclusion of any link in* $S(N)$, *then* $S_\vee(N, A)$ *is equal to* $S(N)$ *($\overline{S_\vee(N, A)}$ is empty).*

Definition 6 (Principal switching [11,12]). *Let A be a formula occurrence. We say that a switching $S_p^\wedge$ (S_v^p) is principal for $eA^\wedge$ (eA_v) when it chooses the edges satisfying the following restrictions:*

1. $\frac{A_{p_1} \; A_{p_2}}{A_c}$ *is a contraction link and a premise A_{p_i} is the formula occurrence A ($\frac{A_p}{A_{c_1} \; A_{c_2}}$ is an expansion link and a conclusion A_{c_i} is the formula occurrence A): the switching chooses the edge with A.*
2. $\frac{X_{p_1} \; X_{p_2}}{X_c}$ *is a contraction link and only one premise X_{p_i} belongs to $eA^\wedge$ (eA_v): the switching links the conclusion with the premise which is not in $eA^\wedge$ (eA_v).*
3. $\frac{X_p}{X_{c_1} \; X_{c_2}}$ *is an expansion link and only one conclusion X_{c_i} belongs to $eA^\wedge$ (eA_v): $S_p^\wedge$ (S_v^p) selects the edge which has the conclusion that is not in $eA^\wedge$ (eA_v).*

Lemma 3. *The north (south) empire of A exists and is given by the two following equivalent conditions:*

1. *$\bigcap_S S^\wedge(N, A)$ ($\bigcap_S S_v(N, A)$), where S ranges over all switches of N;*
2. *the smallest set of formula occurrences of N closed under the following conditions:*

   (a) *$A \in eA^\wedge$ ($A \in eA_v$);*

   (b) *if $\frac{X}{Y}$ is a simple link and $Y \in eA^\wedge$, then $X \in eA^\wedge$ (if $Y \neq A$ and $Y \in eA_v$, then $X \in eA_v$);*

   (c) *if $\frac{X \; Y}{Z}$ is a conjunctive focussing link and $Z \in eA^\wedge$, then $X, Y \in eA^\wedge$ (if $Z \neq A$ and $Z \in eA_v$, then $X, Y \in eA_v$);*

   (d) *if $\frac{X}{Y \; Z}$ is a disjunctive defocussing link and $Y \in eA^\wedge$ or $Z \in eA^\wedge$, then $X \in eA^\wedge$ (if $Y \neq A \neq Z$ and $Y \in eA_v$ or $Z \in eA_v$, then $X \in eA_v$);*

   (e) *if $\frac{X_p}{X_{c_1} \; X_{c_2}}$ is an expansion link and $X_{c_1}, X_{c_2} \in eA^\wedge$, then $X_p \in eA^\wedge$ (if $X_{c_1} \neq A \neq X_{c_2}$ and $X_{c_1}, X_{c_2} \in eA_v$, then $X_p \in eA_v$);*

   (f) *if $\frac{X_{p_1} \; X_{p_2}}{X_c}$ is a contraction link and $X_c \in eA^\wedge$, then $X_{p_1}, X_{p_2} \in eA^\wedge$ (if $X_c \neq A$ and $X_c \in eA_v$, then $X_{p_1}, X_{p_2} \in eA_v$);*

   (g) *if $\frac{X}{Y}$ is a simple link, $X \neq A$ and $X \in eA^\wedge$, then $Y \in eA^\wedge$ (if $X \in eA_v$, then $Y \in eA_v$);*

   (h) *if $\frac{X \; Y}{Z}$ is a conjunctive focussing link, $X \neq A \neq Y$ and $X \in eA^\wedge$ or $Y \in eA^\wedge$, then $Z \in eA^\wedge$ (if $X \in eA_v$ or $Y \in eA_v$, then $Z \in eA_v$);*

   (i) *if $\frac{X}{Y \; Z}$ is a disjunctive defocussing link, $X \neq A$ and $X \in eA^\wedge$, then $Y, Z \in eA^\wedge$ (if $X \in eA_v$, then $Y, Z \in eA_v$);*

   (j) *if $\frac{X_p}{X_{c_1} \; X_{c_2}}$ is an expansion link, $X_p \neq A$ and $X_p \in eA^\wedge$, then $X_{c_1}, X_{c_2} \in eA^\wedge$ (if $X_p \in eA_v$, then $X_{c_1}, X_{c_2} \in eA_v$);*

   (k) *if $\frac{X_{p_1} \; X_{p_2}}{X_c}$ is a contraction link, $X_{p_1} \neq A \neq X_{p_2}$ and $X_{p_1}, X_{p_2} \in eA^\wedge$, then $X_c \in eA^\wedge$ (if $X_{p_1}, X_{p_2} \in eA_v$, then $X_c \in eA_v$).*

Proof. We will prove the case for $eA^\wedge$ according to [5] (the case for eA_v is similar)

I $2 \subseteq 1$: we show that 1 is closed under conditions defining 2. Its immediate $A \in \bigcap_S S^\wedge(N, A)$ ($S^\wedge(N, A)$ contains A for every switching S). If $B_1 \in \bigcap_S S^\wedge(N, A)$ and in all switches there is an edge (B_1, B_2), then we conclude $B_2 \in \bigcap_S S^\wedge(N, A)$ (imperialistic lemma [12]). So the construction is also closed under conditions 2b, 2c, 2d, 2g, 2h and 2i. Conditions 2e and 2k are also simple. Now suppose that 1 does not respect 2f. Then there is a contraction link $\frac{X_{p_1} \quad X_{p_2}}{X_c}$ such that $X_c \in \bigcap_S S^\wedge(N, A)$, but $X_{p_i} \notin \bigcap_S S^\wedge(N, A)$, for $i = 1$ or $i = 2$. Consider the first one: $X_{p_1} \notin S^\wedge(N, A)$ for some S. Since $X_c \in S^\wedge(N, A)$, then $(X_{p_2}, X_p) \in S(N, A)$ and so $X_{p_2} \in S^\wedge(N, A)$. Once $\overline{S^\wedge(N, A)}$ is not empty, A must be premise of a link whose one conclusion is A' and $A' \in \overline{S^\wedge(N, A)}$. Let π be the path between X_{p_1} and A' in $\overline{S^\wedge(N, A)}$. Since $(X_{p_1}, X_c) \notin S(N)$, this edge does not belong to π. Consider now a switch S' like S, except that $(X_{p_1}, X_c) \in S'(N)$. Note that π is in $S'(N)$ too and $X_c \in S'^\wedge(N, A)$ (because $X_c \in \bigcap_S S^\wedge(N, A)$). Then we may extend π and get a path between A' and A without the edge (A, A') in $S'(N)$: we obtain a cycle in $S'(N)$, which is a contradiction. Therefore 1 is closed under 2f. For similar reason, we conclude that 1 is closed under 2j too.

II $1 \subseteq 2$: let $S_p^\wedge$ a principal switching for $eA^\wedge$. We will prove $S_p^\wedge(N, A) \subseteq 2$. $S_p^\wedge(N, A) \cap 2 \neq \emptyset$, because both contain A. But definition 6 ensures that it is impossible to leave $eA^\wedge$ once we are in $S_p^\wedge(N, A)$. Since $\bigcap_S S^\wedge(N, A) \subseteq S_p^\wedge(N, A)$, we conclude that $\bigcap_S S^\wedge(N, A) \subseteq eA^\wedge$. $\qquad\square$

Corollary 1. $S_p^\wedge = eA^\wedge$ and $S_\vee^p = eA_\vee$.

Corollary 2. Let A be a premise and B a conclusion. Then $eA_\vee = eB^\wedge = N$.

Lemma 4. $eA^\wedge$ and $eA_\vee$ are the largest sub-N-Graphs which contains A as a lower and upper door, respectively.

Proof. The proof uses the same argument presented in [5] (see Proposition 2). $\qquad\square$

Fig. 3 illustrates some concepts about empires. For example, in the N-Graph on left, we have $eA^\wedge = \{A, \ A \vee C, \ C, \ C \wedge Z, \ Z, \ \neg A \wedge Z\}$ (formulas in green), and $eA_\vee = \{A, \ \bot, \ \neg A, \ \neg A \wedge Z\}$ (formulas in yellow). The formula occurrence in red belongs to both empires. We can see that there is no sub-N-Graph which contains A as conclusion (premise) and is larger than $eA^\wedge$ ($eA_\vee$), as both conclusions of a expansion link are needed to add its premise (condition 2e). For the second N-Graph we have the same color scheme for A, and here we can not have the conclusion of the contraction link because we need both premises (condition 2k).

4.2 Nesting Lemmas

Lemma 5 (Nesting of empires I [12]). Let A and B be distinct formula occurrences in a N-Graph. If $A \in eB^\wedge$ and $B \notin eA^\wedge$, then $eA^\wedge \subsetneq eB^\wedge$.

Lemma 6 (Nesting of empires II [12]). Let A and B be distinct formula occurrences in a N-Graph. If $A \notin eB^\wedge$ and $B \notin eA^\wedge$, then $eA^\wedge \cap eB^\wedge = \emptyset$.

Proof (I and II). Construct a principal switching $S_p^\wedge$ for $eB^\wedge$ with some additional details:

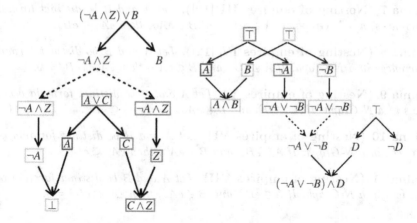

Fig. 3. N-Graphs for $(\neg A \wedge Z) \vee B$, $A \vee C \vdash B$, $C \wedge Z$ and $\vdash A \wedge B$, $(\neg A \vee \neg B) \wedge D$, $\neg D$.

I *contraction link whose conclusion belongs to* $eB^\wedge$: if the conclusion is not in $eA^\wedge$, then we proceed as we do for a principal switching for $eA^\wedge$ (if only one premise is in $eA^\wedge$, choose the other premise);

II *expansion link whose premise belongs to* $eB^\wedge$: if the premise is not in $eA^\wedge$, then we proceed as we do for a principal switching for $eA^\wedge$ (if only one conclusion is in $eA^\wedge$, choose the other conclusion);

III *if A is a premise of a link whose conclusion A' is in* $eB^\wedge$: then we choose the edge (A, A').

First suppose $A \in eB^\wedge$. We try to go from A to B without passing through (A, A'). Since $S_p^\wedge$ is principal for $eB^\wedge$ and $A \in eB^\wedge$, all formulas in the path from A to B belong to $eB^\wedge$. But $B \notin eA^\wedge$ and sometime we leave $eA^\wedge$. By construction 2 of lemma 3, there are only two ways of leaving $eA^\wedge$ without passing through (A, A'): passing through a contraction link whose only one premise belongs to $eA^\wedge$, or passing through an expansion link whose only one conclusion belongs to $eA^\wedge$; but, steps I and II avoid this cases, respectively.

Therefore it is impossible to leave $eA^\wedge$ in $S_p^\wedge(N, B)$, unless $(A, A') \in S_p^\wedge(N, B)$. This implies $S_p^\wedge(N, A) \subsetneq S_p^\wedge(N, B)$. Since $eA^\wedge \subset S_p^\wedge(N, A)$ and $eB^\wedge = S_p^\wedge(N, B)$, we conclude $eA^\wedge \subsetneq eB^\wedge$.

Now suppose $A \notin eB^\wedge$. I and II ensure we do not have any edges between $eA^\wedge$ and $\overline{eA^\wedge}$[3] in $eB^\wedge$, except perhaps for (A, A'). But now $A \notin eB^\wedge$ and therefore $A \notin S_p^\wedge(N, B)$. So $(A, A') \notin S_p^\wedge(N, B)$. Since $eB^\wedge = S_p^\wedge$ and $B \notin eA^\wedge$, no formula of $eA^\wedge$ belongs to $eB^\wedge$ and thus $eA^\wedge \cap eB^\wedge = \emptyset$.   □

From these two previous lemmas we have nesting lemmas 7 and 8 for south empires too (the proofs are similar to the previous ones) and from these four nesting lemmas, it is possible to proof nesting lemmas between north and south (9, 10 and 11).

---

[3] $\overline{eA^\wedge}$ represents the set of all formula occurrenes which are not in $eA^\wedge$.

Lemma 7 (Nesting of empires III [12]). *Let A and B be distinct formula occurrences in a N-Graph. If $A \in eB_\vee$ and $B \notin eA_\vee$, then $eA_\vee \subsetneq eB_\vee$.*

Lemma 8 (Nesting of empires IV [12]). *Let A and B be distinct formula occurrences in a N-Graph. If $A \notin eB_\vee$ and $B \notin eA_\vee$, then $eA_\vee \cap eB_\vee = \emptyset$.*

Lemma 9 (Nesting of empires V). *Let A and B be distinct formula occurrences in a N-Graph. If $A \in eB^\wedge$ and $B \notin eA_\vee$, then $eA_\vee \subsetneq eB^\wedge$.*

Lemma 10 (Nesting of empires VI). *Let A and B be distinct formula occurrences in a N-Graph. If $A \in eB_\vee$ and $B \notin eA^\wedge$, then $eA^\wedge \subsetneq eB_\vee$.*

Lemma 11 (Nesting of empires VII). *Let A and B be distinct formula occurrences in a N-Graph. If $A \notin eB^\wedge$ and $B \notin eA_\vee$, then $eA_\vee \cap eB^\wedge = \emptyset$.*

5 Whole Empires

We defined the *whole empire of A* as the union of the north and the south empires of A. Now we use the north and south empires properties to find new ones about whole empires.

Lemma 12. *wA is a sub-N-graph.*

Proof. Once we proved that $eA^\wedge$ and $eA_\vee$ are N-graphs (lemma 4) and $eA^\wedge \cap eA_\vee = \{A\}$, we get $wA = eA^\wedge \cup eA_\vee$ is a sub-N-graph by lemma 1.   □

Corollary 3. *Let A be a premise and B a conclusion. Then $wA = wB = N$ (by corollary 2).*

Lemma 13 (Nesting of whole empires I). *Let A and B be distinct occurrences. If $A \notin wB$ and $B \notin wA$, then $wA \cap wB = \emptyset$.*

Proof. Since $A \notin wB$ and $B \notin wA$, we get: $A \notin eB^\wedge$, $A \notin eB_\vee$, $B \notin eA^\wedge$ and $B \notin eA_\vee$. We apply nesting of empires lemmas:

1. if $B \notin eA^\wedge$ and $A \notin eB^\wedge$, then $eA^\wedge \cap eB^\wedge = \emptyset$ (by lemma 6);
2. if $B \notin eA_\vee$ and $A \notin eB^\wedge$, then $eA_\vee \cap eB^\wedge = \emptyset$ (by lemma 11);
3. uniting 1 and 2 and applying the distributive law: $eB^\wedge \cap (eA^\wedge \cup eA_\vee) = \emptyset$;
4. if $B \notin eA^\wedge$ and $A \notin eB_\vee$, then $eA^\wedge \cap eB_\vee = \emptyset$ (by lemma 11);
5. if $B \notin eA_\vee$ and $A \notin eB_\vee$, then $eA_\vee \cap eB_\vee = \emptyset$ (by lemma 8);
6. uniting 4 and 5 and applying the distributive law: $eB_\vee \cap (eA^\wedge \cup eA_\vee) = \emptyset$;
7. uniting 3 and 6, the distributive law: $(eA^\wedge \cup eA_\vee) \cap (eB^\wedge \cup eB_\vee) = \emptyset$.

□

Lemma 14 (Nesting of whole empires II). *Let A and B be distinct occurrences. If $A \in wB$ and $B \notin wA$, then $wA \subsetneq wB$.*

Proof. Once $B \notin wA$, we have $B \notin eA^\wedge$ and $B \notin eA_\vee$. For $A \in wB$ we get $A \in eB^\wedge$ or $A \in eB_\vee$. We will prove the lemma for $A \in eB^\wedge$ (the case for south is analogous):

1. if $A \in eB^\wedge$ and $B \notin eA^\wedge$, then $eA^\wedge \subsetneq eB^\wedge$ (by lemma 5);
2. if $A \in eB^\wedge$ and $B \notin eA_\vee$, then $eA_\vee \subsetneq eB^\wedge$ (by lemma 9);
3. if $eA^\wedge \subsetneq eB^\wedge$ and $eA_\vee \subsetneq eB^\wedge$, then $eA^\wedge \cup eA_\vee \subsetneq eB^\wedge$;
4. if $eA^\wedge \cup eA_\vee \subsetneq eB^\wedge$, then $eA^\wedge \cup eA_\vee \subsetneq eB^\wedge \cup eB_\vee$.

$\square$

Definition 7 ($\ll$). *Let A and B be formula occurrences of N. We say $A \ll B$ iff $wA \subsetneq wB$.*

It is immediate that $\ll$ is a strict ordering of formula occurrences of N which are not premises neither conclusions, since we have for any domain set X and any subset Q of $P(X)^4$, $(\subseteq, Q)$ is a poset. Maximal formulas with regard to $\ll$ will split N. Given that the whole empires of premises and conclusions are always equal to N by corollary 3, we are not interested in these formulas. So they are not in the domain of $\ll$. One may easily verify that if there are no contraction and extension links, for all formula A of N, $wA = N$ and so any formula would be maximal. The next two following lemmas show how these links act on $\ll$.

Lemma 15. *Let $l = \frac{\cdots X \cdots}{\cdots Y \cdots}$ be a link such that there is a formula-occurrence A which $X \in wA$ and $Y \notin wA$. Then $A \ll Y$.*

Proof. Once $X \in eA^\wedge \cup eA_\vee$, we have two cases. If $X \in eA^\wedge$, then since $Y \notin eA^\wedge$, l must be a contraction link and its other premise does not belong to $eA^\wedge$ (contruction 2 in lemma 3). Therefore Y is a conclusion of a contraction link and this implies $X \in eY^\wedge$ (by 2f in lemma 3). So $eA^\wedge \cap eY^\wedge \neq \emptyset$. If $A \notin eY^\wedge$, then we will have $eA^\wedge \cap eY^\wedge = \emptyset$ (by lemma 6): a contradiction. Thus $A \in eY^\wedge$ and, by lemma 14, we conclude $wA \subsetneq wY$. The case for $X \in eA_\vee$ is analogous.

$\square$

Next lemma is similar, but for expansion link:

Lemma 16. *Let $l = \frac{\cdots X \cdots}{\cdots Y \cdots}$ be a link such that there is a formula-occurrence A which $X \notin wA$ and $Y \in wA$. Then $A \ll X$.*

6   Sequentialization

We saw in Sections 4 and 5 how to define empires for proof graphs with switchable defocussing links (expansion) and proved some properties. Now we will show a new proof of sequentialization for this proof-graphs (N-Graphs without "$\rightarrow$" connective). Without loss of generality, we assume $\top$ as $A \vee \neg A$ and $\bot$ as $A \wedge \neg A$, where the formula A belongs to the premise or conclusion of the link.

Theorem 1 (Sequentialization for fragment without "$\rightarrow$" connective). *Given a N-Graph derivation N without "$\rightarrow$" connective, there is a sequent calculus derivation $SC(N)$ of $A_1, \ldots, A_n \vdash B_1, \ldots, B_m$ in the classical sequent calculus whose occurrences of formulas $A_1, \ldots, A_n$ and $B_1, \ldots, B_m$ are in one-to-one correspondence with the elements of $PREMIS(N)$ and $CONC(N)$, respectively.*

---

[4] $P(X)$ is the power set of X.

Proof. We proceed by induction on the number of links of N.

1. N *does not have any link (it has only one vertex v labelled with A)*: this case is immediate. $SC(N)$ is $A \vdash A$.

2. N *has only one link*: since N is a N-Graph, then this link is not a contraction neither an expansion. This case is simple, once there is a simple mapping between links and sequent calculus rules, which makes the construction of $SC(N)$ immediate (completeness proof [17,18]). For example, in case $\wedge - I$:

$$\frac{A \vdash A \quad B \vdash B}{A, B \vdash A \wedge B} \wedge -R$$

3. N *has an initial expansion link*: the induction hypothesis has built a derivation Π ending with $\mathbf{A}, \mathbf{A}, \ldots, A_n \vdash B_1, \ldots, B_m$. Then $SC(N)$ is achieved by left contraction:

$$\frac{\begin{array}{c} \Pi \\ \mathbf{A}, \mathbf{A}, \ldots, A_n \vdash B_1, \ldots, B_m \end{array}}{\mathbf{A}, \ldots, A_n \vdash B_1, \ldots, B_m} LC$$

4. N *has a final contraction link*: here the induction hypothesis has built a derivation Π ending with $A_1, \ldots, A_n \vdash \mathbf{B}, \mathbf{B}, \ldots, B_m$. Hence $SC(N)$ is obtained by right contraction:

$$\frac{\begin{array}{c} \Pi \\ A_1, \ldots, A_n \vdash \mathbf{B}, \mathbf{B}, \ldots, B_m \end{array}}{A_1, \ldots, A_n \vdash \mathbf{B}, \ldots, B_m} RC$$

5. N *has more than one link, but does not have initial expansion link neither a final contraction link*: this case is more complicated and is similar to that one in MLL$^-$ in which all terminal links are $\otimes$. Yet here we have an additional challenge: the split node is in the middle of the proof. Choose a formula occurrence A which is maximal with respect to $\ll$. We claim that $wA = eA^\wedge \cup eA_\vee = N$. That is, A labels the split node.
   Suppose not. Then let Z be a formula occurrence such that $Z \in N - (eA^\wedge \cup eA_\vee)$ and $S_p^\wedge$ be a principal switching for $eA^\wedge$. Given that $Z \notin eA^\wedge$, the path ρ from A to Z in $S_p^\wedge(N)$ passes through a conclusion A' of A. Let $A_\vee$ be the last node which belongs to $eA_\vee$ in ρ and W the next one in ρ (i.e. $W \notin eA_\vee$). There are two cases for the edge incident to $A_\vee$ and W:

   (a) $(A_\vee, W)$ *belongs to a contraction link whose other premise is not in $eA_\vee$*: according to lemma 15 we have $A \ll W$, contradicting the maximality of A in $\ll$.

   (b) $(W, A_\vee)$ *belongs to an expansion link whose other conclusion is not in $eA_\vee$*: we apply lemma 16 and conclude $A \ll W$ here too. We contradict our choice again.

   Thus $wA = N$. Let $\Gamma_1, \Gamma_2, \Delta_1, \Delta_2$ be sets of formula occurrences such that: $\Gamma_1 \cup \Gamma_2 = \{A_1, \ldots, A_n\}$, $\Delta_1 \cup \Delta_2 = \{B_1, \ldots, B_m\}$ and $\Gamma_1 \cap \Gamma_2 = \Delta_1 \cap \Delta_2 = \emptyset$. Since $eA^\wedge$ is a N-Graph and A is a lower door, the induction hypothesis built $SC(eA^\wedge)$ ending with $\Gamma_1 \vdash \Delta_1, A$. Once $eA_\vee$ is a N-Graph and A is an upper

door, the induction hypothesis made $SC(eA_\lor)$ ending with $A, \Gamma_2 \vdash \Delta_2$. So $SC(N)$ is achieved by cut rule:

$$\frac{SC(eA^\land) \qquad SC(eA_\lor)}{\dfrac{\Gamma_1 \vdash \Delta_1, A \qquad A, \Gamma_2 \vdash \Delta_2}{\Gamma_1, \Gamma_2 \vdash \Delta_1, \Delta_2}}$$

$\square$

7 Conclusion

With N-Graphs, the structural links are based on the sequent calculus, but the logical links emulate the rules of ND. Sequent calculus (classical and linear) have only introduction rules. On the other hand, natural deduction and N-Graphs present elimination rules, so we need to adapt the notion of empire from proof-nets to account for multiple-conclusion ND. This was done with south empires.

Moving the definitions from sequent calculus to ND according to this way we have been able to formulate a new and rather general method of performing surgical cuts on proofs in multiple conclusion ND, producing subnets for classical logic. In N-Graphs the split nodes, maximal with regards to the ordering induced by the union of the empires, may be located anywhere in the proof, not only as terminal node representing a conclusion.

We show an example in Fig. 3: every initial link is defocussing unswitchable and every final link is focussing unswitchable in both N-Graphs. Their maximal nodes are highlighted and they split the proofs into two also correct proofs. It is illustrated in Fig. 4 for the N-Graph on the left (Fig. 3). The same procedure could be applied to any of the two maximal nodes in the N-Graph on the right.

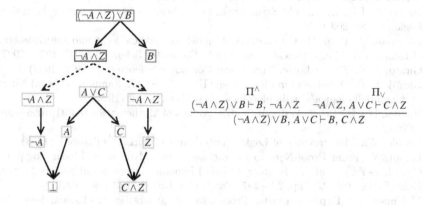

Fig. 4. Example of how to cut using the maximal node

Acknowledgements. We should like to thank Gianluigi Bellin for his very useful comments and suggestions on an earlier paper dealing with the same topic, as well as for his advice on the occasion of a visit to our group in April 2013.

References

1. Alves, G.V., de Oliveira, A.G., de Queiroz, R.J.G.B.: Proof-graphs: a thorough cycle treatment, normalization and subformula property. Fundamenta Informaticae 106, 119–147 (2011)
2. Alves, G.V., de Oliveira, A.G., de Queiroz, R.J.G.B.: Transformations via Geometric Perspective Techniques Augmented with Cycles Normalization. In: Ono, H., Kanazawa, M., de Queiroz, R. (eds.) WoLLIC 2009. LNCS, vol. 5514, pp. 84–98. Springer, Heidelberg (2009)
3. Alves, G.V., de Oliveira, A.G., de Queiroz, R.J.G.B.: Towards normalization for proof-graphs. In: Logic Cooloquium, Bulletin of Symbolic Logic 2005, Torino, United States of America, vol. 11, pp. 302–303 (2005)
4. Andrade, L., Carvalho, R., de Oliveira, A., de Queiroz, R.: Linear Time Proof Verification on N-Graphs: A Graph Theoretic Approach. In: Libkin, L., Kohlenbach, U., de Queiroz, R. (eds.) WoLLIC 2013. LNCS, vol. 8071, pp. 34–48. Springer, Heidelberg (2013)
5. Bellin, G., Van de Wiele, J.: Subnets of Proof-nets in MLL⁻. Advances in Linear Logic. In: Girard, J.-Y., Lafont, Y., Regnier, L. (eds.) London Math Soc. Lect. Notes Series, vol. 222, pp. 249–270. Cambridge University Press (1995)
6. Blute, R.F., Cockett, J.R.B., Seely, R.A.G., Trimble, T.H.: Natural deduction and coherence for weakly distributive categories. Journal of Pure and Applied Algebra 113(3), 229–296 (1996)
7. Carbone, A.: Interpolants, Cut Elimination and Flow Graphs for the Propositional Calculus. Annals of Pure and Applied Logic 83, 249–299 (1997)
8. Cruz, M.Q., de Oliveira, A.G., de Queiroz, R.J.G.B., de Paiva, V.: Intuitionistic N-graphs. Logic Journal of the IGPL (2013) (Print)
9. Danos, V., Regnier, L.: The Structure of Multiplicatives. Archive for Mathematical Logic 28, 181–203 (1989)
10. Führman, C., Pym, D.: On categorical models of classical logic and the Geometry of Interaction. Mathematical Structures in Computer Science 17, 957–1027 (2007)
11. Girard, J.-Y.: Linear Logic. Theoretical Computer Science 50, 1–102 (1987)
12. Girard, J.-Y.: Quantifiers in Linear Logic II. In: Corsi, G., Sambin, G. (eds.) Nuovi problemi della logica e della filosofia della scienza, vol. 2 (1991)
13. Hughes, D.J.D.: Proofs Without Syntax. Annals of Mathematics 164(3), 1065–1076 (2006)
14. Kneale, W.: The Province of Logic. Contemporary British Philosophy (1958)
15. Lafont, Y.: From Proof-Nets to Interaction Nets. Advances in Linear Logic. In: Girard, J.-Y., Lafont, Y., Regnier, L. (eds.) London Mathematical Society Lecture Notes Series, vol. 222, pp. 225–247. Cambridge University Press (1995)
16. McKinley, R.: Expansion nets: Proof-nets for propositional classical logic. In: Fermüller, C.G., Voronkov, A. (eds.) LPAR-17. LNCS, vol. 6397, pp. 535–549. Springer, Heidelberg (2010)
17. de Oliveira, A.G.: Proofs from a Geometric Perspective. PhD Thesis, Universidade Federal de Pernambuco (2001)

18. de Oliveira, A.G., de Queiroz, R.J.G.B.: Geometry of Deduction via Graphs of Proof. In: de Queiroz, R. (ed.) Logic for Concurrency and Synchronisation, pp. 3–88. Kluwer (2003)
19. Robinson, E.: Proof Nets for Classical Logic. Journal of Logic and Computation 13, 777–797 (2003)
20. Shoesmith, D.J., Smiley, T.J.: Multiple-Conclusion Logic. Cambridge University Press, London (1978)
21. Statman, R.: Structural Complexity of Proofs. PhD thesis, Stanford (1974)
22. Ungar, A.M.: Normalization, Cut-elimination and the Theory of Proofs. In: CSLI Lecture Notes, Center for the Study of Language and Information, vol. 28 (1992)

A   N-Graphs Dealing with Cycles

In a multiple conclusion calculus we have rules with more than one conclusion ($\lor - elimination$) as well as rules with more than one premise ($\land - introduction$). This allows the existence of cycles and makes soundness difficult to prove. In order to deal with cycles, the proposed solution for N-Graphs [17,18] adopts some ideas from Shoesmith & Smiley [20] and Ungar [22] for classical logic and the simplicity of Danos & Regnier's solution for MLL$^-$. The main idea is to distinguish a link that represents a logical operation from the one that represents a structural operation. The second kind of links is represented by contraction and expansion links. A proof may contain cycles[5], which shall be controlled by these two structural links.

An invalid cycle occurs when a conjunction is made on terms generated by the same disjunction (Fig. 5). A valid cycle happens when a disjunction yields to the same formula twice and they are contracted to a single occurrence. The latter is similar to the $\lor - elimination$ rule in ND.

Fig. 5. Proof-graphs for $A \lor A \vdash A \land A$: a sound one is on the right

de Oliveira proposes grouping conclusions into equivalence classes using the contraction link, and grouping assumptions similarly to single conclusion calculi with the use of the expansion link. In the sequent calculus it is done by right and left contractions, respectively. In Fig. 5, at the rightmost proof-graph, the cycle involving $A \lor A$ is opened and closed only by disjunctive links, and therefore is

---

[5] Sometimes we talk about "cycles" when in fact "semicycles" are meant (i.e. the direction of edges is not relevant).

valid. In the other hand, to allow the conjunction of a formula with itself $(A \wedge A)$, we need two instances of A, which may be joined by an expansion so we can complete the proof of $A \vee A \vdash A \wedge A$. The cycle that includes $A \wedge A$ in Fig. 5 is also valid because it is opened and closed only by conjunctive links. Here the importance of the switchable links is revealed, as we can see the necessity of such structural links where the geometry and the semantic are contradictory in order to create valid cycles in a proof.

Interrogative Dependencies and the Constructive Content of Inquisitive Proofs

Ivano Ciardelli*

ILLC, University of Amsterdam
Science Park 107
1098 XG Amsterdam

Abstract. This paper shows how dichotomous inquisitive semantics gives rise to a general notion of entailment that unifies standard declarative entailment with answerhood and interrogative dependency, the relation holding when an answer to a question determines an answer to another. We investigate the associated logic, presenting a new completeness proof based on an explicit canonical model construction. On the way to this proof, we establish a new result, the *resolution theorem*, which shows that inquisitive proofs have a natural computational interpretation. We conclude arguing that, as a logic of dependencies, inquisitive logic has certain theoretical and practical advantages over related systems.

Keywords: Inquisitive logic, logic of questions, dependence logic.

1   Introduction

Inquisitive semantics [9,1,14,4, a.o.] pursues a semantic framework that encompasses both information and issues, thus reflecting the primary function of language as a tool for information exchange. The most standard logical incarnation of the framework is *basic inquisitive semantics*, InqB [1,5], obtained associating the connectives with the natural algebraic operations in the space of inquisitive meanings [14]. Recently, a close relative of InqB has been investigated, the system $InqD_\pi$ of *dichotomous inquisitive semantics* [10,3], whose syntax enforces a strict distinction of formulas into declaratives and interrogatives.

In the present paper, we take a closer look at the logic that arises from this system. We point out that this logic subsumes standard declarative entailment, answerhood, and interrogative dependencies as three particular cases of a unique, cross-categorial entailment relation. We provide a new completeness proof, more explicit and better suited to generalizations than the one given in [3]. On our way to this proof, we establish a new result which brings out how inquisitive proofs may be seen as encoding methods for computing interrogative dependencies. Finally, we look at $InqD_\pi$ from the perspective of dependence logic, pointing out certain advantages of the inquisitive logical setup.

---

* I am indebted to Lucas Champollion, Jeroen Groenendijk, Lauri Hella, Floris Roelofsen and Fan Yang for discussions on the ideas presented here. Financial support from the Dutch Organization for Scientific Research (NWO) is gratefully acknowledged.

U. Kohlenbach et al. (Eds.): WoLLIC 2014, LNCS 8652, pp. 109–123, 2014.
© Springer-Verlag Berlin Heidelberg 2014

2   Dichotomous Inquisitive Semantics

This section provides a minimal introduction to dichotomous inquisitive semantics. For a comprehensive exposition and proofs, the reader is referred to [3].

Unlike *basic* inquisitive semantics InqB, which is based on a standard propositional language, *dichotomous* inquisitive semantics $InqD_\pi$ enriches a propositional language, whose formulas are called *declaratives*, with a new syntactic category of *interrogative* formulas. Given a set of atoms $\mathcal{P}$, the set $\mathcal{L}_!$ of declaratives and the set $\mathcal{L}_?$ of interrogatives are defined recursively as follows:[1]

Definition 1 (Syntax).

1. *for any* $p \in \mathcal{P}$, $p \in \mathcal{L}_!$
2. $\bot \in \mathcal{L}_!$
3. *if* $\alpha_1, \ldots, \alpha_n \in \mathcal{L}_!$, *then* $?\{\alpha_1, \ldots, \alpha_n\} \in \mathcal{L}_?$
4. *if* $\varphi, \psi \in \mathcal{L}_\circ$, *then* $\varphi \wedge \psi \in \mathcal{L}_\circ$, *where* $\circ \in \{!, ?\}$
5. *if* $\varphi \in \mathcal{L}_! \cup \mathcal{L}_?$ *and* $\psi \in \mathcal{L}_\circ$, *then* $\varphi \to \psi \in \mathcal{L}_\circ$, *where* $\circ \in \{!, ?\}$

We will also make use of some abbreviations. We will write $\neg\varphi$ for $\varphi \to \bot$ and $\varphi \leftrightarrow \psi$ for $(\varphi \to \psi) \wedge (\psi \to \varphi)$. Moreover, for α and β declaratives, we will write $\alpha \vee \beta$ for $\neg(\neg\alpha \wedge \neg\beta)$, and $?\alpha$ for a *polar interrogative* $?\{\alpha, \neg\alpha\}$.

Throughout the paper, we let α, β, γ range over declaratives, μ, ν, λ over interrogatives, and φ, ψ, χ over the whole language. Moreover, Γ ranges over sets of declaratives, Λ over sets of interrogatives, and Φ over arbitrary sets.

The semantics is based on possible world models for propositional logic.[2]

Definition 2 (Models). *A model for* $InqD_\pi$ *is a pair* $M = \langle \mathcal{W}, V \rangle$, *where:*

- $\mathcal{W}$ *is a set whose elements we refer to as* possible worlds;
- $V : \mathcal{W} \to \wp(\mathcal{P})$ *is a* valuation map, *yielding for each* $w \in \mathcal{W}$ *the set* $V(w)$ *of atoms true at* w.

Usually, semantics is synonymous with truth-conditions. However, our language now contains interrogatives as well. We do not lay out the meaning of an interrogative by specifying what a state of affairs has to be like to make it true, but rather by specifying what information is needed to resolve it. Thus, the natural evaluation points for interrogatives are not possible worlds, but rather bodies of information. These as referred to as *information states* and modeled formally by identifying them with the set of worlds compatible with the information.

Definition 3 (Information states).
An information state *in a model* M *is a set* $s \subseteq \mathcal{W}$ *of possible worlds.*

---

[1] Our language is richer than the one in [3], which does not allow interrogatives as antecedents of an implication. This enrichment is essential: as we will see, interrogative dependencies are expressed precisely by implications among two interrogatives.

[2] In [3], a fixed model ω is assumed, consisting of *all* propositional valuations. Since any possible world model is embeddable in ω, this difference is not an essential one.

To retain uniform semantic notions, InqD_π lifts the interpretation of *all* sentences in the language to the level of information states. The semantics is thus given in the form of a relation of *support* between information states and formulas. Intuitively, for a declarative being supported in a state s amounts to being *established* in s, while for an interrogative it amounts to being *resolved* in s.

Definition 4 (Support). *Let M be a model and s an information state in M.*

1. $M, s \models p \iff p \in V(w)$ *for all worlds* $w \in s$
2. $M, s \models \bot \iff s = \emptyset$
3. $M, s \models\, ?\{\alpha_1, \ldots, \alpha_n\} \iff M, s \models \alpha_1$ *or* $\ldots$ *or* $M, s \models \alpha_n$
4. $M, s \models \varphi \wedge \psi \iff M, s \models \varphi$ *and* $M, s \models \psi$
5. $M, s \models \varphi \rightarrow \psi \iff$ *for any* $t \subseteq s$, *if* $M, t \models \varphi$ *then* $M, t \models \psi$

A first, crucial feature of the semantics is that support is *persistent*.

Fact 1 (Persistence). *If* $M, s \models \varphi$ *and* $t \subseteq s$, *then* $M, t \models \varphi$.

Second, although our semantics is defined in terms of support, we can recover *truth* at worlds by defining it as support at the corresponding singleton state.

Definition 5 (Truth).
φ *is true at a world w in M, notation $M, w \models \varphi$, in case $M, \{w\} \models \varphi$.*

Computing the support clauses for singleton states, we find that the connectives all get their standard truth-conditional clauses. Moreover, persistence implies that a world makes a formula true iff it is contained in some supporting state.

Fact 2. $M, w \models \varphi \iff w \in s$ *for some state s such that $M, s \models \varphi$.*

In general, truth conditions do not determine support conditions. For instance, the polar interrogatives $?p$ and $?q$ are both true everywhere, but clearly, in general they have different support conditions. However, the semantics of *declaratives* is still completely determined by truth-conditions: for, a declarative is supported in a state iff it is true at all the worlds in the state.

Fact 3. *For any declarative α:* $M, s \models \alpha \iff (M, w \models \alpha$ *for all* $w \in s)$

To any formula φ, we associate a set $\mathcal{R}(\varphi)$ of declaratives that we call the *resolutions* of φ. This set is defined recursively as follows.

Definition 6 (Resolutions).

- $\mathcal{R}(p) = \{p\}$
- $\mathcal{R}(\bot) = \{\bot\}$
- $\mathcal{R}(?\{\alpha_1, \ldots, \alpha_n\}) = \{\alpha_1, \ldots, \alpha_n\}$
- $\mathcal{R}(\varphi \wedge \psi) = \{\alpha \wedge \beta \mid \alpha \in \mathcal{R}(\varphi) \text{ and } \beta \in \mathcal{R}(\psi)\}$
- $\mathcal{R}(\varphi \rightarrow \psi) = \{\bigwedge_{\alpha \in \mathcal{R}(\varphi)} \alpha \rightarrow f(\alpha) \mid f : \mathcal{R}(\varphi) \rightarrow \mathcal{R}(\psi)\}$

It is easy to see that a declarative α has itself as unique resolution: $\mathcal{R}(\alpha) = \{\alpha\}$. As for an interrogative μ, we may think of its resolutions of as syntactically generated answers to it.[3] Indeed, the next fact says that establishing some resolution is a necessary and sufficient condition to resolve an interrogative.

Fact 4. *For any M, s and φ, $M, s \models \varphi \iff M, s \models \alpha$ for some $\alpha \in \mathcal{R}(\varphi)$*

As a corollary, we get the following normal form result: every formula φ is equivalent to a basic interrogative having the resolutions of φ as constituents.

Corollary 1 (Normal form). *For any φ, $\varphi \equiv \, ?\mathcal{R}(\varphi)$.*

In terms of resolutions we define the notion of *presupposition* of an interrogative.

Definition 7 (Presupposition of an interrogative).
The presupposition of an interrogative μ is the declarative $\pi_\mu = \bigvee \mathcal{R}(\mu)$.

Since the interrogative operator has the same truth conditions as a disjunction, it follows from Corollary 1 that μ and π_μ have the same truth conditions. Incidentally, this tells us how truth should be read for interrogatives: an interrogative μ is true in w iff some resolution to μ is true in w.

The notion of resolution may be generalized to sets of formulas as follows.

Definition 8 (Resolutions of a set).
The set $\mathcal{R}(\Phi)$ of resolutions of a set Φ contains those sets Γ of declaratives s.t.:

- *for all $\varphi \in \Phi$ there is an $\alpha \in \Gamma$ such that $\alpha \in \mathcal{R}(\varphi)$*
- *for all $\alpha \in \Gamma$ there is a $\varphi \in \Phi$ such that $\alpha \in \mathcal{R}(\varphi)$*

That is, a resolution of Φ is a set of declaratives which is obtained by replacing every formula in Φ by one or more of its resolutions. Since a declarative has itself as unique resolution, we obtain a resolution of Φ by keeping all the declaratives in Φ, and substituting each interrogative by one or more resolutions. Fact 4 generalizes to sets: writing $M, s \models \Phi$ for '$M, s \models \varphi$ for all $\varphi \in \Phi$', we have:

Fact 5. *For any M, s and Φ, $M, s \models \Phi \iff M, s \models \Gamma$ for some $\Gamma \in \mathcal{R}(\Phi)$*

We end this section by taking a closer look at the behavior of implication. First, if the antecedent is declarative, the clause amounts to the following simpler one:

$$M, s \models \alpha \to \varphi \iff M, s \cap \{w \in \mathcal{W} \mid M, w \models \alpha\} \models \varphi$$

That is, the conditional $\alpha \to \varphi$ is established (resolved) in s iff φ is established (resolved) in the state that results from augmenting s with the assumption that α is true. For conditional *declaratives*, this delivers a standard material implication, as predicted by Fact 3 together with the fact that truth-conditions are standard.

---

[3] Our *resolutions* are a more general version of the *basic answers* in the interrogative frameworks of Hintikka [11,12] and Wisniewski [16]. We use the term *resolutions* as a reminder that this is a specific technical notion, sufficient for the present purposes. Our notion of *presupposition* of a question is also shared with the mentioned theories.

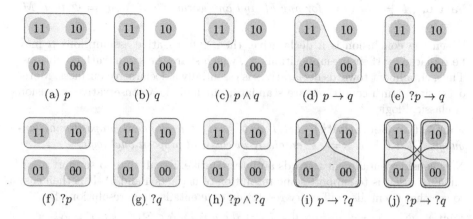

Fig. 1. The meanings of some simple sentences. 11 represents a world where both p and q are true, 10 a world where p is true and q is false, etc. For simplicity, only *maximal* supporting states are depicted. Notice that, by Fact 3, the declaratives on the top row have exactly one maximal supporting state: the set of all worlds where they are true.

At the same time, the clause also delivers conditional interrogatives like $p \to ?q$, which is resolved precisely when one of $p \to q$ and $p \to \neg q$ is established.

Now consider the case of an interrogative antecedent μ. If the consequent is a declarative α, the whole conditional is declarative, thus fully truth-conditional; hence, μ may simply be replaced by its presupposition: $\mu \to \alpha \equiv \pi_\mu \to \alpha$.

If the consequent is itself an interrogative ν, on the other hand, the clause says that $\mu \to \nu$ is resolved in s in case, if we extend s so as to resolve μ, the resulting state resolves ν. So, we can resolve $\mu \to \nu$ if we can resolve ν *conditionally* on having a resolution of μ, i.e., if our information is such that a resolution of μ determines a resolution of ν. E.g., the conditional interrogative $?p \to ?q$ is resolved precisely in case at least one of the following declaratives is established:

1. $(p \to q) \land (\neg p \to q) \ \equiv q$ 3. $(p \to \neg q) \land (\neg p \to q) \ \equiv q \leftrightarrow \neg p$
2. $(p \to q) \land (\neg p \to \neg q) \equiv q \leftrightarrow p$ 4. $(p \to \neg q) \land (\neg p \to \neg q) \equiv \neg q$

which correspond precisely to the four ways in which a resolution to $?p$ may determine a resolution to $?q$. Thus, such a conditional interrogative asks for enough information to establish a certain *interrogative dependency*.

3 Entailment and Interrogative Dependencies

Entailment in InqD_π is defined in the natural way, as preservation of support.

Definition 9 (Entailment).
$\Phi \models \psi \iff$ *for any model M and state s, if $M, s \models \Phi$ then $M, s \models \psi$.*

To see what this notion captures, consider first entailment towards a declarative. Fact 3 implies that, in this case, only truth-conditional content matters.

Fact 6. $\Phi \models \alpha \iff$ *for any M and any world w : if $M, w \models \Phi$ then M,* $w \models \alpha$.

When the conclusion is a declarative, then, interrogative assumptions μ may be replaced by their presuppositions π_μ, which share the same truth conditions. Thus, entailment towards declaratives is essentially a declarative business. Moreover, since truth-conditions are standard, our logic is a conservative extension of classical logic.

Fact 7 (Conservativity). *Let Γ, α be part of a standard propositional language. Then $\Gamma \models \alpha \iff \Gamma$ entails α in classical propositional logic.*

Now consider entailment towards an interrogative. Recall that to support a formula, or a set, is to support some resolution of it (facts 4 and 5). It is then easy to see that Φ entails ψ iff every resolution of Φ entails some resolution of ψ.

Fact 8. $\Phi \models \psi \iff$ *for all $\Gamma \in \mathcal{R}(\varphi)$ there is an $\alpha \in \mathcal{R}(\psi)$ s.t. $\Gamma \models \alpha$.*

Decomposing Φ into a set Γ of declaratives and a set Λ of interrogatives, and assuming ψ is an interrogative μ, this tells us that $\Gamma, \Lambda \models \mu$ holds iff any resolution of all interrogatives in Λ, together with Γ, entails some resolution of μ; that is, if *given Γ, any resolution of the interrogatives in Λ determines some resolution of μ*. To illustrate this, consider the following example of valid entailment:

$$p \leftrightarrow q \wedge r, \ ?q \wedge ?r \models ?p$$

Given the declarative $p \leftrightarrow q \wedge r$, any resolution of the conjunctive question $?q \wedge ?r$ determines a resolution of $?p$: for instance, the resolution $q \wedge r$ determines the resolution p, the resolution $q \wedge \neg r$ determines the resolution $\neg p$, and so on. Thus, an interrogative dependency is captured as a particular case of entailment in InqD_π, involving an interrogative conclusion and some interrogative assumptions.

Now that we know that interrogative dependencies are a case of entailment, it should no longer surprise us that they are internalized in the language as implications, such as $?p \rightarrow ?q$. As we have seen, the resolutions of such formulas embody precisely the different possible ways in which the consequent may be determined by the antecedent. Notice that the specific *way* in which an interrogative is determined by another may in turn be one of the variables on which the resolution of a certain interrogative conclusion depends. For instance, the following entailment captures the fact that a resolution of $?q$ is determined once a resolution to $?p$ is given and a specific dependency of $?q$ on $?p$ is established.

$$?p, \ ?p \rightarrow ?q \models ?q$$

How about the case in which we have an interrogative conclusion and *no* interrogative assumption? Since a set of declaratives Γ is the only resolution of itself, Fact 8 has the following corollary: Γ entails an interrogative μ iff it establishes some particular resolution of μ, i.e., in case it settles μ in a particular way.

Fact 9. *If Γ is a set of declaratives, $\Gamma \models \psi \iff \Gamma \models \alpha$ for some $\alpha \in \mathcal{R}(\psi)$.*

Summing up, then, inquisitive entailment brings under the same umbrella three crucial and seemingly independent notions of a logic of information and issues: *standard declarative entailment, answerhood,* and *interrogative dependency.*

4   Derivation System

A natural deduction system for InqD_π is described in the table below. In this system, the standard connectives—conjunction, implication, and falsum—are all assigned their standard inference rules. Thus, the core logical features of the connectives are preserved when these connectives are generalized to interrogatives.[4]

What does *not* generalize, by contrast, is the double negation axiom: for, double negation elimination is valid only for formulas whose semantics is truth-conditional, i.e., for which being supported amounts to being true at every world.

Fact 10 (Double negation characterizes truth-conditionality).
$\neg\neg\varphi \to \varphi$ *is valid iff for all* M, s: $M, s \models \varphi \iff (M, w \models \varphi$ *for all* $w \in s)$

Conjunction				Implication

$$\frac{\varphi \quad \psi}{\varphi \wedge \psi} \qquad \frac{\varphi \wedge \psi}{\varphi} \quad \frac{\varphi \wedge \psi}{\psi} \qquad \qquad \frac{\begin{array}{c}[\varphi]\\ \vdots \\ \psi\end{array}}{\varphi \to \psi} \qquad \frac{\varphi \quad \varphi \to \psi}{\psi}$$

Interrogative   **Falsum**

$$\frac{\alpha_i}{?\{\alpha_1,\ldots,\alpha_n\}} \qquad \frac{\begin{array}{ccc}[\alpha_1] & & [\alpha_n]\\ \vdots & & \vdots \\ \varphi & \cdots & \varphi \quad ?\{\alpha_1,\ldots,\alpha_n\}\end{array}}{\varphi} \qquad \frac{\bot}{\varphi}$$

Kreisel-Putnam axiom   **Double negation**
$$(\alpha \to ?\{\beta_1,\ldots,\beta_n\}) \to ?\{\alpha \to \beta_1,\ldots,\alpha \to \beta_n\} \qquad \neg\neg\alpha \to \alpha$$

The rules for the interrogative operator are simply the usual ones for a disjunction. This is hardly surprising, since the semantics of ? is disjunctive. The introduction rule says that if we have established some α_i, then we have resolved $?\{\alpha_1,\ldots,\alpha_n\}$. The elimination rule says that if we can infer φ from the assumption that α_i is established for each i, then we can infer φ from the assumption that $?\{\alpha_1,\ldots,\alpha_n\}$ is resolved.[5] Finally, the last component of the system is the

---

[4] We refer to the introduction and elimination rule for a connective o as (oi) and (oe).

[5] The standard rules for negation and disjunction, which are derived connectives in our system, are admissible, with one caveat: a disjunction may only be eliminated towards a declarative. This restriction marks the difference between $\vee$ and ? and prevents unsound derivations such as $p \vee \neg p \vdash ?p$.

Kreisel-Putnam axiom, which distributes an implication over an interrogative consequent, provided the antecedent is a declarative.[6]

Definition 10.
We write $P : \Phi \vdash \psi$ if P is a proof with conclusion ψ whose set of undischarged assumptions is included in Φ. We write $\Phi \vdash \psi$ if some proof $P : \Phi \vdash \psi$ exists. We say φ and ψ are provably equivalent, notation $\varphi \dashv\vdash \psi$, if $\varphi \vdash \psi$ and $\psi \vdash \varphi$.

As usual, proving soundness is a tedious but straightforward matter.

Theorem 1 (Soundness). *If $\Phi \vdash \psi$ then $\Phi \models \psi$.*

Next, the normal form result of Corollary 1 is provable in the system. For the lengthy but straightforward inductive proof, the reader is referred to [3].

Lemma 1. *For any φ, $\varphi \dashv\vdash ?\mathcal{R}(\varphi)$.*

As a corollary, $(?i)$ ensures that a formula is derivable from any of its resolutions.

Corollary 2. *If $\alpha \in \mathcal{R}(\varphi)$, then $\alpha \vdash \varphi$.*

The following theorem, central to the completeness proof, says that derivability shares the property of entailment expressed by Fact 8: from Φ we can derive ψ iff from any specific resolution Γ of Φ we can derive some resolution α of ψ.

Theorem 2 (Resolution theorem).
$\Phi \vdash \psi \iff$ *for all $\Gamma \in \mathcal{R}(\Phi)$ there exists some $\alpha \in \mathcal{R}(\psi)$ s.t. $\Gamma \vdash \alpha$.*

The proof is given in the appendix. The inductive proof of the left-to-right direction of the theorem is constructive, providing an effective procedure that, given a proof $P : \Phi \vdash \psi$ and a resolution Γ of Φ, produces a new proof $Q : \Gamma \vdash \alpha$ of a specific resolution α of ψ. This procedure is illustrated in the next section.

Let us now turn to prove completeness. In [3], completeness was proved by reducing it to the completeness of classical propositional logic. Here, by contrast, we will provide an explicit canonical model construction. This strategy has the merit of generalizing to extensions of the language, such as *inquisitive epistemic logic* [6], for which a reduction strategy would not be viable [2].

The worlds in our canonical model will be *complete theories of declaratives*, i.e., sets Γ of declaratives which are (i) closed under deduction of declaratives; (ii) consistent; and (iii) complete: for any declarative α, either α or $\neg\alpha$ is in Γ.

Definition 11 (Canonical model).
The canonical model for InqD_π is the model $M^c = \langle \mathcal{W}^c, V^c \rangle$, where:

- *the elements of $\mathcal{W}^c$ are the complete theories of declaratives;*
- *$V^c : \mathcal{W}^c \to \wp(\mathcal{P})$ is the map $V^c(\Gamma) = \{p \in \mathcal{P} \mid p \in \Gamma\}$*

---

[6] This axiom is tightly related to the axiom $(\neg\varphi \to \psi \vee \chi) \to (\neg\varphi \to \psi) \vee (\neg\varphi \to \chi)$ first investigated in the context of intermediate logics in [13], whence the name.

The bridge between derivability and semantics is given by the following *support lemma*, stating that support at a state S in the canonical model amounts to derivability from the intersection $\bigcap S$ of all the theories in S (where $\bigcap \emptyset$ is the set of all formulas). The inductive proof is given in the appendix.

Lemma 2 (Support lemma). *For any S and φ, $M^c, S \models \varphi \iff \bigcap S \vdash \varphi$.*

As shown in the appendix, the support lemma allows us to use the canonical model to give counterexamples to entailment, and thus to prove completeness.

Theorem 3 (Completeness theorem). *If $\Phi \models \psi$, then $\Phi \vdash \psi$.*

5   Computational Content of Inquisitive Proofs

In this section we illustrate with an example the effective procedure described in the proof of the resolution theorem, and discuss the significance of this procedure. Consider again the example of valid entailment given in section 3:

$$p \leftrightarrow q \wedge r, \; ?q \wedge ?r \models ?p$$

Let $\Phi = \{p \leftrightarrow q \wedge r, \; ?q \wedge ?r\}$. By Fact 8, the validity of this entailment implies that any specific resolution Γ of Φ is bound to entail some resolution α of $?p$. The proof of the resolution theorem, then, tells us how to use a proof $P : \Phi \vdash ?p$ to produce a proof of some specific $\alpha \in \mathcal{R}(\psi)$ from a given $\Gamma \in \mathcal{R}(\Phi)$.

Below we have a proof $P : \Phi \vdash ?p$ in our deduction system. Subproofs involving only classical propositional logic have been omitted and denoted P_1, P_2, P_3.

$$
\cfrac{
  \cfrac{
    \cfrac{?q \wedge ?r}{?r}\,(\wedge e) \qquad
    \cfrac{\cfrac{[q]\;\;[r]\;\;p \leftrightarrow q \wedge r}{p}\,(P_1)}{?p}\,(?i) \qquad
    \cfrac{\cfrac{[\neg r]\;\;p \leftrightarrow q \wedge r}{\neg p}\,(P_2)}{?p}\,(?i)
  }{?p}\,(?e) \qquad
  \cfrac{\cfrac{[\neg q]\;\;p \leftrightarrow q \wedge r}{\neg p}\,(P_3)}{?p}\,(?i)
}{?p}\,(?e)
$$

Suppose now that we are given the specific resolution $\Gamma = \{p \leftrightarrow q \wedge r, \; q \wedge r\}$ of Φ, that is, suppose $?q \wedge ?r$ is resolved to $q \wedge r$, and let us see how we can use P to build a proof Q from Γ which yields either p or $\neg p$, whichever is entailed by Γ. The procedure builds the proof Q inductively on the structure of P, as follows:

1. replace assumptions of $?q \wedge ?r$ in P by the corresponding resolution $q \wedge r$;
2. where $(\wedge e)$ was used in P to get $?q$ and $?r$ from $?p \wedge ?r$, use the same rules to obtain q and r from $q \wedge r$; this gives us two proofs $Q_1 : \Gamma \vdash q$ and $Q_2 : \Gamma \vdash r$;
3. the proofs P_1, P_2 and P_3, which only involve declaratives, are left unchanged;
4. the step of $?$-introduction right below these proofs is simply erased;
5. where in P the rule $(?e)$ is used to eliminate from $?r$, take the proof $Q_2 : \Gamma \vdash r$ obtained above and plug it in place of the assumption of r in $P_1 : \Gamma \cup \{q, r\} \vdash p$. In this way we obtain a proof $Q_3 : \Gamma \cup \{q\} \vdash p$;

6. similarly, where $(?e)$ was used to eliminate from $?q$, take the proof $Q_1 : \Gamma \vdash q$ that we have obtained and plug it in place of the assumption of q in Q_3.

The outcome is the following proof $Q : \Gamma \vdash p$, showing that Γ determines the resolution p of $?p$.

$$\cfrac{\dfrac{q \wedge r}{q}\ (\wedge e) \qquad \dfrac{q \wedge r}{r}\ (\wedge e) \qquad p \leftrightarrow q \wedge r}{p}\ (P_1)$$

Thus, the proof $P : \Phi \vdash ?p$ above does more than just witnessing the validity of the entailment: it encodes the *specific way* in which a resolution of $?p$ may be obtained, given $p \leftrightarrow q \wedge r$, from a resolution of $?q \wedge ?r$, that is, it provides a *method* for turning a resolution of $?q \wedge ?r$ into a corresponding resolution of $?p$.

Summing up, an inquisitive proof concluding with an interrogative μ builds up a way of resolving μ, dependent on resolutions to certain interrogative assumptions. Hence, while the declarative fragment of the logic is completely classical, when it comes to interrogatives, inquisitive proofs have a distinctive constructive content. This resonates with the observation that, in the realm of interrogatives, the double negation axiom is no longer valid. Indeed, our natural deduction system is essentially a system for intuitionistic logic—where ? plays the role of disjunction—with certain syntactic restrictions on formulas, and two extra ingredients: declarative double negation, and the Kreisel-Putnam axiom.[7]

6   Inquisitive Logic as a Logic of Dependencies

We have seen how inquisitive logic captures interrogative dependencies as a case of entailment. In recent years, the logical study of dependency, sparked first by the study of bound variables in first-order logic, has received increasing attention, leading to the development of *dependence logic* [15,7,8, a.o.]. Lately, it has been realized [17] that propositional inquisitive semantics and propositional dependence logic are close relatives: they have the same semantic structures, identical expressive power, and differ only in their repertoire of logical operators.

Propositional dependence logic starts with dependence atoms $=(p_1, \ldots, p_n, q)$, expressing the fact that the value of an atom q is determined by the values of $p_1, \ldots, p_n$. This is a particular kind of interrogative dependency, which may be expressed in inquisitive semantics by means of the formula $?p_1 \wedge \ldots \wedge ?p_n \rightarrow ?q$. So, in inquisitive semantics, the dependence *atom* actually has a *complex* logical structure, consisting of an implication, a determining antecedent—a conjunction

---

[7] The syntactic restrictions are only needed to keep the language dichotomous, and are not essential: we could replace the operator ? by a binary inquisitive disjunction $\vee$, and allow the connectives to apply unrestrictedly. We would then end up with the system InqB of *basic inquisitive semantics* [1,5] , based on a simple propositional language. The results we have seen—in particular, the resolution theorem—translate, *mutatis mutandis*, to the setting of InqB. Thus, the properties described in this paper pertain to inquisitive logic at large, rather than just to the specific system $InqD_\pi$.

of atomic polar interrogatives—and a determined consequent—an atomic polar interrogative. I would like to argue here that, for a propositional logic of dependencies, the inquisitive repertoire of logical operations is a mathematically natural choice as well as a practically advantageous one.

First, once we recognize that dependencies are captured as a special case of entailment, it seems natural to internalize them in the language by means of an implication operation, especially since, perhaps surprisingly, such an operator exists and enjoys a natural semantics and a standard proof-theory.

One practical advantage of expressing dependencies by means of an implication, in combination with other operators, is that it becomes immediately clear that there is no need to restrict this operation to atomic polar interrogatives, or to polar interrogatives... or even to interrogatives. Implication makes sense in a much broader context, including the context of declaratives, in which we recover standard material implication. This extra generality allows us to express easily a wide spectrum of dependencies. For instance, the implication $?\{p, q\} \to ?\{r, s\}$ is supported in case every resolution to $?\{p, q\}$ determines a resolution to $?\{r, s\}$, that is, in case establishing either of p and q determines one of r and s.

Second, if we look at the set of *inquisitive propositions*—the non-empty, downward closed sets of states, which are the meaning objects of both inquisitive and propositional dependence logic—and we consider them ordered by entailment, we find that this space has a natural algebraic structure, namely, it forms a Heyting algebra [14]; the connectives of inquisitive semantics perform precisely the natural operations in this algebra: they express *meet*, *join*, and (algebraic) *implication*, which are responsible for the logical properties of these connectives.

As a concrete advantage, this principled treatment of connectives yields the well-behaved logical calculus described in this paper, and the one described in [1,5] for the system InqB. Indeed, when it comes to proof theory, the dependence atoms are not ruled by simple logical laws: to axiomatize their logic, syntactically involved rules are needed, which essentially amount to an implicit stipulation of their resolutions [17]. By contrast, in inquisitive logic, formulas expressing dependencies are complex objects, built up by several logical operations—implication, conjunction, and question mark/inquisitive disjunction—each of which has natural, even standard inference rules. Taken together, these simple rules determine the complex logical properties manifested by dependence formulas. This lends further support to the view that the operations expressed by these connectives are the natural building blocks in the given semantic setting.

Conclusions I would like to close recapitulating the three main conclusions drawn in this paper: first, the relation of interrogative dependency is a particular case of entailment, involving interrogative conclusions and interrogative assumptions. Second, as a case of entailment, dependencies are internalized in the logical language as implications among interrogatives. Lastly, the associated logical calculus is a specific kind of constructive logic, whose proofs encode dependencies, and provide methods to compute a resolution of the interrogative conclusion from resolutions of the interrogative assumptions.

Appendix

Proof of Theorem 2 (Resolution Theorem)

Let us start from the left-to-right direction of the theorem: if Φ derives ψ, then any resolution Γ of Φ derives some resolution α of ψ.

The proof is a constructive one: given a proof $P : \Phi \vdash \psi$ and a resolution Γ of Φ, we show how to use P and Γ to build a proof $Q : \Gamma \vdash \alpha$ of some resolution α of ψ. The construction proceeds by induction on the structure of the proof P. We distinguish a number of cases depending on the last rule applied in P.

- ψ is an undischarged assumption, $\psi \in \Phi$. In this case, any resolution Γ of Φ contains a resolution α of ψ by definition, so $\Gamma \vdash \alpha$.
- ψ is an axiom. If ψ is an instance of the double negation axiom, then it is a declarative, and the claim is trivially true. If ψ is an instance of the Kreisel-Putnam axiom, of the form $(\beta \to ?\{\gamma_1, \ldots, \gamma_n\}) \to ?\{\beta \to \gamma_1 \ldots, \beta \to \gamma_n\}$, take $\alpha = \bigwedge_{1 \le i \le n}((\beta \to \gamma_i) \to (\beta \to \gamma_i))$: α is a resolution of ψ and, being a classical tautology, we have $\Gamma \vdash \alpha$ for any set Γ whatsoever, in particular for any $\Gamma \in \mathcal{R}(\Phi)$.
- $\psi = \chi \wedge \xi$ was obtained by $(\wedge i)$ from χ and ξ. Take any resolution Γ of Φ. Since the set of undischarged assumptions above both χ and ξ is included in Φ, by induction hypothesis from Γ we can deduce a resolution β of χ and a resolution γ of ξ. But then, by applying a conjunction introduction rule, from Γ we can deduce $\beta \wedge \gamma$, which is a resolution of ψ.
- $\psi = \chi \to \xi$ was obtained by $(\to i)$. Then the immediate subproof of P is a proof of ξ from the set of assumptions $\Phi \cup \{\chi\}$. Take any resolution Γ of Φ. Suppose $\alpha_1, \ldots, \alpha_n$ are the resolutions of χ. For any $1 \le i \le n$, then, $\Gamma \cup \{\alpha_i\}$ is a resolution of $\Phi \cup \{\chi\}$, whence by induction hypothesis we have a proof $Q_i : \Gamma \cup \{\alpha_i\} \vdash \beta_i$ for some resolution β_i of ξ. But then, extending Q_i with an implication introduction, we derive $\alpha_i \to \beta_i$ from Γ. And since this is the case for any $1 \le i \le n$, from Γ we can derive $(\alpha_1 \to \beta_1) \wedge \ldots \wedge (\alpha_n \to \beta_n)$, where $\{\beta_1, \ldots, \beta_n\} \subseteq \mathcal{R}(\xi)$, which is a resolution of $\chi \to \xi = \psi$.
- $\psi = ?\{\alpha_1, \ldots, \alpha_n\}$ was obtained by $(?i)$ from α_i. Thus, the immediate subproof of P is a proof of α_i from Φ. Take any resolution Γ of Φ. By induction hypothesis, from Γ we can then derive a resolution of α_i, that is to say, we can derive α_i, since a declarative is the only resolution of itself. So, the induction hypothesis gives us a proof of α_i from Γ, which is what we needed, since α_i is a resolution of ψ.
- ψ was obtained by $(\wedge e)$ from $\psi \wedge \chi$. Then the immediate subproof of P is a proof of $\psi \wedge \chi$ from Φ. Take a resolution Γ of Φ. By induction hypothesis there is a proof $Q : \Gamma \vdash \alpha$ for some resolution α of $\psi \wedge \chi$. By definition, such a resolution is of the form $\alpha = \beta \wedge \gamma$ where β is a resolution of ψ and γ is a resolution of χ. Extending Q with a rule of conjunction elimination, then, we have a proof of the resolution β of ψ from Γ. Of course, the argument is analogous if ψ was obtained by $(\wedge e)$ from a conjunction $\chi \wedge \psi$.
- ψ was obtained by $(\to e)$ from χ and $\chi \to \psi$. Then the immediate subproofs of P are a proof of χ from Φ, and a proof of $\chi \to \psi$ from Φ. Consider a

resolution Γ of Φ. By induction hypothesis we have a proof $Q_1 : \Gamma \vdash \beta$ where $\beta \in \mathcal{R}(\chi)$, and a proof $Q_2 : \Gamma \vdash \gamma$, where $\gamma \in \mathcal{R}(\chi \to \psi)$. Now, if $\mathcal{R}(\chi) = \{\beta_1, \ldots, \beta_n\}$, then $\beta = \beta_i$ for some i, and (by definition of the resolutions of an implication) $\gamma = (\beta_1 \to \gamma_1) \wedge \ldots \wedge (\beta_n \to \gamma_n)$ where $\{\gamma_1, \ldots, \gamma_n\} \subseteq \mathcal{R}(\psi)$. Now, extending Q_2 with an application of $(\wedge e)$ we get a proof of $\beta_i \to \gamma_i$ from Γ. Finally, putting together this proof with $Q_1 : \Gamma \vdash \beta_i$ and applying $(\to e)$, we obtain the resolution γ_i of $\mathcal{R}(\psi)$.

- ψ was obtained by $(?e)$ from a basic interrogative $?\{\beta_1, \ldots, \beta_m\}$. Then the immediate subproofs of P are a proof $P_0 : \Phi \vdash ?\{\beta_1, \ldots, \beta_m\}$ and, for $1 \leq i \leq n$ a proof $P_i : \Phi \cup \{\beta_i\} \vdash \psi$. Now consider a resolution Γ of Φ. By induction hypothesis we have a proof $Q_0 : \Gamma \vdash \beta$ for some $\beta \in \mathcal{R}(?\{\beta_1, \ldots, \beta_m\})$. Moreover, for any $1 \leq i \leq n$, since $\Gamma \cup \{\beta_i\}$ is a resolution of $\Phi \cup \{\beta_i\}$, by induction hypothesis we have a proof $Q_i : \Gamma \cup \{\beta_i\} \vdash \alpha_i$ where $\alpha_i \in \mathcal{R}(\psi)$. Now since β is a resolution of $?\{\beta_1, \ldots, \beta_m\}$, by definition $\beta = \beta_i$ for some i. But then, substituting any undischarged assumption of β_i in Q_i with an occurrence of the proof Q_0 with conclusion β_i, we obtain a proof of α_i from Γ, which is what we needed, since $\alpha_i \in \mathcal{R}(\psi)$.

- ψ was obtained by $(\bot e)$. This means that the immediate subproof of P is a proof of $\bot$ from Φ. Take any resolution Γ of Φ. By induction hypothesis, from Γ we can prove a resolution of $\bot$, that is, since a declarative is the only resolution of itself, we can prove $\bot$. But then, by an application of $(\bot e)$, from Γ we can obtain any formula, in particular any of the resolutions of ψ (crucially, the set of resolutions of a formula is always non-empty).

This completes the description of the inductive procedure to construct the required proof Q, and thus proves the left-to-right direction of the theorem. In order to establish the converse, let us make a detour through the following lemma.

Lemma 3. *If $\Phi \not\vdash \psi$ then there exists a $\Gamma \in \mathcal{R}(\Phi)$ such that $\Gamma \not\vdash \psi$.*

Proof. First assume Φ is finite. By induction on the number of formulas in Φ, we prove that *for any ψ, if $\Phi \not\vdash \psi$ there is some $\Gamma \in \mathcal{R}(\Phi)$ such that $\Gamma \not\vdash \psi$.*

If $\Phi = \emptyset$, the claim is trivially true. Now make the inductive hypothesis that the claim is true for sets of n formulas, and let us consider a set Φ of $n + 1$ formulas. Then Φ is of the form $\Psi \cup \{\chi\}$ for some set Ψ of n formulas and some formula χ. Now consider a formula ψ such that $\Psi, \chi \not\vdash \psi$. By Lemma 1, we must also have $\Psi, ?\mathcal{R}(\chi) \not\vdash \psi$ whence, by the ?-introduction rule, we must have $\Psi, \alpha \not\vdash \psi$ for some $\alpha \in \mathcal{R}(\chi)$. By the rules for implication, we must then have $\Psi \not\vdash \alpha \to \psi$. So, by induction hypothesis there is a $\Gamma \in \mathcal{R}(\Psi)$ such that $\Gamma \not\vdash \alpha \to \psi$. Finally, again by the rules for implication we have $\Gamma, \alpha \not\vdash \psi$, which proves the claim since $\Gamma \cup \{\alpha\}$ is a resolution of $\Psi \cup \{\chi\}$.

Our inductive proof is thus complete, and the claim is proved for the case in which Φ is finite. The conclusion can then be extended to the infinite case by an argument using König's lemma. For the details, we refer to [2].

Proof of theorem 2, continued. What remained to be shown is the right-to-left direction of the theorem: if every resolution of Φ derives some resolution of ψ, then $\Phi \vdash \psi$. Contrapositively, suppose $\Phi \nvdash \psi$. By Lemma 3 there exists a $\Gamma \in \mathcal{R}(\Phi)$ such that $\Gamma \nvdash \psi$. Now, since for any $\alpha \in \mathcal{R}(\psi)$ we have $\alpha \vdash \psi$ (Corollary 2), Γ cannot derive any $\alpha \in \mathcal{R}(\psi)$, otherwise it would derive ψ as well. So, Γ is a resolution of Φ which does not derive any resolution of ψ.   □

Lemma 4. *For any state $S \subseteq \mathcal{W}^c$ and declarative α, $\bigcap S \vdash \alpha \iff \alpha \in \bigcap S$*

Proof. If $\alpha \in \bigcap S$ then obviously $\bigcap S \vdash \alpha$. For the converse, suppose $\bigcap S \vdash \alpha$. For any $\Gamma \in S$ we have $\bigcap S \subseteq \Gamma$, so also $\Gamma \vdash \alpha$. But then, because Γ is closed under deduction of declaratives, we must have $\alpha \in \Gamma$. So, $\alpha \in \bigcap S$.

Proof of Lemma 2 (Support lemma)
The proof goes by induction on the complexity of φ. The straightforward cases for atoms, falsum, and conjunction are omitted.

Implication Suppose $\bigcap S \vdash \varphi \to \psi$. Take any $T \subseteq S$: if $T \models \varphi$ then by induction hypothesis $\bigcap T \vdash \varphi$. Since $T \subseteq S$, we have $\bigcap T \supseteq \bigcap S$, and since $\bigcap S \vdash \varphi \to \psi$, also $\bigcap T \vdash \varphi \to \psi$. By $(\to e)$, from $\bigcap T \vdash \varphi \to \psi$ and $\bigcap T \vdash \varphi$ it follows $\bigcap T \vdash \psi$, which by induction hypothesis implies $T \models \psi$. So, every substate of S that supports φ also supports ψ, that is, $S \models \varphi \to \psi$.

Viceversa, suppose $\bigcap S \nvdash \varphi \to \psi$. By the introduction rule for implication, this means that $\bigcap S, \varphi \nvdash \psi$. Now by Lemma 3 there is a a resolution of $(\bigcap S) \cup \{\varphi\}$ which does not derive ψ. Since $\bigcap S$ is a set of declaratives, this resolution must include a set of the form $(\bigcap S) \cup \{\alpha\}$ where α is a resolution of φ. Hence, there must exist a resolution α of φ such that $\bigcap S, \alpha \nvdash \psi$.

Now let $T = \{\Gamma \in S \mid \alpha \in \Gamma\}$. First, by definition we have $\alpha \in \bigcap T$, whence $\bigcap T \vdash \varphi$ by Corollary 2. By induction hypothesis we then have $T \models \varphi$. Now, if we can show that $\bigcap T \nvdash \psi$ we are done. For then, the induction hypothesis gives $T \nvDash \psi$, which means that T is a substate of S that supports φ but not ψ, which shows that $S \nvDash \varphi \to \psi$.

So, we are left to show that $\bigcap T \nvdash \psi$. Towards a contradiction, suppose that $\bigcap T \vdash \psi$. Since $\bigcap T$ is a set of declaratives, Theorem 2 tells us that $\bigcap T \vdash \beta$ for some resolution β of ψ, which by Lemma 4 amounts to $\beta \in \bigcap T$. So, for any $\Gamma \in T$ we have $\beta \in \Gamma$ and thus also $\alpha \to \beta \in \Gamma$, since Γ is closed under deduction of declaratives and $\beta \vdash \alpha \to \beta$. Now consider any $\Gamma \in S - T$: this means that $\alpha \notin \Gamma$; then since Γ is complete we have $\neg \alpha \in \Gamma$, whence $\alpha \to \beta \in \Gamma$, again because Γ is closed under deduction of declaratives and $\neg \alpha \vdash \alpha \to \beta$. This would mean, then, that $\alpha \to \beta \in \Gamma$ for any $\Gamma \in S$, whether $\Gamma \in T$ or $\Gamma \in S - T$. We can then conclude $\alpha \to \beta \in \bigcap S$, whence $\bigcap S, \alpha \vdash \beta$. And since β is a resolution of ψ we also have $\bigcap S, \alpha \vdash \psi$. But this is a contradiction, since by assumption α is such that $\bigcap S, \alpha \nvdash \psi$.

Interrogative operator If $S \models ?\{\alpha_1, \ldots, \alpha_n\}$, then $S \models \alpha_i$ for some i, so by induction hypothesis we have $\bigcap S \vdash \alpha_i$ and by ?-introduction also $\bigcap S \vdash ?\{\alpha_1, \ldots, \alpha_n\}$. Conversely, suppose $\bigcap S \vdash ?\{\alpha_1, \ldots, \alpha_n\}$. Since $\bigcap S$ is a set of declaratives, Theorem 2 implies $\bigcap S \vdash \alpha_i$ for some $1 \leq i \leq n$. By induction hypothesis we then have $S \models \alpha_i$, and thus also $S \models ?\{\alpha_1, \ldots, \alpha_n\}$.   □

Proof of Theorem 3 (Completeness theorem)
Suppose $\Phi \not\vdash \psi$. By Theorem 2, there is a resolution Θ of Φ which does not derive any resolution of ψ. Let $\mathcal{R}(\psi) = \{\alpha_1, \ldots, \alpha_n\}$: for each i, since $\Theta \not\vdash \alpha_i$, the set $\Theta \cup \{\neg\alpha_i\}$ is consistent, and thus extendible to a complete theory of declaratives $\Gamma_i \in W^c$. Let $S = \{\Gamma_1, \ldots, \Gamma_n\}$ be the set of theories obtained for $1 \leq i \leq n$. It is easy to verify that, by construction, $S \models \Phi$ but $S \not\models \psi$, whence $\Phi \not\models \psi$. $\qquad\square$

References

1. Ciardelli, I.: Inquisitive Semantics and Intermediate Logics. MSc thesis, University of Amsterdam (2009)
2. Ciardelli, I.: Modalities in the realm of questions: axiomatizing inquisitive epistemic logic (2014) (Manuscript, submitted for publication)
3. Ciardelli, I., Groenendijk, J., Roelofsen, F.: On the semantics and logic of declaratives and interrogatives. Synthese (2013), doi:10.1007/s11229-013-0352-7
4. Ciardelli, I., Groenendijk, J., Roelofsen, F.: Inquisitive semantics: a new notion of meaning. Language and Linguistics Compass 7(9), 459–476 (2013)
5. Ciardelli, I., Roelofsen, F.: Inquisitive Logic. Journal of Philosophical Logic 40(1), 55–94 (2011)
6. Ciardelli, I., Roelofsen, F.: Inquisitive dynamic epistemic logic. Synthese (2014), doi:10.1007/s11229-014-0404-7
7. Galliani, P.: The dynamics of imperfect information. PhD thesis, University of Helsinki (2012)
8. Galliani, P., Väänänen, J.: On dependence logic. arXiv preprint arXiv:1305.5948 (2013)
9. Groenendijk, J.: Inquisitive Semantics: Two Possibilities for Disjunction. In: Bosch, P., Gabelaia, D., Lang, J. (eds.) TbiLLC 2007. LNCS, vol. 5422, pp. 80–94. Springer, Heidelberg (2009)
10. Groenendijk, J.: Erotetic languages and the inquisitive hierarchy. In: a Festschrift for Martin Stokhof (2011)
11. Hintikka, J.: Inquiry as inquiry: A logic of scientific discovery. Kluwer Academic Publishers (1999)
12. Hintikka, J.: Socratic epistemology: explorations of knowledge-seeking by questioning. Cambridge University Press (2007)
13. Kreisel, G., Putnam, H.: Eine Unableitbarkeitsbeweismethode für den intuitionistischen Aussagenkalkül. Archiv für mathematische Logik und Grundlagenforschung 3(3-4), 74–78 (1957)
14. Roelofsen, F.: Algebraic foundations for the semantic treatment of inquisitive content. Synthese 190(1), 79–102 (2013)
15. Väänänen, J.: Dependence Logic: A New Approach to Independence Friendly Logic. Cambridge University Press (2007)
16. Wiśniewski, A.: The logic of questions as a theory of erotetic arguments. Synthese 109(1), 1–25 (1996)
17. Yang, F.: On extensions and variants of dependence logic: a study of intuitionistic connectives in the team semantics setting. PhD thesis, University of Helsinki (2014)

Modal Logics of Uncertainty with Two-Layer Syntax: A General Completeness Theorem

Petr Cintula[1] and Carles Noguera[1,2]

[1] Institute of Computer Science, Academy of Sciences of the Czech Republic,
Pod Vodárenskou věží 271/2, 182 07 Prague, Czech Republic
cintula@cs.cas.cz
[2] Institute of Information Theory and Automation, Academy of Sciences of the
Czech Republic, Pod Vodárenskou věží 4, 182 08 Prague, Czech Republic
noguera@utia.cas.cz

Abstract. Modal logics with two syntactical layers (both governed by classical logic) have been proposed as logics of uncertainty following Hamblin's seminal idea of reading the modal operator $\Box\varphi$ as 'probably φ', meaning that the probability of φ is bigger than a given threshold.

An interesting departure from that (classical) paradigm has been introduced by Hájek with his *fuzzy probability logic* when, while still keeping classical logic as interpretation of the lower syntactical layer, he proposed to use Łukasiewicz logic in the upper one, so that the truth degree of $\Box\varphi$ could be directly identified with the probability of φ. Later, other authors have used the same formalism with different kinds of uncertainty measures and other pairs of logics, allowing for a treatment of uncertainty of vague events (i.e. also changing the logic in the lower layer).

The aim of this paper is to provide a general framework for two-layer modal logics that encompasses all the previously studied two-layer modal fuzzy logics, provides a general axiomatization and a semantics of measured Kripke frames, and prove a general completeness theorem.

1 Introduction

Two-layer modal logics restrict the usage of modalities by employing a two-layer syntax with: (i) non-modal formulae, (ii) atomic modal formulae obtained by applying the modality operator(s) only to non-modal ones, and (iii) complex modal formulae built from the atomic ones. Classical logic governs the behavior of formulae of both modal and non-modal layers. This formalism has been usefully exploited by Fagin, Halpern and many others (see e.g. [8,17]), in modal logics of uncertainty following Hamblin's seminal idea of reading the modal operator $\Box\varphi$ as 'probably φ' [18], meaning that the probability of φ is bigger than a given threshold.

On the other hand, Mathematical Fuzzy Logic (MFL) studies many-valued logics of *comparable truth*, i.e. logics with an intended algebraic semantics (typically over the real interval $[0, 1]$) where all truth degrees are comparable because they are linearly ordered [3]. This feature makes fuzzy logics amenable to treat some aspects of vagueness [22].

U. Kohlenbach et al. (Eds.): WoLLIC 2014, LNCS 8652, pp. 124–136, 2014.
© Springer-Verlag Berlin Heidelberg 2014

Also uncertainty, albeit conceptually clearly separated from vagueness, has been addressed inside MFL. Hamblin's main idea was rediscovered and reformulated in the setting of many-valued logics by Hájek and Harmancová in [16] and was later developed in Hájek's monograph [15]. The main intuition was that one could use probability to determine the truth degree of statements such as 'tomorrow it will probably rain' or 'the probability that tomorrow it will rain is high'. Indeed, one takes classical logic and its formulae φ to describe crisp events, introduces a new modal operator $\square$ which can be applied on them to create atomic modal formulae $\square\varphi$ which may be read as 'probably φ' (or 'the probability of φ is high'), and finally these atomic modal formulae are combined by using the connectives of Łukasiewicz logic Ł [2,20,3]. This yields a two-layer syntax where modalities are never nested and can only be applied to formulae of classical logic. Hájek proposed a calculus in that syntax consisting of:

- axioms of classical propositional logic for non-modal formulae and axioms of Łukasiewicz logic Ł for modal ones,
- *modus ponens* rules for both non-modal and modal formulae,
- additional axioms:
  A1 $\square\varphi \to_Ł (\square(\varphi \to \psi) \to_Ł \square\psi)$
  A2 $\square\neg\varphi \leftrightarrow_Ł \neg_Ł\square\varphi$
  A3 $\square(\varphi \vee \psi) \leftrightarrow_Ł [(\square\varphi \to_Ł \square(\varphi \wedge \psi)) \to_Ł \square\psi]$
- a modal rule of necessitation $\varphi \vdash \square\varphi$.[1]

The resulting logic is called FP(Ł). Its semantical counterpart are *probability Kripke frames*, that is, structures $\mathbf{F} = \langle W, \mathbf{2}, [0,1]_Ł, \mu \rangle$, where W is a set of possible worlds, $\mathbf{2}$ is the Boolean algebra of two elements, $[0,1]_Ł$ is the standard Łukasiewicz algebra[2] and μ is a finitely additive probability measure. The idea is that non-modal (classical) formulae are evaluated in $\mathbf{2}$, μ is the interpretation of $\square$, and modal formulae are interpreted in $[0,1]_Ł$. This is formally achieved by the notion of *Kripke model* over a probability Kripke frame $\mathbf{F}$, i.e. $\mathbf{M} = \langle \mathbf{F}, \langle e_w \rangle_{w \in W} \rangle$ where:

- e_w is a classical evaluation of non-modal formulae, for each $w \in W$,
- $\{w \mid e_w(\varphi) = 1\}$ is in the domain of μ, for each non-modal formula φ.

The truth value of non-modal formulae in each world $w \in W$ is interpreted by the corresponding classical evaluation e_w, the truth value of an atomic modal formula $\square\varphi$ is uniformly defined as $\|\square\varphi\|_{\mathbf{M}} = \mu(\{w \mid e_w(\varphi) = 1\})$, and truth values of complex modal formulae are obtained by means of the operations in $[0,1]_Ł$. This gives a semantics for both modal and non-modal formulae of FP(Ł).

---

[1] All modal logics we consider in the paper are global modal logics in the sense that necessitation is a rule of the system, not just an admissible rule of the form: 'if φ is a theorem, then so $\square\varphi$'. This is the traditional approach in this research area. The question whether it would make sense to study also their local variants is beyond the scope of this paper.

[2] I.e. the real unit interval endowed with Łukasiewicz operations: $a \to_Ł b = \min\{1, 1 - a + b\}$, $a \leftrightarrow_Ł b = \min\{a \to_Ł b, b \to_Ł a\}$, and $\neg_Ł a = 1 - a$.

The corresponding completeness theorem then says that, given finite sets Φ and Γ of resp. non-modal and modal formulae and a modal formula γ, the following are equivalent:

- $\Phi, \Gamma \vdash_{\text{FP}(\text{Ł})} \gamma$
- $||\gamma||_{\mathbf{M}} = 1$ for all Kripke models $\mathbf{M}$, satisfying Γ and Φ, over all probability Kripke frames.

Several works have followed this idea with variations. In [13] Godo, Esteva and Hájek replaced Łukasiewicz logic on the second layer by the logic ŁΠ,[3] but kept classical logic for non-modal formulae. The logic ŁΠ, with its expanded language, enabled them to deal with conditional probability. Flaminio and Montagna also considered conditional probability in [11], and Godo and Marchioni investigated coherent conditional probabilities in [14]. Marchioni also proposed a class of *logics of uncertainty* in [19] with different kinds of measures besides probability (e.g. possibility and necessity measures, see [6]) to quantify the uncertainty of events. In all of these works classical logic has been kept as the underlying logic for non-modal formulae.

However, if one wants to deal with uncertainty and vagueness at once, i.e. with the probability of vague events, as in 'tomorrow it will probably rain *heavily*', the two-layer paradigm can still be useful provided that the underlying classical logic is substituted by a fuzzy logic. This idea has been also investigated in some works, as [9] where finite Łukasiewicz systems $Ł_n$ are taken as the logics of vague events. Other recent works along these lines are surveyed in [10]. There has even been a first attempt at an abstract theory of two-layer modal fuzzy logics in [21], but it is restricted to the family of *core fuzzy logics* which has become too narrow to contain the current scope of MFL (see [3]).

The primary aim of this paper is to provide a new general framework for two-layer modal fuzzy logics that encompasses the current state of the art and paves the way for future development. Actually, we will obtain much more than this. Indeed, we show how one can construct a modal logic (with arbitrary modalities, not necessarily read as probability) over a wide class of *non-classical logics* (satisfying certain weak technical requirements). Therefore, we need not assume that the starting logic is fuzzy, and we can develop a general theory of two-layer modal logics covering the known examples, showing how the methods used in the literature can lead to completeness results using very few properties of the underlying logics.

As a semantics, we propose particular kinds of *measured Kripke Frames* and prove corresponding completeness theorems from which the previous ones in the literature follow as particular instances. We will see that the strength of the completeness we obtain depends on the strength of the completeness of the initial logics. The proof is based on Hájek's idea from [15] of translating formulae and proofs of the two-layer modal logic to formulae and proofs of the upper logic.

---

[3] ŁΠ is the expansion of Ł with two new connectives $\&_\Pi$ and $\to_\Pi$ whose standard interpretation is, for each $a, b \in [0, 1]$, $a \&_\Pi b = a \cdot b$ (product of real numbers), $a \to_\Pi b = b/a$ if $a > b$, and $a \to_\Pi b = 1$ if $a \leq b$ (see [7]).

The paper is organized as follows: after this introduction, Section 2 gives the logical framework of the paper, with the minimal assumptions we require for the underlying propositional logics; then Section 3 gives the basic definitions regarding syntax and semantics of two-layer modal logics and, finally, Sections 4 and 5 give their corresponding forms of completeness theorems.

2   The Logical Framework

This section presents the necessary basic definitions and notation (for further information on Algebraic Logic notions see [12,5]). The definitions of a propositional language $\mathcal{L}$, the free term algebra $\boldsymbol{Fm}_\mathcal{L}$ over a denumerable set of generators (propositional variables), and finitary Hilbert-style proof systems are as usual. Let us introduce the notion of propositional logic that we use in this paper.

Convention 1. *Let $\mathcal{L}$ be a propositional language containing binary connectives $\rightarrow$ and $\wedge$, a defined binary connective $\varphi \leftrightarrow \psi = (\varphi \rightarrow \psi)\wedge(\psi \rightarrow \varphi)$, and a nullary connective $\bar{1}$.*[4] *In this paper a propositional logic L in propositional language $\mathcal{L}$ is a finitary lattice-conjunctive weakly implicative logic with unit (as studied in [4]).*

In more details, this means that L is identified with the provability relation $\vdash_L$ on $\boldsymbol{Fm}_\mathcal{L}$ given by a finitary Hilbert-style system such that the following formulae (rules respectively) are theorems (resp. derived rules) of L:[5]

$$\vdash_L \varphi \rightarrow \varphi \qquad \varphi, \varphi \rightarrow \psi \vdash_L \psi \qquad \varphi \rightarrow \psi, \psi \rightarrow \chi \vdash_L \varphi \rightarrow \chi \qquad \varphi \vdash_L \bar{1} \rightarrow \varphi$$

$$\vdash_L \varphi \wedge \psi \rightarrow \varphi \qquad \vdash_L \varphi \wedge \psi \rightarrow \psi \qquad \chi \rightarrow \varphi, \chi \rightarrow \psi \vdash_L \chi \rightarrow \varphi \wedge \psi \qquad \vdash_L \bar{1}$$

$$\{\varphi_i \leftrightarrow \psi_i \mid i \leq n\} \vdash_L c(\varphi_1, \ldots, \varphi_n) \leftrightarrow c(\psi_1, \ldots, \psi_n) \qquad \text{for every } n\text{-ary } c \in \mathcal{L}.$$

We recall now the basics of semantics.[6] Let us fix a logic L in a language $\mathcal{L}$; then $\mathcal{L}$-algebras are algebras with signature $\mathcal{L}$ and homomorphisms from $\boldsymbol{Fm}_\mathcal{L}$ to an $\mathcal{L}$-algebra $\boldsymbol{A}$ are called $\boldsymbol{A}$-*evaluations*. For an $\mathcal{L}$-algebra $\boldsymbol{A}$ we define the set $F_{\boldsymbol{A}} = \{x \mid x \wedge^{\boldsymbol{A}} \bar{1}^{\boldsymbol{A}} = \bar{1}^{\boldsymbol{A}}\}$.[7] We say that $\boldsymbol{A}$ is an L-algebra, $\boldsymbol{A} \in \mathbb{L}$ in symbols, if

- for each $\Gamma \cup \{\varphi\} \subseteq Fm_\mathcal{L}$ such that $\Gamma \vdash_L \varphi$, we have that for each $\boldsymbol{A}$-evaluation e, if $e[\Gamma] \subseteq F_{\boldsymbol{A}}$, then $e(\varphi) \in F_{\boldsymbol{A}}$,
- for each $x, y \in A$, if $x \leftrightarrow^{\boldsymbol{A}} y \in F_{\boldsymbol{A}}$, then $x = y$.

---

[4] Also the connectives $\rightarrow$, $\wedge$, and $\bar{0}$ might be definable from primitive ones.

[5] These theorems and rules capture, in a very elementary way, the minimal reasonable behavior of basic connectives we need to assume for our results.

[6] The reader familiar with the notion of algebraizable logics will note that our logics are algebraizable in the sense of [1] with a truth definition given by the single equation $x \wedge \bar{1} \approx \bar{1}$, and thus the following definitions and results are absolutely standard.

[7] $F_{\boldsymbol{A}}$ is usually called the *filter* of $\boldsymbol{A}$ and its members are the *designated elements* of the algebra, i.e. those that represent the notion of truth to be preserved in correct derivations.

It can be easily shown that $\mathbb{L}$ is a quasivariety and the equivalent algebraic semantics of L in the sense of [1], thus, in particular, L is strongly complete with respect to the semantics given by $\mathbb{L}$, i.e. the following are equivalent:

- $\Gamma \vdash_L \varphi$
- $e(\varphi) \in F_{\boldsymbol{A}}$ for each $\boldsymbol{A} \in \mathbb{L}$ and each $\boldsymbol{A}$-evaluation e such that $e[\Gamma] \subseteq F_{\boldsymbol{A}}$.

Of course some logics are strongly complete w.r.t. more specific classes of algebras.[8] Interestingly enough, in some prominent logics (e.g. Łukasiewicz logic) one can obtain a completeness result w.r.t. one intended algebra ($[0,1]_{\text{Ł}}$ in the case of Łukasiewicz) only when restricting to derivation from *finite* theories. Therefore we distinguish two forms of strong completeness. Namely, given a subclass $\mathbb{K} \subseteq \mathbb{L}$, we say that L has *(finite) strong $\mathbb{K}$-completeness*, S$\mathbb{K}$C (or FS$\mathbb{K}$C resp.) whenever for each (finite) theory $\Gamma \cup \{\varphi\}$ holds that $\Gamma \vdash_L \varphi$ iff for each $\boldsymbol{A} \in \mathbb{K}$ and each $\boldsymbol{A}$-evaluation e, $e(\varphi) \in F_{\boldsymbol{A}}$ whenever $e[\Gamma] \subseteq F_{\boldsymbol{A}}$.

3   Syntax and Semantics of Two-Layer Modal Logics

Let us fix two logics L_1 and L_2 in languages $\mathcal{L}_1$ and $\mathcal{L}_2$ and *modal language* $\mathfrak{L}$ (a non-empty set of modalities together with their arities) such that $\mathfrak{L} \cap (\mathcal{L}_1 \cup \mathcal{L}_2) = \emptyset$. We define three kinds of formulae of a two-level language $Fm^{Var}_{\mathcal{L}_2(\mathcal{L}_1)}$ over a set of variables *Var*:

- non-modal formulae from $Fm^{Var}_{\mathcal{L}_1}$,
- atomic modal formulae of the form $\Box(\varphi_1, \ldots, \varphi_n)$, for $\varphi_1, \ldots, \varphi_n \in Fm^{Var}_{\mathcal{L}_1}$ and an n-ary modality $\Box \in \mathfrak{L}$.
- modal formulae resulting from atomic ones by connectives from $\mathcal{L}_2$.[9]

We use the following notational conventions:

	non-modal	modal
formulae	$\varphi, \psi, \ldots$	$\gamma, \delta, \ldots$
sets of formulae	$\Phi, \Psi, \ldots$	$\Gamma, \Delta, \ldots$

We can extend any substitution[10] σ of $Fm^{Var}_{\mathcal{L}_1}$ first to atomic modal formulae as: $\sigma\Box(\varphi_1, \ldots, \varphi_n) = \Box(\sigma\varphi_1, \ldots, \sigma\varphi_n)$, and then to all modal formulae inductively, i.e., for each n-ary connective $c \in \mathcal{L}_2$ we set $\sigma c(\gamma_1, \ldots, \gamma_n) = c(\sigma\gamma_1, \ldots, \sigma\gamma_n)$.

---

[8] E.g. in classical logic $\mathbb{L}$ is the class of all Boolean algebras, but classical logic is also strongly complete w.r.t. the Boolean algebra of two elements; also the Łukasiewicz logic is strongly complete w.r.t. the class of all linearly ordered algebras from $\mathbb{L}$, where the order is given by the semi-lattice conjunction $\wedge$.

[9] As illustrative examples, observe that given a non-modal formula φ the following are *not* modal formulae in our sense: $\varphi, \varphi \to \psi, \Box\Box\varphi, \varphi \to \Box\varphi$. Also note that we do not exclude the possibility that $\mathcal{L}_1$ and $\mathcal{L}_2$ may share some connectives (e.g. $\to, \wedge, \bar{1}$, and $\leftrightarrow$), and hence they occur in both levels; the context will always settle any possible ambiguity.

[10] By substitution we mean an endomorphism on $\boldsymbol{Fm}^{Var}_{\mathcal{L}_1}$, i.e., a mapping $\sigma \colon Fm^{Var}_{\mathcal{L}_1} \to Fm^{Var}_{\mathcal{L}_1}$ such that for each n-ary connective $c \in \mathcal{L}_1$ and each formulae $\varphi_1, \ldots, \varphi_n$, we have $\sigma c(\varphi_1, \ldots, \varphi_n) = c(\sigma\varphi_1, \ldots, \sigma\varphi_n)$.

Definition 1. *The* minimal L_2-modal logic over logic L_1 *(denoted by $L_2(L_1)$)* *has formulae* $Fm_{\mathcal{L}_2(\mathcal{L}_1)}^{Var}$ *and an axiomatic system consisting of*

- *the axioms and rules of L_1 for non-modal formulae,*
- *axioms and rules of L_2 for modal formulae,*
- *and the following congruence rule:*

$$\varphi_1 \leftrightarrow \psi_1, \ldots, \varphi_n \leftrightarrow \psi_n \vdash \Box(\varphi_1, \ldots, \varphi_n) \leftrightarrow \Box(\psi_1, \ldots, \psi_n) \qquad \text{(CONGR)}$$

A modal rule is a tuple $\langle \Phi, \gamma \rangle$ for some finite set of non-modal premises Φ and a modal conclusion γ (if Φ is empty we call it modal axiom *instead). An L_2-modal logic over a logic L_1 is an extension of $L_2(L_1)$ by a set of modal rules.*

We understand rules as schemata, i.e., for each substitution σ on $Fm_{\mathcal{L}_1}^{Var}$, if $\langle \Phi, \gamma \rangle$ is a modal rule, then $\langle \sigma[\Phi], \sigma\gamma \rangle$ is also a modal rule.

Definition 2. *Let L be an L_2-modal logic over a logic L_1 given by a set of modal rules R. A proof in L from a set of non-modal premises Φ and modal premises Γ is a sequence of formulae where each element is either an axiom of L_i,[11] a modal axiom, an element of $\Phi \cup \Gamma$, or it follows from the previous formulae in the sequence using a rule of L_i, the rule (CONGR) or an additional modal rule from R. We use the symbol $\vdash_L$ to denote the provability relation in L.*

One can imagine that the proof consists of three separate parts: proving non-modal formulae, application of the modal rules on proved non-modal formulae, and proving modal formulae. Note that provability of non-modal formulae is fully determined by the logic L_1 and non-modal premises, i.e., we have: $\Phi, \Gamma \vdash_L \varphi$ iff $\Phi \vdash_{L_1} \varphi$.

Observe that Hájek's logic FP(Ł) (recall its definition in the introduction) is an Ł-modal logic over classical logic (not the minimal one, as it will be obvious when we introduce the semantics), as the congruence rule is easily derivable in its proof system.

Let us fix sets $\mathbb{K}_i$ of L_i-algebras for $i \in \{1, 2\}$. Now we define the general notion of Kripke frame, which nicely illuminates the different nature of both layers: the algebras from $\mathbb{K}_1$ are used to compute the *local* truth values of non-modal formulae in particular possible words,[12] the algebras from $\mathbb{K}_2$ are used to compute the *global* truth degrees of modal formulae in a frame, where the truth degrees of atomic formulae are obtained by *measuring* the collection of truth degrees in all possible worlds. Note that this measure is *partial*, i.e., not necessarily all possible collections of truth values in all possible worlds are measurable.

---

[11] When we speak about using an axiom γ of L_2 we formally assume the existence of a mapping τ from variables to atomic modal formulae and an axiom $\varphi(v_1, \ldots, v_n)$ in $Fm_{\mathcal{L}_2}^{Var}$ such that $\gamma = \varphi(\tau(v_1), \ldots, \tau(v_n))$; analogously for rules.

[12] In order to prove our general completeness theorems we have to allow, unlike in the traditional approach, possibly different algebras in different possible words.

Definition 3. *A* $\mathbb{K}_1$-*based* $\mathbb{K}_2$-*measured Kripke* $\mathfrak{L}$-*frame is defined as a tuple* $\mathbf{F} = \langle W, \langle \boldsymbol{A}_w \rangle_{w \in W}, \boldsymbol{B}, \langle \mu^\square \rangle_{\square \in \mathfrak{L}} \rangle$ *where* W *is a set (of possible worlds),* $\boldsymbol{A}_w \in \mathbb{K}_1$ *for each* $w \in W$, $\boldsymbol{B} \in \mathbb{K}_2$, *and for each* n-*ary* $\square \in \mathfrak{L}$, $\mu^\square$ *is a* partial *mapping*

$$\mu^\square \colon \big(\prod_{w \in W} A_w \big)^n \to B.$$

Next we define Kripke models in the expected way: by providing evaluations of non-modal formulae in all possible worlds. Note, however, that we also require that any collection of truth degrees of a formula in all possible worlds (or n formulae in the case of n-ary modalities) has be measurable. This assumption will ensure that the truth degrees of modal formulae are always defined.

Definition 4. *Let* $\mathbf{F} = \langle W, \langle \boldsymbol{A}_w \rangle_{w \in W}, \boldsymbol{B}, \langle \mu^\square \rangle_{\square \in \mathfrak{L}} \rangle$ *be a* $\mathbb{K}_1$-*based* $\mathbb{K}_2$-*measured Kripke* $\mathfrak{L}$-*frame. For each* $w \in W$, *take an evaluation* $e_w \colon Fm_{\mathcal{L}_1}^{Var} \to A_w$ *and let* $\mathbf{M}$ *be the tuple* $\langle \mathbf{F}, \langle e_w \rangle_{w \in W} \rangle$. *Given an* $\mathcal{L}_1$-*formula* φ, *we define* $\varphi^{\mathbf{M}} \in \prod_{w \in W} A_w$ *as the function* $\varphi^{\mathbf{M}}(w) = e_w(\varphi)$.

We say say that $\mathbf{M}$ *is a* Kripke model $\mathbf{M}$ *over* $\mathbf{F}$ *if for each* n-*ary* $\square \in \mathfrak{L}$ *and non-modal formulae* $\varphi_1, \ldots, \varphi_n$, *the tuple* $\langle \varphi_1^{\mathbf{M}}, \ldots, \varphi_n^{\mathbf{M}} \rangle$ *belongs to the domain of* $\mu^\square$.

The truth value of an atomic modal formula is defined as $\|\square(\varphi_1, \ldots, \varphi_n)\|_{\mathbf{M}} = \mu^\square(\varphi_1^{\mathbf{M}}, \ldots, \varphi_n^{\mathbf{M}})$. *The truth value of non-atomic modal formulae is computed by using operations from* $\boldsymbol{B}$, *i.e., for each* n-*ary connective* $c \in \mathcal{L}_2$ *and each modal formulae* $\gamma_1, \ldots, \gamma_n$, *we have* $\|c(\gamma_1, \ldots, \gamma_n)\|_{\mathbf{M}} = c^{\boldsymbol{B}}(\|\gamma_1\|_{\mathbf{M}}, \ldots, \|\gamma_n\|_{\mathbf{M}})$.

We say that $\mathbf{M}$ *is a model of a non-modal formula* φ *if* $e_w(\varphi) \in F_{\boldsymbol{A}_w}$ *for each* $w \in W$; *we say that it is a model of a modal formula* γ *if* $\|\gamma\|_{\mathbf{M}} \in F_{\boldsymbol{B}}$.

We conclude this section by defining Kripke frames for a given logic.

Definition 5. *A measured Kripke* $\mathfrak{L}$-*frame* $\mathbf{F}$ *is a frame for an* L_2-*modal logic over a logic* L_1 *if all its additional modal rules are valid in all Kripke models over* $\mathbf{F}$, *i.e. for any rule and any Kripke model over* $\mathbf{F}$ *which is a model of its premises is also a model of its conclusion.*

We can show that Hájek's probability Kripke frames are exactly the $\{\mathbf{2}\}$-based $\{[0,1]_{\text{Ł}}\}$-measured frames for FP(Ł). Clearly any probability Kripke frame is a $\{\mathbf{2}\}$-based $\{[0,1]_{\text{Ł}}\}$-measured frame. We need to show that a Kripke frame $\mathbf{F} = \langle W, \mathbf{2}, [0,1]_{\text{Ł}}, \mu \rangle$ is frame of FP(Ł) iff μ finitely additive probability measure. Both directions are relatively easy to prove, e.g. a Kripke model $\mathbf{M}$ is a model of axiom A3 iff $\mu(\varphi^{\mathbf{M}} \vee \psi^{\mathbf{M}}) = \mu(\varphi^{\mathbf{M}}) + \mu(\psi^{\mathbf{M}}) - \mu(\varphi^{\mathbf{M}} \wedge \psi^{\mathbf{M}})$ (this is equivalent with finite additivity because necessitation and axiom $A2$ entails that $\mu(\overline{0}^{\mathbf{M}}) = 0$).

4   Strong Completeness Theorem

In this section we state and prove the first of our main results: the strong completeness of an L_2-modal logic over a logic L_1.

Theorem 1. *Let* L *be an* L_2-*modal logic over a logic* L_1 *such that* L_i *has the* SK_iC *for* $i \in \{1, 2\}$. *Then the following are equivalent for each non-modal theory* Φ *and modal theory* $\Gamma \cup \{\gamma\}$:

- $\Phi, \Gamma \vdash_L \gamma$
- *for each* $\mathbb{K}_1$-*based* $\mathbb{K}_2$-*measured Kripke frame* **F** *for* L *and each Kripke model* **M** *over* **F**, *if* **M** *is a model of* $\Phi \cup \Gamma$, *then it is a model of* γ.

The same equivalence holds if L_1 *has the* FSK_1C *only, but only restricted to finite sets* Φ.

The proof of soundness is straightforward: the only non-trivial part is to show the soundness of the rule (CONGR), which easily follows from the properties of algebraic semantics (the behavior of $\leftrightarrow$) and the fact that the interpretations of modalities are (partial) functions.[13] We prove the completeness in a series of lemmata. The first step is to translate the modal logic L into the logic L_2.

Definition 6. *Take a new set of variables* $Var_m = \{p^{\square}_{\varphi_1,...,\varphi_n} \mid \square$ *an* n-*ary modality and* $\varphi_1, \ldots, \varphi_n$ *non-modal formulae*}. *We define the translation of modal formulae into* $Fm^{Var_m}_{\mathcal{L}_2}$-*formulae in the following way:*

- $(\square(\varphi_1, \ldots, \varphi_n))^* = p^{\square}_{\varphi_1,...,\varphi_n}$, *for any atomic modal formula* $\square(\varphi_1, \ldots, \varphi_n)$
- $c(\gamma_1, \ldots, \gamma_n)^* = c(\gamma_1^*, \ldots, \gamma_n^*)$, *for any* n-*ary connective* c *in* $\mathcal{L}_2$ *and any modal formulae* $\gamma_1, \ldots, \gamma_n$

Finally, given a set Γ *of modal formulae and a set* Φ *of non-modal formulae we define (we write* '$\Phi \vdash \Psi$' *for* '$\Phi \vdash \psi$ *for each* $\psi \in \Psi$'):

- $\Gamma^* = \{\gamma^* \mid \gamma \in \Gamma\}$
- $\Phi^* = \{\gamma^* \mid$ *there is a set* Ψ *s.t.* $\Phi \vdash_{L_1} \Psi$ *and* $\langle \Psi, \gamma \rangle$ *is a modal rule of* L$\}$.

Observe that Φ^* consists of *-translations of conclusions of modal rules whose premises are provable from Φ (in particular Φ^* always contains *-translations of all modal axioms).

Lemma 1. *For each set of non-modal formulae* Φ *and each set of modal formulae* $\Gamma \cup \{\gamma\}$ *we have:* $\Phi, \Gamma \vdash_L \gamma$ *iff* $\Phi^*, \Gamma^* \vdash_{L_2} \gamma^*$.

Proof. From left to right: we take any proof of γ from $\Phi \cup \Gamma$ in L and show that if we omit all non-modal formulae from the proof and replace each modal formula δ by δ^* we obtain a proof of γ^* from $\Phi^* \cup \Gamma^*$ in L_2. Indeed, if δ is an axiom of L_2 or element of Γ the claim is obvious; if δ is a consequence of some modal rule then $\delta^* \in \Phi^*$; and finally if δ follows, by some rule of L_2, from premises Δ, then so does δ^* from premises Δ^*.

From right to left: we take any proof of γ^* from $\Phi^* \cup \Gamma^*$ in L_2 and construct a new sequence of formulae. Note that for any $Fm^{Var_m}_{\mathcal{L}_2}$-formula A there is unique modal formula δ such that $\delta^* = A$, i.e., we can assume that the proof consists of formulae of the form δ^*. We distinguish the following two cases:

---

[13] The partially of the interpretation of modalities should not come as a surprise, taking into account the original probabilistic motivation of our work.

– if $\delta^* \in \Phi^*$ i.e., there is a modal rule $\langle \Psi, \delta \rangle$ such that $\Phi \vdash_{L_1} \Psi$: then we replace δ^* by δ preceded by a sequence of all proofs (in arbitrary order) of elements of Ψ from Φ.

– otherwise we just replace δ^* by δ.

Clearly, the resulting sequence is a proof of γ from $\Phi \cup \Gamma$ in L. □

For the next step of the proof of Theorem 1, let us assume that $\Phi, \Gamma \nvdash_L \gamma$; thus, by the previous lemma, $\Phi^*, \Gamma^* \nvdash_{L_2} \gamma^*$. Therefore, using the assumption of S$\mathbb{K}_2$C of L_2, we obtain a $\mathbb{K}_2$-algebra B and a B-evaluation e such that $e[\Phi^* \cup \Gamma^*] \subseteq F_B$ and $e(\gamma^*) \notin F_B$.

Let us take $W = \{\varphi \mid \Phi \nvdash_{L_1} \varphi\}$. For each $\varphi \in W$ we have, by S$\mathbb{K}_1$C of L_1, a $\mathbb{K}_1$-algebra A_φ and an A_φ-evaluation e_φ s.t. $e_\varphi[\Phi] \subseteq F_{A_\varphi}$ and $e_\varphi(\varphi) \notin F_{A_\varphi}$. Note that if Φ is finite, then it is sufficient to assume FS$\mathbb{K}_1$C of L_1.

Consider the structure $\mathbf{F} = \langle W, \langle A_\varphi \rangle_{\varphi \in W}, B, \langle \mu^{\square} \rangle_{\square \in \mathfrak{L}} \rangle$, where for each n-ary $\square \in \mathfrak{L}$ the corresponding partial mapping is defined as:

$$\mu^{\square}(\langle \langle p_\varphi^1 \rangle_{\varphi \in W}, \ldots, \langle p_\varphi^n \rangle_{\varphi \in W} \rangle) = \begin{cases} e((\square(\pi_1, \ldots, \pi_n))^*) & \text{if there are } \pi_i \text{s s.t.} \\ & \forall \varphi \in W, \ p_\varphi^i = e_\varphi(\pi_i), \\ \text{undefined} & \text{otherwise.} \end{cases}$$

Next we prove that we have indeed constructed a Kripke frame.

Lemma 2. $\mathbf{F}$ *is a* $\mathbb{K}_1$*-based* $\mathbb{K}_2$*-measured Kripke frame.*

Proof. We only need to show that the definition of each function $\mu^{\square}$ is sound. Consider a measurable tuple $\langle \langle p_\varphi^1 \rangle_{\varphi \in W}, \ldots, \langle p_\varphi^n \rangle_{\varphi \in W} \rangle$ such that, for each i and each φ we have $p_\varphi^i = e_\varphi(\pi_i) = e_\varphi(\pi_i')$. To complete the proof we need to show that $e((\square(\pi_1, \ldots, \pi_n))^*) = e((\square(\pi_1', \ldots, \pi_n'))^*)$.

Our assumption entails, for each i, $\Phi \vdash_{L_1} \pi_i \leftrightarrow \pi_i'$ (otherwise $\pi_i \leftrightarrow \pi_i' \in W$ and $e_{\pi_i \leftrightarrow \pi_i'}(\pi_i \leftrightarrow \pi_i') \notin F_{A_{\pi_i \leftrightarrow \pi_i'}}$ thus $e_{\pi_i \leftrightarrow \pi_i'}(\pi_i) \neq e_{\pi_i \leftrightarrow \pi_i'}(\pi_i')$, a contradiction) and so, due to the rule (CONGR):

$$(\square(\pi_1, \ldots, \pi_n))^* \leftrightarrow (\square(\pi_1', \ldots, \pi_n'))^* \in \Phi^*.$$

This entails that $e((\square(\pi_1, \ldots, \pi_n))^* \leftrightarrow (\square(\pi_1', \ldots, \pi_n'))^*) \in F_B$ and, therefore, $e((\square(\pi_1, \ldots, \pi_n))^*) = e((\square(\pi_1', \ldots, \pi_n'))^*)$. □

Thus $\mathbf{M}_0 = \langle \mathbf{F}, \langle e_\varphi \rangle_{\varphi \in W} \rangle$ is a Kripke model over $\mathbf{F}$; indeed the definability condition is obviously satisfied: for each n-ary $\square \in \mathfrak{L}$ and non-modal formulae $\chi_1, \ldots, \chi_n$, the element $\langle \chi_1^{\mathbf{M}_0}, \ldots, \chi_n^{\mathbf{M}_0} \rangle = \langle \langle e_\varphi(\chi_1) \rangle_{\varphi \in W}, \ldots, \langle e_\varphi(\chi_n) \rangle_{\varphi \in W} \rangle$ obviously belongs to the domain of $\mu^{\square}$.

Note that $\mu^{\square}$ is defined *only* for arguments which can be uniformy described by values of some non-modal formule in the fixed system of evaluations by $\langle e_\varphi \rangle_{\varphi \in W}$; thus it is almost always undefined. This peculiar feature of $\mathbf{F}$ is used in the next lemma, which shows that all possible Kripke models over $\mathbf{F}$ can be effectively described by a substitution and the model $\mathbf{M}_0$. The substitution also determines the relation between truth value of a modal formula and its *-translation.

Lemma 3. *For each Kripke model* $\mathbf{M} = \langle \mathbf{F}, \langle \hat{e}_\varphi \rangle_{\varphi \in W} \rangle$ *there is an* $\mathcal{L}_1$*-substitution* σ *such that for each* $\varphi \in W$*, non-modal* ψ*, and modal* δ*:*

$$\hat{e}_\varphi(\psi) = e_\varphi(\sigma\psi) \qquad and \qquad ||\delta||_\mathbf{M} = e((\sigma\delta)^*).$$

Proof. Let v be an atomic $\mathcal{L}_1$-formula and take the atomic modal formula $\Box(v, \ldots, v)$. By definition of Kripke model, $\langle \langle \hat{e}_\varphi(v) \rangle_{\varphi \in W}, \ldots, \langle \hat{e}_\varphi(v) \rangle_{\varphi \in W} \rangle$ has to be in the domain of $\mu^\Box$. Therefore, since the function is defined only in such case, there has to be an $\mathcal{L}_1$-formula $\sigma(v)$ such that for each $\varphi \in W$ $\hat{e}_\varphi(v) = e_\varphi(\sigma(v))$ (actually, there might be many formulae $\sigma(v)$ satisfying this condition; just choose the first one in any fixed enumeration). We extend σ to an $\mathcal{L}_1$-substitution and show the first claim by induction. The base case is clear. Take $\psi = c(\chi_1, \ldots, \chi_n)$. Then $\hat{e}_\varphi(c(\chi_1, \ldots, \chi_n)) = c^{A_\varphi}(\hat{e}_\varphi(\chi_1), \ldots, \hat{e}_\varphi(\chi_n)) = c^{A_\varphi}(e_\varphi(\sigma\chi_1), \ldots, e_\varphi(\sigma\chi_n)) = e_\varphi(c(\sigma\chi_1, \ldots, \sigma\chi_n)) = e_\varphi(\sigma c(\chi_1, \ldots, \chi_n))$.

Now we prove the second claim first for atomic modal formulae; assume that $\delta = \Box(\chi_1, \ldots, \chi_n)$. Then we have:

$$
\begin{aligned}
||\delta||_\mathbf{M} &= \mu^\Box(\langle \hat{e}_\varphi(\chi_1) \rangle_{\varphi \in W}, \ldots, \langle \hat{e}_\varphi(\chi_n) \rangle_{\varphi \in W}) \\
&= \mu^\Box(\langle e_\varphi(\sigma\chi_1) \rangle_{\varphi \in W}, \ldots, \langle e_\varphi(\sigma\chi_n) \rangle_{\varphi \in W}) \\
&= e((\Box(\sigma\chi_1, \ldots, \sigma\chi_n))^*) \\
&= e((\sigma\delta)^*).
\end{aligned}
$$

The rest is a simple induction. $\qquad\qquad\qquad\qquad\qquad\qquad\qquad\qquad\qquad\quad \Box$

Having described all possible Kripke models over $\mathbf{F}$ we can easily prove that $\mathbf{F}$ is a Kripke frame for the logic L.

Lemma 4. $\mathbf{F}$ *is a Kripke frame for the logic* L.

Proof. Consider a Kripke model $\mathbf{M} = \langle \mathbf{F}, \langle \hat{e}_\varphi \rangle_{\varphi \in W} \rangle$ and a modal rule $\langle \Psi, \delta \rangle$ such that for each $\varphi \in W$ we have $\hat{e}_\varphi[\Psi] \subseteq F_{A_\varphi}$. We show that $\Phi \vdash_{L_1} \sigma[\Psi]$, where σ is the substitution determined by the previous lemma: indeed, otherwise we would have $\sigma\psi \in W$ for some $\psi \in \Psi$ and so $e_{\sigma\psi}(\sigma\psi) \notin F_{A_{\sigma\psi}}$ and, by the previous lemma, $\hat{e}_{\sigma\psi}(\psi) \notin F_{A_{\sigma\psi}}$, a contradiction. Due to structurality we also have a modal rule $\langle \sigma[\Psi], \sigma\delta \rangle$ in L and thus $(\sigma\delta)^* \in \Phi^*$ and so $||\delta||_\mathbf{M} = e((\sigma\delta)^*) \in F_B$ (the equality is due to the previous lemma). $\qquad\qquad\qquad\qquad\qquad\quad \Box$

Let us again consider the Kripke model $\mathbf{M}_0 = \langle \mathbf{F}, \langle e_\varphi \rangle_{\varphi \in W} \rangle$ which we now know is a model over a Kripke frame for L. Consider any world φ and observe that from the definition of e_φ we know that $e_\varphi[\Phi] \subseteq F_{A_\varphi}$. Furthermore observe that for each modal formula δ we have $||\delta||_{\mathbf{M}_0} = e(\delta^*)$ (due to Lemma 3) and so for each $\gamma' \in \Gamma$ we have $||\gamma'||_{\mathbf{M}_0} \in F_B$ while $||\gamma||_{\mathbf{M}_0} \notin F_B$. Thus we have established that $\mathbf{M}_0$ is a model of $\Phi \cup \Gamma$ but not of γ. This concludes the proof of Theorem 1.

5 Finite Strong Completeness Theorem

As mentioned in the introduction several prominent logics enjoy only the *finite* strong completeness theorem with respect to their intended semantics, including the Łukasiewicz logic featuring in the prototypical example of Hájek's probability logic FP(Ł). Theorem 1 shows that if this is the case for L_1, it is still not a huge problem: we only need to restrict ourselves to finite sets of non-modal premises. A more serious problem arises when the logic L_2 does not enjoy the strong completeness (as explained right after the next theorem).

The next theorem shows that we can obtain finite strong completeness for an L_2-modal logic over L_1 even if both logics satisfy only the finite strong completeness but at the price of adding some further conditions. One can easily observe that Hájek's probability logic FP(Ł) satisfies these conditions, thus the completeness of FP(Ł) is a corollary (recalling that Hájek's probability Kripke frames are in fact **2**-based $[0, 1]_{Ł}$-measured frames for FP(Ł)).

Theorem 2. *Let* L *be an* L_2-*modal logic over a logic* L_1 *such that* L_i *has* FSK_iC, $\mathbb{L}_1$ *is locally finite and* L *has finitely many additional modal rules. Then the following are equivalent for each* finite *non-modal theory* Φ *and each* finite *modal theory* $\Gamma \cup \{\gamma\}$:

- $\Phi, \Gamma \vdash_L \gamma$
- *for each* $\mathbb{K}_1$-*based* $\mathbb{K}_2$-*measured Kripke frame* **F** *for* L *and each Kripke model* **M** *over* **F**, *if* **M** *is a model of* $\Phi \cup \Gamma$, *then it is a model of* γ.

Recall that, in the proof of Theorem 1, the SK_2C of L_2 was used to obtain a witness showing that $\Phi^*, \Gamma^* \nvdash_{L_2} \gamma^*$. Now we only have FSK_2C so we need the set $\Phi^* \cup \Gamma^*$ to be finite. The finiteness of Γ^* follows from the finiteness of Γ. The analog is however not true for Φ^*: recall that Φ^* consists of $*$-translations of conclusions of modal rules whose premises are provable from Φ. To force this set to be finite we need a different translation $*$. Note that for our purposes we can safely assume that the set of propositional variables *Var* is finite (containing only those occurring in $\Phi \cup \Gamma \cup \{\gamma\}$), therefore our assumption of local finiteness entails that there are only finitely many non-equivalent non-modal formulae. After we define the new translation and prove the analog of Lemma 1, the proof of the present theorem is then the same as the proof of Theorem 1.

Definition 7. *Given a non-formula* φ, *we denote the class of its equivalent formulae as* $[\varphi] = \{\psi \mid \vdash_{L_1} \varphi \leftrightarrow \psi\}$. *Take a new set of variables* $Var_m = \{p^{\square}_{[\varphi_1], \ldots, [\varphi_n]} \mid \square$ *an* n-*ary modality and* $\varphi_1, \ldots, \varphi_n$ *non-modal formulae*}. *Now we define the translation of modal formulae into* $Fm^{Var_m}_{\mathcal{L}_2}$-*formulae as:*

- $(\square(\varphi_1, \ldots, \varphi_n))^* = p^{\square}_{[\varphi_1], \ldots, [\varphi_n]}$ *for any* n-*ary modality* $\square$
- $c(\gamma_1, \ldots, \gamma_n)^* = c(\gamma_1^*, \ldots, \gamma_n^*)$, *where* c *is any* n-*connective from* $\mathcal{L}_2$.

Finally, for any pair of sets Γ *and* Φ *of formulae we define:*

- $\Gamma^* = \{\gamma^* \mid \gamma \in \Gamma\}$
- $\Phi^* = \{\gamma^* \mid$ *there is a set* Ψ *s.t.* $\Phi \vdash_{L_1} \Psi$ *and* $\langle \Psi, \gamma \rangle$ *is a modal rule of* L}.

Lemma 5. *For each set of non-modal formulae Φ and each set of modal formulae $\Gamma \cup \{\gamma\}$ we have: $\Phi, \Gamma \vdash_{\mathrm{L}} \gamma$ iff $\Phi^*, \Gamma^* \vdash_{\mathrm{L}_2} \gamma^*$.*

Furthermore Φ^* is a finite set whenever the following conditions are met:

- *Var is finite,*
- *$\mathbb{L}_1$ is locally finite, and*
- *L has only finitely many additional modal rules (up to substitutions).*

Proof. The proof of the left-to-right direction is the same as in the proof of Lemma 1. The converse direction is a bit more complicated; let us first fix a 'pseudo preimage' of $*$, i.e., any function σ assigning to each $Fm_{\mathcal{L}_2}^{Var_m}$-formula A a modal formula σA such that $(\sigma A)^* = A$. This function has the following useful properties:

- if $\langle X, A \rangle$ is an instance of some rule of L_2, then so is $\langle \sigma[X], \sigma A \rangle$.
- $\Phi \vdash_{\mathrm{L}} \sigma \delta^* \leftrightarrow \delta$ for each modal formula δ (indeed, we prove it by induction: if $\delta = \Box(\varphi_1, \ldots, \varphi_n)$, then $\delta^* = p_{[\varphi_1], \ldots, [\varphi_n]}^{\Box}$ and so $\sigma \delta^* = \Box(\varphi'_1, \ldots, \varphi'_n)$ for some $\varphi'_1, \ldots, \varphi'_n$ such that $\vdash_{\mathrm{L}_1} \varphi_i \leftrightarrow \varphi'_i$; the rule (CONGR) completes the proof of the base case and the induction step is simple).

Next let us take any proof $A_1, \ldots, A_n$ of γ^* from $\Phi^* \cup \Gamma^*$ and construct a new sequence of formulae by distinguishing three cases:

- $A_i \in \Phi^*$ i.e., there is a modal rule $\langle \Psi, \delta \rangle$ such that $\Phi \vdash_{\mathrm{L}_1} \Psi$ and $A_i = \delta^*$: we replace A_i by σA_i preceded by a sequence of proofs (in arbitrary order) of all elements of Ψ from Φ, the formula δ and a proof of the formula $\delta \to \sigma A_i$.
- $A_i \in \Gamma^*$ i.e., there is a modal formula $\delta \in \Gamma$ such that and $A_i = \delta^*$: we replace A_i by σA_i preceded by δ and a proof of the formula $\delta \to \sigma A_i$.
- Otherwise we just replace A by σA_i.

Then we add a proof of the formula $\sigma A_n \to \gamma$ and the formula γ. Clearly, the resulting sequence is a proof of γ from $\Psi \cup \Gamma$ in the logic L.

It remains to be proved that Φ^* is finite whenever the three conditions are met. From the local finiteness of $\mathbb{L}_1$ and the finiteness of Var we know that there are only finitely many non-equivalent formulae in L_1. Observe that for each modal rule $\langle \Psi, \gamma \rangle$ the set of formulae

$$\{(\rho\gamma)^* \mid \rho \text{ is a substitution such that } \Phi \vdash_{\mathrm{L}_1} \rho[\Psi]\}$$

is finite. Since Φ^* is a union of finitely many such sets, it has to be finite. $\Box$

Acknowledgments. The authors are indebted to the anynomous referees for their corrections and remarks. Both authors were supported by the grant GAP202/10/1826 of the Czech Science Foundation. Petr Cintula also acknowledges the support of RVO 67985807. Carles Noguera was also supported by the FP7-PEOPLE-2009-IRSES project MaToMUVI (PIRSES-GA-2009-247584).

References

1. Blok, W.J., Pigozzi, D.L.: Algebraizable Logics. Memoirs of the American Mathematical Society, vol. 396. American Mathematical Society, Providence (1989), http://orion.math.iastate.edu/dpigozzi/
2. Cignoli, R., D'Ottaviano, I.M., Mundici, D.: Algebraic Foundations of Many-Valued Reasoning. Trends in Logic, vol. 7. Kluwer, Dordrecht (1999)
3. Cintula, P., Hájek, P., Noguera, C. (eds.): Handbook of Mathematical Fuzzy Logic (in 2 volumes). Studies in Logic, Mathematical Logic and Foundations, vol. 37, 38. College Publications, London (2011)
4. Cintula, P., Noguera, C.: A general framework for mathematical fuzzy logic. In: Cintula, P., Hájek, P., Noguera, C. (eds.) Handbook of Mathematical Fuzzy Logic - Volume 1. Studies in Logic, Mathematical Logic and Foundations, vol. 37, pp. 103–207. College Publications, London (2011)
5. Czelakowski, J.: Protoalgebraic Logics. Trends in Logic, vol. 10. Kluwer, Dordrecht (2001)
6. Dubois, D., Prade, H.: Possibility theory. Plenum Press, New-York (1988)
7. Esteva, F., Godo, L.: Putting together Łukasiewicz and product logic. Mathware and Soft Computing 6(2-3), 219–234 (1999)
8. Fagin, R., Halpern, J.Y., Megiddo, N.: A logic for reasoning about probabilities. Information and Computation 87(1-2), 78–128 (1990)
9. Flaminio, T., Godo, L.: A logic for reasoning about the probability of fuzzy events. Fuzzy Sets and Systems 158(6), 625–638 (2006)
10. Flaminio, T., Godo, L., Marchioni, E.: Reasoning about uncertainty of fuzzy events: An overview. In: Cintula, P., Fermüller, C., Godo, L. (eds.) Understanding Vagueness: Logical, Philosophical, and Linguistic Perspectives. Studies in Logic, vol. 36, pp. 367–400. College Publications, London (2011)
11. Flaminio, T., Montagna, F.: A logical and algebraic treatment of conditional probability. Archive for Mathematical Logic 44(2), 245–262 (2005)
12. Font, J.M., Jansana, R., Pigozzi, D.L.: A survey of Abstract Algebraic Logic. Studia Logica 74(1-2), 13–97 (2003), Special Issue on Abstract Algebraic Logic II
13. Godo, L., Esteva, F., Hájek, P.: Reasoning about probability using fuzzy logic. Neural Network World 10(5), 811–823 (2000), Special issue on SOFSEM 2000
14. Godo, L., Marchioni, E.: Coherent conditional probability in a fuzzy logic setting. Logic Journal of the Interest Group of Pure and Applied Logic 14(3), 457–481 (2006)
15. Hájek, P.: Metamathematics of Fuzzy Logic. Trends in Logic, vol. 4. Kluwer, Dordrecht (1998)
16. Hájek, P., Harmancová, D.: Medical fuzzy expert systems and reasoning about beliefs. In: Wyatt, J.C., Stefanelli, M., Barahona, P. (eds.) AIME 1995. LNCS, vol. 934, pp. 403–404. Springer, Heidelberg (1995)
17. Halpern, J.Y.: Reasoning About Uncertainty. MIT Press (2005)
18. Hamblin, C.L.: The modal 'probably'. Mind 68, 234–240 (1959)
19. Marchioni, E.: Possibilistic conditioning framed in fuzzy logics. International Journal of Approximate Reasoning 43(2), 133–165 (2006)
20. Mundici, D.: Advanced Lukasiewicz Calculus and MV-Algebras. Trends in Logic, vol. 35. Springer, New York (2011)
21. Rusnok, P.: Probability in formal fuzzy logic, Master thesis, Czech Technical University in Prague (2008)
22. Smith, N.J.: Vagueness and Degrees of Truth. Oxford University Press, Oxford (2009)

Ancestral Logic: A Proof Theoretical Study

Liron Cohen[1] and Arnon Avron[2]

[1] School of Mathematical Sciences, Tel-Aviv University, Israel
liron.cohen@math.tau.ac.il
[2] School of Computer Science, Tel Aviv University, Israel
aa@tau.ac.il

Abstract. Many efforts have been made in recent years to construct formal systems for mechanizing mathematical reasoning. A framework which seems particularly suitable for this task is *ancestral logic* – the logic obtained by augmenting first-order logic with a transitive closure operator. While the study of this logic has so far been mostly model-theoretical, this work is devoted to its proof theory (which is much more relevant for the task of mechanizing mathematics). We develop a Gentzen-style proof system TC_G which is sound for ancestral logic, and prove its equivalence to previous systems for the reflexive transitive closure operator by providing translation algorithms between them. We further provide evidence that TC_G indeed encompasses all forms of reasoning for this logic that are in practice. The central rule of TC_G is an induction rule which generalizes that of Peano Arithmetic (PA). In the case of arithmetics we show that the ordinal number of TC_G is ε_0.

1 Introduction

In light of recent advances in the field of automated reasoning, formal systems for mechanizing mathematical reasoning are attracting a lot of interest (see, e.g., [10,5,6,15]). Most of these systems go beyond first-order logic (FOL), because the latter is too weak for this task: one cannot even give in it a categorical characterization of the most basic concept of mathematics - the natural numbers. Using second-order logic (SOL) for this task, however, has many disadvantages. SOL has doubtful semantics, as it is based on debatable ontological commitments. Moreover, it does not seem satisfactory that dealing with basic notions (such as the natural numbers) requires using the strong notions involved in SOL, such as quantifying over all subsets of infinite sets. In addition, SOL is difficult to deal with from a proof-theoretical point of view.

The above considerations imply that the most suitable framework for mechanizing mathematical reasoning should be provided by some logic between FOL and SOL. A framework that seems particularly suitable for this task is *ancestral logic* – the logic obtained by augmenting FOL with the concept of transitive closure of a given relation. Indeed, ancestral logic provides a suitable framework for the formalization of mathematics as it is appropriate for defining fundamental abstract formulations of transitive relations that occur commonly in basic mathematics (see, e.g., [2,16,17]).

U. Kohlenbach et al. (Eds.): WoLLIC 2014, LNCS 8652, pp. 137–151, 2014.
© Springer-Verlag Berlin Heidelberg 2014

Most of the works on ancestral logic have so far been carried out in the context of finite model theory (see, e.g., [7]). Clearly, the focus on finite structures renders these works irrelevant for the task of formalizing mathematics. Moreover, most of this research has been dedicated to model theory, whereas for mechanizing mathematics we need useful *proof systems*.

This work provides a proof-theoretical study of ancestral logic. In [2] a formal proof system for ancestral logic was suggested. Therein it was stated that: "a major research task here is to find out what other rules (if any) should be added in order to make the system 'complete' in some reasonable sense". In this work we provide an answer to this question. We show that the system proposed in [2] is too weak, as it fails to prove certain fundamental properties of the transitive closure operator. We then take further steps towards a useful proof system for ancestral logic by proposing a stronger system, TC_G, which is sound for this logic and encompasses all forms of reasoning for this logic that are used in practice. TC_G is proven to be equivalent to systems previously suggested in the literature for the reflexive transitive closure, in the sense that there are translation algorithms between them that preserve provability. We further investigate the proof theoretical method of constructive consistency proofs and show that in the case of arithmetics the ordinal number of the system TC_G is ε_0.

2   Logics with a Transitive Closure Operator

In mathematics, the transitive closure of a binary relation R is defined as the minimal transitive relation that contains R. In general, the transitive closure operator, TC, is not first-order definable (see, e.g., [8,1]). Thus, we present ancestral logic, which is the logic obtained by augmenting FOL with a transitive closure operator[1]. Below are the corresponding formal definitions of a first-order language augmented by a transitive closure operator, and its semantics.

In this paper σ denotes a first-order signature with equality. A structure for a first-order language based on σ is an ordered pair $M = \langle D, I \rangle$, where D is a non-empty set of elements (the domain) and I is an interpretation function on σ. To avoid confusion regarding parentheses, we use (,) for parentheses in a formal language, and [,] for parentheses in the metalanguage.

Definition 1. *Let σ be a signature for a first-order language with equality, and let $M = \langle D, I \rangle$ be a structure for σ and v an assignment in M.*

- *The language $L_{TC}(\sigma)$ is defined as the first-order language based on σ, with the addition of the TC operator defined by: for any formula φ in $L_{TC}(\sigma)$, x, y distinct variables, and s, t terms, $(TC_{x,y}\varphi)(s,t)$ is a formula in $L_{TC}(\sigma)$. The free occurrences of x and y in φ are bound in this formula.*
- *The pair $\langle M, v \rangle$ is said to satisfy $(TC_{x,y}\varphi)(s,t)$ if there exist $a_0, ..., a_n \in D$ $(n > 0)$ such that $v[s] = a_0$, $v[t] = a_n$, and φ is satisfied by M and $v[x := a_i, y := a_{i+1}]$[2] for $0 \le i \le n-1$.*
  The logic obtained is called Ancestral Logic and it is denoted by $\mathcal{L}_{TC}$.

---

[1] Such logics are also sometimes called Transitive Closure Logic.

[2] $v[x := a]$ denotes the x-variant of v which assigns to x the element a from D.

In the semantics presented here, $(TC_{x,y}\varphi)(s,t)$ requires that there should be at least one φ-step between s and t. However, another well studied form of the transitive closure operator [11,12,14] is the reflexive form, RTC.

Definition 2. *Let σ be a first-order signature, and let $M = \langle D, I \rangle$ be a structure for σ and v an assignment in M.*

- *The language $L_{RTC}(\sigma)$ is defined as $L_{TC}(\sigma)$ with TC replaced by RTC.*
- *The pair $\langle M, v \rangle$ is said to satisfy $(RTC_{x,y}\varphi)(s,t)$ if $s = t$ or there exist $a_0, ..., a_n \in D$ $(n > 0)$ such that $v[s] = a_0$, $v[t] = a_n$, and φ is satisfied by M and $v[x := a_i, y := a_{i+1}]$ for $0 \leq i \leq n-1$.*
  Similarly, the obtained logic is denoted by $\mathcal{L}_{RTC}$.

Using equality, the two forms of the transitive closure operator are definable in terms of each other. The reflexive transitive closure operator is definable using the non-reflexive form by

$$(RTC_{x,y}\varphi)(s,t) := (TC_{x,y}\varphi)(s,t) \vee s = t,$$

while the non-reflexive TC operator is definable, for example, by

$$(TC_{x,y}\varphi)(s,t) := \exists z \left(\varphi \left\{ \frac{s}{x}, \frac{z}{y} \right\} \wedge (RTC_{x,y}\varphi)(z,t) \right)$$

where z is a fresh variable.[3]

One difference between the two forms is the ability to define quantifiers. The existential quantifier can be defined using the TC operator [2], however it cannot be defined using the RTC operator, as we prove below.

Proposition 1. *The existential quantifier is not definable in the quantifier-free fragment of $\mathcal{L}_{RTC}$.*

Proof. Take σ to consist of a constant symbol 0 and a unary predicate symbol P. It can be easily shown by induction that each quantifier-free sentence ψ in $\mathcal{L}_{RTC}^{\sigma}$ is logically equivalent to one of the following sentences: $P(0)$, $\neg P(0)$, $0 = 0$, or $0 \neq 0$. Since $\exists x P(x)$ is clearly not logically equivalent to any of these four sentences, we conclude that the existential quantifier cannot be defined in the quantifier-free fragment of $\mathcal{L}_{RTC}$. $\qquad\square$

The concept of the transitive closure operator is embedded in our understanding of the natural numbers. Therefore, it is only natural to explore the expressive power of various first-order languages for arithmetic augmented by the TC operator. Let 0 be a constant symbol and s a unary function symbol. It is known that in $\mathcal{L}_{TC}^{\{0,s\}}$ together with the standard axioms for the successor function, the following sentence categorically characterize the natural numbers:

$$\forall x \, (x = 0 \vee (TC_{w,u}(s(w) = u))(0, x)) \tag{1}$$

---

[3] $\varphi\left\{\frac{t_1}{x_1}, ..., \frac{t_n}{x_n}\right\}$ denotes the formula obtained from φ by substituting t_i for each free occurrence of x_i in φ, assuming that $t_1, ..., t_n$ are free for $x_1, ...x_n$ in φ.

In [2] it was also shown that all recursive functions and relations are definable in $\mathcal{L}_{TC}^{\{0,s,+\}}$, where $+$ is a binary function symbol. This implies that the upward Lï¿œewenheim-Skolem theorem fails for ancestral logic, and that ancestral logic is finitary, i.e. the compactness theorem fails for it. Moreover, ancestral logic is not even arithmetic, thus any formal deductive system which is sound for it is incomplete.

3   Gentzen-Style Proof Systems for Ancestral Logic

Ideally, we would like to have a consistent, sound, and complete axiomatic system for ancestral logic. However, since there could be no sound and complete system for ancestral logic, one should instead look for useful and effective *partial* formal systems that are still adequate for formalizing mathematical reasoning. The systems defined in this section are extensions of Gentzen's system for classical first-order logic with equality, $\mathcal{LK}_=$ [9].

In what follows the letters Γ, Δ represent finite (possibly empty) multisets of formulas, φ, ψ, ϕ arbitrary formulas, x, y, z, u, v, w variables, and r, s, t terms. For convenience, we shall denote a sequent of the form $\Gamma \Rightarrow \{\varphi\}$ by $\Gamma \Rightarrow \varphi$, and employ other standard abbreviations, such as Γ, Δ instead of $\Gamma \cup \Delta$. To improve readability, in some derivations we omit the context from the sequents.

In [11,12,14] two equivalent Hilbert-style systems for ancestral logic in which the reflexive transitive closure operator, RTC, was taken as primitive were suggested. Below is a Gentzen-style proof system for the RTC operator which is equivalent to the Hilbert-style systems presented in the original papers.

Definition 3. *The system* RTC_G *is defined by adding to* $\mathcal{LK}_=$ *the axiom*

$$\Gamma \Rightarrow \Delta, (RTC_{x,y}\varphi)(s,s) \tag{2}$$

and the following inference rules:

$$\frac{\Gamma \Rightarrow \Delta, \varphi\left\{\frac{s}{x}, \frac{t}{y}\right\}}{\Gamma \Rightarrow \Delta, (RTC_{x,y}\varphi)(s,t)} \tag{3}$$

$$\frac{\Gamma \Rightarrow \Delta, (RTC_{x,y}\varphi)(s,r) \quad \Gamma \Rightarrow \Delta, (RTC_{x,y}\varphi)(r,t)}{\Gamma \Rightarrow \Delta, (RTC_{x,y}\varphi)(s,t)} \tag{4}$$

$$\frac{\Gamma, \psi(x), \varphi(x,y) \Rightarrow \Delta, \psi\left\{\frac{y}{x}\right\}}{\Gamma, \psi\left\{\frac{s}{x}\right\}, (RTC_{x,y}\varphi)(s,t) \Rightarrow \Delta, \psi\left\{\frac{t}{x}\right\}} \tag{5}$$

In all three rules we assume that the terms which are substituted are free for substitution and that no forbidden capturing occurs. In Rule (5) x *should not occur free in* Γ *and* Δ, *and* y *should not occur free in* Γ, Δ *and* ψ.

Rule (5) is a generalized induction principle which states that if t is a φ-descendant of s (or equal to it), then if s has some property which is passed down from one object to another if they are φ-related, then t also has that property.[4]

We next show that RTC_G is adequate for RTC, in the sense that it does give the RTC operator the intended meaning of the reflexive transitive closure, and can derive all fundamental rules concerning the RTC operator that have been suggested in the literature (as far as we know).

Proposition 2. *The following rules are derivable in* RTC_G:[5]

$$\frac{\Gamma \Rightarrow \Delta, \varphi\left\{\frac{s}{x}, \frac{r}{y}\right\} \quad \Gamma \Rightarrow \Delta, (RTC_{x,y}\varphi)(r,t)}{\Gamma \Rightarrow \Delta, (RTC_{x,y}\varphi)(s,t)}$$

$$\frac{\Gamma \Rightarrow \Delta, (RTC_{x,y}\varphi)(s,r) \quad \Gamma \Rightarrow \Delta, \varphi\left\{\frac{r}{x}, \frac{t}{y}\right\}}{\Gamma \Rightarrow \Delta, (RTC_{x,y}\varphi)(s,t)} \tag{6}$$

$$\frac{\Gamma \Rightarrow \Delta, (RTC_{x,y}\varphi)(s,t)}{\Gamma \Rightarrow \Delta, s = t, \exists z\left((RTC_{x,y}\varphi)(s,z) \wedge \varphi\left\{\frac{z}{x}, \frac{t}{y}\right\}\right)}$$

$$\frac{\Gamma \Rightarrow \Delta, (RTC_{x,y}\varphi)(s,t)}{\Gamma \Rightarrow \Delta, s = t, \exists z\left(\varphi\left\{\frac{s}{x}, \frac{z}{y}\right\} \wedge (RTC_{x,y}\varphi)(z,t)\right)} \tag{7}$$

$$\frac{\Gamma \Rightarrow \Delta, (RTC_{x,y}\varphi)(s,t)}{\Gamma \Rightarrow \Delta, (RTC_{y,x}\varphi)(t,s)} \quad \frac{(RTC_{x,y}\varphi)(s,t), \Gamma \Rightarrow \Delta}{(RTC_{y,x}\varphi)(t,s), \Gamma \Rightarrow \Delta} \tag{8}$$

$$\frac{\Gamma \Rightarrow \Delta, (RTC_{x,y}\varphi)(s,t)}{\Gamma \Rightarrow \Delta, \left(RTC_{u,v}\varphi\left\{\frac{u}{x}, \frac{v}{y}\right\}\right)(s,t)} \quad \frac{(RTC_{x,y}\varphi)(s,t), \Gamma \Rightarrow \Delta}{\left(RTC_{u,v}\varphi\left\{\frac{u}{x}, \frac{v}{y}\right\}\right)(s,t), \Gamma \Rightarrow \Delta} \tag{9}$$

$$\frac{\Gamma, \varphi \Rightarrow \Delta, \psi}{\Gamma, (RTC_{x,y}\varphi)(s,t) \Rightarrow \Delta, (RTC_{x,y}\psi)(s,t)} \tag{10}$$

$$\frac{(RTC_{x,y}\varphi)(s,t), \Gamma \Rightarrow \Delta}{(RTC_{u,v}(RTC_{x,y}\varphi)(u,v))(s,t), \Gamma \Rightarrow \Delta} \tag{11}$$

$$\frac{\varphi\left\{\frac{s}{x}\right\}, \Gamma \Rightarrow \Delta}{(RTC_{x,y}\varphi)(s,t), \Gamma \Rightarrow s = t, \Delta} \quad \frac{\varphi\left\{\frac{t}{y}\right\}, \Gamma \Rightarrow \Delta}{(RTC_{x,y}\varphi)(s,t), \Gamma \Rightarrow s = t, \Delta} \tag{12}$$

Conditions:

- In all the rules we assume that the terms which are substituted are free for substitution and that no forbidden capturing occurs.

---

[4] For other works on sequent systems with induction see, e.g., [13,18].

[5] These rules are counterparts of the Hilbert-style rules suggested in [11,12,14].

- In (7) z should not occur free in Γ, Δ and $\varphi \left\{ \frac{s}{x}, \frac{t}{y} \right\}$.
- In (9) the conditions are the usual ones concerning the α-rule.
- In (10) x, y should not occur free in Γ, Δ.
- In (11) u, v should not occur free in φ.
- In (12) y should not occur free in Γ, Δ or s in the left rule, and x should not occur free in Γ, Δ or t in the right rule.

In [2] a Gentzen-style system for the non-reflexive transitive closure operator was presented. Therein it was stated that: "a major research task here is to find out what other rules (if any) should be added in order to make the system 'complete' in some reasonable sense". In this section we answer this (two part) research question. First we show that the system in [2] is too weak for ancestral logic, as it fails to prove certain fundamental properties of the transitive closure operator. Then we present a stronger variation of the system which encompasses all forms of reasoning for ancestral logic that are used in practice.

Below is the proof system for the TC operator suggested in [2].

Definition 4. *The system TC'_G is defined by adding to $\mathcal{LK}_=$ the following inference rules:*

$$\frac{\Gamma \Rightarrow \Delta, \varphi \left\{ \frac{s}{x}, \frac{t}{y} \right\}}{\Gamma \Rightarrow \Delta, (TC_{x,y}\varphi)(s,t)} \tag{13}$$

$$\frac{\Gamma \Rightarrow \Delta, (TC_{x,y}\varphi)(s,r) \quad \Gamma \Rightarrow \Delta, (TC_{x,y}\varphi)(r,t)}{\Gamma \Rightarrow \Delta, (TC_{x,y}\varphi)(s,t)} \tag{14}$$

$$\frac{\Gamma, \psi(x), \varphi(x,y) \Rightarrow \Delta, \psi \left\{ \frac{y}{x} \right\}}{\Gamma, \psi \left\{ \frac{s}{x} \right\}, (TC_{x,y}\varphi)(s,t) \Rightarrow \Delta, \psi \left\{ \frac{t}{x} \right\}} \tag{15}$$

The same restrictions on the rules in RTC_G apply here.

While all fundamental rules concerning RTC that have been suggested in the literature (as far as we know) are derivable in RTC_G, as shown in Prop. 2, in TC'_G this is not the case. There are fundamental properties of the TC operator which are unprovable in TC'_G.

Proposition 3. *The following valid sequents are unprovable in TC'_G:*

$$(TC_{x,y}\varphi)(s,t) \Rightarrow \varphi \left\{ \frac{s}{x}, \frac{t}{y} \right\}, \exists z \left((TC_{x,y}\varphi)(s,z) \wedge \varphi \left\{ \frac{z}{x}, \frac{t}{y} \right\} \right)$$

$$(TC_{x,y}\varphi)(s,t) \Rightarrow \varphi \left\{ \frac{s}{x}, \frac{t}{y} \right\}, \exists z \left(\varphi \left\{ \frac{s}{x}, \frac{z}{y} \right\} \wedge (TC_{x,y}\varphi)(z,t) \right) \tag{16}$$

$$(TC_{x,y}\varphi)(s,t) \Rightarrow \varphi \left\{ \frac{s}{x} \right\} \qquad (TC_{x,y}\varphi)(s,t) \Rightarrow \varphi \left\{ \frac{t}{y} \right\} \tag{17}$$

where in (16) z is a fresh variable and in (17) y does not occur free in $\varphi \left\{ \frac{s}{x} \right\}$ in the left sequent, and x does not occur free in $\varphi \left\{ \frac{t}{y} \right\}$ in the right sequent.

Proof. Suppose the above sequents are derivable in TC'_G. It is easy to see that all the rules in TC'_G remain valid and derivable in RTC_G if we replace the operator TC with RTC. Hence, the corresponding sequents for RTC are provable in RTC_G. However, they are obviously not valid, since $(RTC_{x,y}\varphi)(s,s)$ holds for all s and φ. $\qquad\square$

In general, any sequent which is valid only for the TC operator and not for the RTC operator will not be derivable in TC'_G. The next natural question is how should the system TC'_G be altered in order to be able to derive in it all the basic rules for the TC operator that are used in practice. Recall that one of the mathematical definitions of the transitive closure of a relation R is the least transitive relation that contains R. Hence, we generalize TC'_G's induction rule in a way that correlates with the minimality requirement in the definition.

Definition 5. *The system TC_G is obtained from TC'_G by replacing Rule (15) by:*

$$\frac{\Gamma, \varphi(x,y) \Rightarrow \Delta, \phi(x,y) \quad \Gamma, \phi\left\{\tfrac{u}{x}, \tfrac{v}{y}\right\}, \phi\left\{\tfrac{v}{x}, \tfrac{w}{y}\right\} \Rightarrow \Delta, \phi\left\{\tfrac{u}{x}, \tfrac{w}{y}\right\}}{\Gamma, (TC_{x,y}\varphi)(s,t) \Rightarrow \Delta, \phi\left\{\tfrac{s}{x}, \tfrac{t}{y}\right\}} \tag{18}$$

where x, y should not occur free in $\Gamma \cup \Delta$, and u, v, w should not occurr free in Γ, Δ, ϕ and φ.

In what follows, we denote the sequent $\psi\left\{\tfrac{u}{x}, \tfrac{v}{y}\right\}, \psi\left\{\tfrac{v}{x}, \tfrac{w}{y}\right\} \Rightarrow \psi\left\{\tfrac{u}{x}, \tfrac{w}{y}\right\}$ by $Trans_{x,y}[\psi]$. The next theorem proves that TC_G is more adequate for ancestral logic than TC'_G.

Theorem 1. *TC_G is an extension TC'_G and all the sequents from Proposition 3 are provable in it.*

Proof. (Outline) In TC_G Rule (15) is derivable by taking for ϕ in Rule (18) the formula $\psi(x) \to \psi\left\{\tfrac{y}{x}\right\}$, for which $Trans_{x,y}[\phi]$ is clearly provable. To show that the first sequent in (16) is provable in TC_G, take for ϕ in Rule (18) the formula $\varphi(x,y) \vee \exists z\left((TC_{x,y}\varphi)(x,z) \wedge \varphi(z,y)\right)$. The provability of the other sequents from Proposition 3 then easily follows. $\qquad\square$

Proposition 4. *In TC_G all the TC-counterparts of the rules in Proposition 2 are derivable.*

Since each of the two forms of the transitive closure operator can be expressed in terms of the other, it is interesting to explore the connection between RTC_G and TC_G. Let φ be a formula in $\mathcal{L}_{TC}$. Define φ^* to be its $\mathcal{L}_{RTC}$-translation by induction as follows: for each formula φ in first-order language define $\varphi^* := \varphi$, and define $((TC_{x,y}A)(s,t))^*$ to be the formula: $\exists z\left(A^*\left\{\tfrac{s}{x}, \tfrac{z}{y}\right\} \wedge (RTC_{x,y}A^*)(z,t)\right)$. Let ψ be a formula in $\mathcal{L}_{RTC}$. Then ψ' is the formula in $\mathcal{L}_{TC}$ defined by induction as follows: for each formula ψ in first-order language define $\psi' := \psi$, and define $((RTC_{x,y}A)(s,t))'$ to be the formula $(TC_{x,y}A')(s,t) \vee s = t$. We use the standard abbreviations: Γ^* for $\{\varphi^* | \varphi \in \Gamma\}$ and Γ' for $\{\varphi' | \varphi \in \Gamma\}$.

First we show that any theorem of TC_G can be translated into a theorem of RTC_G, and vice versa.

Proposition 5. *The following holds:*

1. $\vdash_{TC_G} \Gamma \Rightarrow \Delta$ *implies* $\vdash_{RTC_G} \Gamma^* \Rightarrow \Delta^*$.
2. $\vdash_{RTC_G} \Gamma \Rightarrow \Delta$ *implies* $\vdash_{TC_G} \Gamma' \Rightarrow \Delta'$.

Note that neither $(\varphi')^*$ nor $(\varphi^*)'$ is syntactically equal to φ. For instance, for $\varphi = (TC_{x,y}P(x,y))(s,t)$, $(\varphi^*)'$ is $\exists z (P(s,z) \wedge ((TC_{x,y}P(x,y))(z,t) \vee z = t))$. However, as the next proposition will show, $(\varphi')^*$ and $(\varphi^*)'$ are provably equivalent to φ.

Proposition 6. *The following holds:*

1. $\vdash_{TC_G} (\varphi^*)' \Rightarrow \varphi$ *and* $\vdash_{TC_G} \varphi \Rightarrow (\varphi^*)'$.
2. $\vdash_{RTC_G} (\varphi')^* \Rightarrow \varphi$ *and* $\vdash_{RTC_G} \varphi \Rightarrow (\varphi')^*$.

Theorem 2. TC_G *and* RTC_G *are equivalent, i.e. the following holds:*

1. $\vdash_{RTC_G} \Gamma \Rightarrow \Delta$ *iff* $\vdash_{TC_G} \Gamma' \Rightarrow \Delta'$.
2. $\vdash_{TC_G} \Gamma \Rightarrow \Delta$ *iff* $\vdash_{RTC_G} \Gamma^* \Rightarrow \Delta^*$.

Proof. Follows immediately from Propositions 5 and 6. □

Next we explore some proof-theoretical properties of the system TC_G. A system is said to be consistent if it does not admit a proof of the absurd, i.e. the empty sequent. In $\mathcal{LK}_=$, as well as in TC_G, formulas never disappear, except in cuts (the only other simplification allowed is contraction, in which a repetition is reduced). From this follows that there can be no cut-free proof of the empty sequent. Thus, by proving a weak version of the cut elimination theorem which states cut admissibility only for proofs ending with the empty sequent, one establishes the consistency of the system.

In [9] Gentzen proved the consistency of PA_G (Gentzen-style system for PA)[6] by providing a constructive method for transforming any proof of the empty sequent into a cut-free proof. A crucial step in the proof is the elimination of all appearances of PA_G's induction rule from the end-piece of the proof.[7] First, all free variables which are not used as eigenvariables in the end-piece of the proof are replaced by constants. Then, any application of the induction rule up to a specific natural number is replaced by a corresponding number of structural

---

[6] It should be noted that Gentzen did not prove full cut elimination for PA_G, only consistency.

[7] The end-piece of a proof consists of all the sequents of the proof encountered if we ascend each path starting from the end-sequent and stop when we arrive to an operational inference rule. Thus the lower sequent of this inference rule belongs to the end-piece, but its upper sequents do not.

inference rules. The transformation is done in the following way. Assume that the following application of PA_G's induction rule appears within an end-piece

$$\vdots P$$

$$\frac{\psi\left\{\frac{a}{x}\right\} \Rightarrow \psi\left\{\frac{s(a)}{x}\right\}}{\psi\left\{\frac{0}{x}\right\} \Rightarrow \psi\left\{\frac{t}{x}\right\}}$$

where P denotes the sub-proof ending with the sequent $\psi\left\{\frac{a}{x}\right\} \Rightarrow \psi\left\{\frac{s(a)}{x}\right\}$. Since all free variables were eliminated, t is a closed term and hence there is a term $s(\ldots(s(0))$ such that $\Rightarrow s(\ldots(s(0)) = t$ is provable in PA_G without essential cuts or induction. Therefore, there is also a proof of $\psi(s(\ldots(s(0))) \Rightarrow \psi(t)$ without essential cuts or induction. Let $P(b)$ be the proof obtained from P by replacing a by b throughout the proof. Replace any occurrence of the induction rule by

$$\frac{\begin{array}{c}\vdots P(0) \\ \psi\left\{\frac{0}{x}\right\} \Rightarrow \psi\left\{\frac{s(0)}{x}\right\}\end{array} \quad \frac{\begin{array}{c}\vdots P(s(0)) \\ \psi\left\{\frac{s(0)}{x}\right\} \Rightarrow \psi\left\{\frac{s(s(0))}{x}\right\}\end{array}}{\psi\left\{\frac{0}{x}\right\} \Rightarrow \psi\left\{\frac{s(s(0))}{x}\right\}} \quad \begin{array}{c}\vdots P(s(s(0))) \\ \psi\left\{\frac{s(s(0))}{x}\right\} \Rightarrow \psi\left\{\frac{s(s(s(0)))}{x}\right\}\end{array}}{\psi\left\{\frac{0}{x}\right\} \Rightarrow \psi\left\{\frac{s(s(s(0)))}{x}\right\}}$$

These consecutive cuts are carried on up to the sequent $\psi\left\{\frac{0}{x}\right\} \Rightarrow \psi\left\{\frac{s(\ldots(s(0))}{x}\right\}$. One more cut on $\psi(s(\ldots(s(0))) \Rightarrow \psi(t)$ results in a proof of $\psi\left\{\frac{0}{x}\right\} \Rightarrow \psi\left\{\frac{t}{x}\right\}$.

Can a similar method be applied to the TC-induction rule? The problem is that Gentzen's transformation of the induction rule uses special features of the natural numbers that generally do not exist in TC_G. To see this, notice that the induction rule (Rule (18)) entails all instances of PA_G's induction rule by taking φ to be $s(x) = y$ and ϕ to be $\psi(x) \rightarrow \psi\left\{\frac{y}{x}\right\}$. However, in the general case φ is an arbitrary formula. Thus, unlike in PA_G, we do not have a "built in" measure for the φ-distance between two arbitrary closed terms s and t. The φ-path from s to t is not known apriori. Moreover, it does not have to be unique.

Unfortunately, this generalization of the induction principle renders this standard method for analyzing PA_G inapplicable. Thus, one should look for useful fragments of TC_G in which cuts can be eliminated from proofs of the empty sequent. One such fragment can be obtained via restricting TC_G's induction rule by allowing only φ's of the form $y = t$, where x is the only free variable in t. In this way we force a deterministic φ-path between any two closed terms, while keeping the system strong enough for the task of mechanizing mathematics, as its restricted induction rule still includes that of PA_G. Exploring this direction will be left for further research.

Another proof-theoretical method which arises from Gentzen's constructive consistency proofs is the assignment of ordinals to proof systems. In Gentzen's method, each system is assigned the least ordinal number needed for its constructive consistency proof. This provides a measure for a complexity of a system which is useful for comparing different proof systems. The constructive consistency proof of PA_G entails that the ordinal number of PA_G is at most ε_0, and another theorem of Gentzen shows that it is exactly ε_0.

Definition 6. *The system TC_A is obtained by augmenting TC_G with the standard axioms for successor, addition, and multiplication, together with the axiom characterizing the natural numbers in ancestral logic (Axiom (1)).*

Proposition 7. *TC_A is equivalent to PA_G.*

Proof. (Outline) TC_A is an extension of PA_G, since Rule (18) entails all instances of PA_G's induction rule. In [17] it was shown how it is possible, using a β-function, to encode in PA_G finite sequences and thus define the TC operator. It is easy to see that the system TC_A is equivalent to PA_G, in the sense that there are provability preserving translation algorithms between them. $\square$

Corollary 1. *The ordinal number of the system TC_A is ε_0.*

4 Conclusions and Further Research

In this paper we reviewed the expressive power of logics augmented by a transitive closure operator and explored their reasoning potential. This work focused on working out this potential by presenting effective sound proof systems for ancestral logic that are strong enough for various mathematical needs. The next goal is to improve the computational efficiency of these systems, in order to make them suitable for mechanization.

We believe that ancestral logic should suffice for most of applicable mathematics. Substantiating this claim by creating formal systems based on ancestral logic and formalizing in them large portions of mathematics, is a further future work. A promising candidate for serving as the basis for such system is the predicative set theory PZF, presented in [3,4], which resembles ZF and is suitable for mechanization. The key elements of PZF are that it uses syntactic safety relations between formulas and sets of variables, and that its underlying logic is ancestral logic, which makes it possible to provide inductive definitions of relations and functions. An important criterion for the adequacy of ancestral logic for the task of formalizing mathematics is the extent to which such formalization can be done in a natural way, as close as possible to real mathematical practice.

Acknowledgments. This research was supported by the Ministry of Science and Technology, Israel.

References

1. Aho, A.V., Ullman, J.D.: Universality of data retrieval languages. In: Proceedings of the 6th ACM SIGACT-SIGPLAN Symposium on Principles of Programming Languages, pp. 110–119. ACM (1979)

2. Avron, A.: Transitive closure and the mechanization of mathematics. In: Kamareddine, F.D. (ed.) Thirty Five Years of Automating Mathematics. Applied Logic Series, vol. 28, pp. 149–171. Springer, Netherlands (2003)

3. Avron, A.: Formalizing set theory as it is actually used. In: Asperti, A., Bancerek, G., Trybulec, A. (eds.) MKM 2004. LNCS, vol. 3119, pp. 32–43. Springer, Heidelberg (2004)

4. Avron, A.: A framework for formalizing set theories based on the use of static set terms. In: Avron, A., Dershowitz, N., Rabinovich, A. (eds.) Pillars of Computer Science. LNCS, vol. 4800, pp. 87–106. Springer, Heidelberg (2008)

5. Campbell, J.J.J.A., Reis, J.C.G.D., Wenzel, P.S.M., Sorge, V.: Intelligent computer mathematics (2008)

6. Constable, R.L., Allen, S.F., Bromley, H.M., Cleaveland, W.R., Cremer, J.F., Harper, R.W., Howe, D.J., Knoblock, T.B., Mendler, N.P., Panangaden, P., Sasaki, J.T., Smith, S.F.: Implementing Mathematics with the Nuprl Proof Development System. Prentice-Hall, Inc., Upper Saddle River (1986)

7. Ebbinghaus, H.-D., Flum, J.: Finite Model Theory, vol. 2. Springer (1995)

8. Fagin, R.: Generalized first-order spectra and polynomial-time recognizable sets (1974)

9. Gentzen, G.: Neue Fassung des Widerspruchsfreiheitsbeweises für die reine Zahlentheorie. Forschungen zur Logik 4, 19–44 (1969); English translation in: Szabo, M.E.: The collected work of Gerhard Gentzen. North-Holland, Amsterdam

10. Kamareddine, F.D.: Thirty five years of automating mathematics, vol. 28. Springer (2003)

11. Martin, R.M.: A homogeneous system for formal logic. The Journal of Symbolic Logic 8(1), 1–23 (1943)

12. Martin, R.M.: A note on nominalism and recursive functions. The Journal of Symbolic Logic 14(1), 27–31 (1949)

13. Momigliano, A., Tiu, A.: Induction and co-induction in sequent calculus. In: Berardi, S., Coppo, M., Damiani, F. (eds.) TYPES 2003. LNCS, vol. 3085, pp. 293–308. Springer, Heidelberg (2004)

14. Myhill, J.: A derivation of number theory from ancestral theory. The Journal of Symbolic Logic 17(3), 192–197 (1952)

15. Rudnicki, P.: An overview of the mizar project. In: Proceedings of the 1992 Workshop on Types for Proofs and Programs, pp. 311–330 (1992)

16. Shapiro, S.: Foundations without Foundationalism: A Case for Second-Order Logic: A Case for Second-Order Logic. Oxford University Press (1991)

17. Smith, P.: Ancestral arithmetic and isaacson's thesis. Analysis 68(297), 1–10 (2008)

18. Tiu, A., Momigliano, A.: Cut elimination for a logic with induction and co-induction. Journal of Applied Logic 10(4), 330–367 (2012); Selected papers from the 6th International Conference on Soft Computing Models in Industrial and Environmental Applications

Appendix

In what follows, for readability, we shall not distinguish between the sequents $\varphi \wedge \psi, \Gamma \Rightarrow \Delta$ and $\varphi, \psi, \Gamma \Rightarrow \Delta$ as they are provable from one another.

Proof of Proposition 2:

- The first rule in (6) (The proof of the second rule in (6) is analogous.): From $\Gamma \Rightarrow \Delta, \varphi \left\{ \frac{s}{x}, \frac{r}{y} \right\}$, using Rule (3), we can deduce $\Gamma \Rightarrow \Delta, (RTC_{x,y}\varphi)(s,r)$. Applying Rule (4) on the last sequent and $\Gamma \Rightarrow \Delta, (RTC_{x,y}\varphi)(r,t)$ entails a proof of $\Gamma \Rightarrow \Delta, (RTC_{x,y}\varphi)(s,t)$.
- The first rule in (7): Consider the following proof, P_1:

$$
\cfrac{
\cfrac{
\Rightarrow (RTC_{x,y}\varphi)(y,y)
}{
s = y \Rightarrow (RTC_{x,y}\varphi)(s,y)
} \quad \varphi\left\{\frac{y}{x},\frac{z}{y}\right\} \Rightarrow \varphi\left\{\frac{y}{x},\frac{z}{y}\right\}
}{
\cfrac{
s = y, \varphi\left\{\frac{y}{x},\frac{z}{y}\right\} \Rightarrow (RTC_{x,y}\varphi)(s,y) \wedge \varphi\left\{\frac{y}{x},\frac{z}{y}\right\}
}{
s = y, \varphi\left\{\frac{y}{x},\frac{z}{y}\right\} \Rightarrow \exists w \left((RTC_{x,y}\varphi)(s,w) \wedge \varphi\left\{\frac{w}{x},\frac{z}{y}\right\}\right)
}
}
$$

The sequent $(RTC_{x,y}\varphi)(s,w), \varphi\left\{\frac{w}{x}\right\} \Rightarrow (RTC_{x,y}\varphi)(s,y)$ is provable in RTC_G using (6). Thus, by applying standard $\mathcal{LK}_=$ rules we can construct a proof, P_2, of $\exists w \left((RTC_{x,y}\varphi)(s,w) \wedge \varphi\left\{\frac{w}{x}\right\}\right), \varphi\left\{\frac{y}{x},\frac{z}{y}\right\} \Rightarrow \exists w \left((RTC_{x,y}\varphi)(s,w) \wedge \varphi\left\{\frac{w}{x},\frac{z}{y}\right\}\right)$. Denote by $A(y)$ the formula $\exists w \left((RTC_{x,y}\varphi)(s,w) \wedge \varphi\left\{\frac{w}{x}\right\}\right) \vee s = y$. From P_1 and P_2 we obtain a proof of the sequent $A(y), \varphi\left\{\frac{y}{x},\frac{z}{y}\right\} \Rightarrow A\left\{\frac{z}{y}\right\}$, from which, using Rule (5), we deduce $A\left\{\frac{s}{y}\right\}, (RTC_{x,y}\varphi)(s,t) \Rightarrow A\left\{\frac{t}{y}\right\}$. Since $\Rightarrow A\left\{\frac{s}{y}\right\}$ is derivable from the equality axiom, applying a cut on it results in the desired end-sequent. The proof of the second rule in (7) is symmetric.
- The left rule in (8): The sequent $\varphi(x,y), (RTC_{y,x}\varphi)(x,s) \Rightarrow (RTC_{y,x}\varphi)(y,s)$ is provable in RTC_G using (6). Thus, we can construct the following proof:

$$
\cfrac{
\cfrac{
\cfrac{
\varphi\left\{\frac{z}{y},\frac{s}{x}\right\} \Rightarrow \varphi\left\{\frac{z}{y},\frac{s}{x}\right\}
}{
\varphi\left\{\frac{z}{y},\frac{s}{x}\right\} \Rightarrow (RTC_{y,x}\varphi)(z,s)
}(3) \quad
\cfrac{
\varphi(x,y), (RTC_{y,x}\varphi)(x,s) \Rightarrow (RTC_{y,x}\varphi)(y,s)
}{
(RTC_{x,y}\varphi)(z,t), (RTC_{y,x}\varphi)(z,s) \Rightarrow (RTC_{y,x}\varphi)(t,s)
}(5)
}{
\varphi\left\{\frac{s}{x},\frac{z}{y}\right\} \wedge (RTC_{x,y}\varphi)(z,t) \Rightarrow (RTC_{y,x}\varphi)(t,s)
}
}{
\exists z \left(\varphi\left\{\frac{s}{x},\frac{z}{y}\right\} \wedge (RTC_{x,y}\varphi)(z,t)\right) \Rightarrow (RTC_{y,x}\varphi)(t,s)
}
$$

The sequent $(RTC_{x,y}\varphi)(s,t) \Rightarrow s = t, \exists z \left(\varphi\left\{\frac{s}{x},\frac{z}{y}\right\} \wedge (RTC_{x,y}\varphi)(z,t)\right)$ is provable in RTC_G using Rule (7) and $s = t \Rightarrow (RTC_{y,x}\varphi)(t,s)$ is provable using Axiom (2). From this, by cuts, we obtain a proof of $(RTC_{x,y}\varphi)(s,t) \Rightarrow (RTC_{y,x}\varphi)(t,s)$. The proof of the right rule is symmetric.
- The left rule in (9): In RTC_G the sequent $s = t \Rightarrow \left(RTC_{u,v}\varphi\left\{\frac{u}{x},\frac{v}{y}\right\}\right)(s,t)$ is provable. By a method similar to the one used in the proof of (8) we get the

provability of $\exists z \left((RTC_{x,y}\varphi)(s,z) \wedge \varphi\left\{\frac{z}{x},\frac{t}{y}\right\}\right) \Rightarrow \left(RTC_{u,v}\varphi\left\{\frac{u}{x},\frac{v}{y}\right\}\right)(s,t).$
The sequent $(RTC_{x,y}\varphi)(s,t) \Rightarrow \left(RTC_{u,v}\varphi\left\{\frac{u}{x},\frac{v}{y}\right\}\right)(s,t)$ is then provable by applying cuts and Rule (7). The proof of the right rule is symmetric.

– Rule (10): Consider the following proof:

$$\dfrac{\dfrac{(RTC_{x,y}\psi)(s,z) \Rightarrow (RTC_{x,y}\psi)(s,z) \quad \dfrac{\varphi \Rightarrow \psi}{\varphi\left\{\frac{z}{x},\frac{u}{y}\right\} \Rightarrow \psi\left\{\frac{z}{x},\frac{u}{y}\right\}}}{(RTC_{x,y}\psi)(s,z),\varphi\left\{\frac{z}{x},\frac{u}{y}\right\} \Rightarrow (RTC_{x,y}\psi)(s,u)} (6)}{(RTC_{x,y}\psi)(s,z),(RTC_{x,y}\varphi)(z,t) \Rightarrow (RTC_{x,y}\psi)(s,t)} (5)$$

It is easy to see that $\varphi\left\{\frac{s}{x},\frac{z}{y}\right\} \Rightarrow (RTC_{x,y}\psi)(s,z)$ is provable. From this and the above proof, we can deduce $\exists z \left(\varphi\left\{\frac{s}{x},\frac{z}{y}\right\} \wedge (RTC_{x,y}\varphi)(z,t)\right) \Rightarrow (RTC_{x,y}\psi)(s,t).$ Clearly, the sequent $s = t \Rightarrow (RTC_{y,x}\psi)(s,t)$ is provable in RTC_G using Axiom (2). Using Rule (7) we get $(RTC_{x,y}\varphi)(s,t) \Rightarrow s = t, \exists z \left(\varphi\left\{\frac{s}{x},\frac{z}{y}\right\} \wedge (RTC_{x,y}\varphi)(z,t)\right),$ and two cuts result in a proof of $(RTC_{x,y}\varphi)(s,t) \Rightarrow (RTC_{x,y}\psi)(s,t).$

– Rule (11): The sequent $(RTC_{x,y}\varphi)(s,u),(RTC_{x,y}\varphi)(u,v) \Rightarrow (RTC_{x,y}\varphi)(s,v)$ is provable in RTC_G using Rule (4), from which, by Rule rule (5) we get $(RTC_{x,y}\varphi)(s,s),(RTC_{u,v}(RTC_{x,y}\varphi)(u,v))(s,t) \Rightarrow (RTC_{x,y}\varphi)(s,t).$ A cut on the axiom $\Rightarrow (RTC_{x,y}\varphi)(s,s)$ results in the desired proof.

– The left rule in (12): From $\varphi\left\{\frac{s}{x}\right\} \Rightarrow$, by standard $\mathcal{LK}_=$ rules, we can derive $\exists z \left(\varphi\left\{\frac{s}{x},\frac{z}{y}\right\} \wedge (RTC_{x,y}\varphi)(z,t)\right) \Rightarrow.$ By Rule (7) we have $(RTC_{x,y}\varphi)(s,t) \Rightarrow s = t, \exists z \left(\varphi\left\{\frac{s}{x},\frac{z}{y}\right\} \wedge (RTC_{x,y}\varphi)(z,t)\right).$ Then, $(RTC_{x,y}\varphi)(s,t) \Rightarrow s = t$ is provable by a cut. The proof of the right rule in (12) is analogous. □

Proof of Theorem 1:

Clearly $Trans_{x,y}[\psi(x) \to \psi\left\{\frac{y}{x}\right\}]$ is provable. Thus, we derive Rule (15) by:

$$\dfrac{\dfrac{\dfrac{\psi(x),\varphi(x,y) \Rightarrow \psi\left\{\frac{y}{x}\right\}}{\varphi(x,y) \Rightarrow \psi(x) \to \psi\left\{\frac{y}{x}\right\}} \quad Trans_{x,y}[\psi(x) \to \psi\left\{\frac{y}{x}\right\}]}{(TC_{x,y}\varphi)(s,t) \Rightarrow \psi\left\{\frac{s}{x}\right\} \to \psi\left\{\frac{t}{x}\right\}}}{\psi\left\{\frac{s}{x}\right\},(TC_{x,y}\varphi)(s,t) \Rightarrow \psi\left\{\frac{t}{x}\right\}} (18)$$

To see that the first sequent in (16) is provable in TC_G, take ϕ to be $\varphi(x,y) \vee \exists z ((TC_{x,y}\varphi)(x,z) \wedge \varphi(z,y)).$ For any two terms r_1,r_2, denote by A_{r_1,r_2} the formula $\exists z ((TC_{x,y}\varphi)(r_1,z) \wedge \varphi(z,r_2)).$ Clearly, $\varphi(x,y) \Rightarrow \varphi(x,y) \vee A_{x,y}$ is provable in TC_G. We show that $Trans_{x,y}[\varphi(x,y) \vee A_{x,y}]$ is also provable. Observe the following sub-proof:

$$\dfrac{(TC_{x,y}\varphi)(u,v),(TC_{x,y}\varphi)(v,a) \Rightarrow (TC_{x,y}\varphi)(u,a) \quad (TC_{x,y}\varphi)(u,a),\varphi(a,w) \Rightarrow A_{u,w}}{\dfrac{(TC_{x,y}\varphi)(u,v),(TC_{x,y}\varphi)(v,a) \wedge \varphi(a,w) \Rightarrow A_{u,w}}{(TC_{x,y}\varphi)(u,v),A_{v,w} \Rightarrow A_{u,w}}}$$

It is easy to see that $(TC_{x,y}\varphi)(u,v),\varphi(v,w)\Rightarrow A_{u,w}$ is provable in TC_G, so we can prove the sequent $(TC_{x,y}\varphi)(u,v),\varphi(v,w)\vee A_{v,w}\Rightarrow\varphi(u,w)\vee A_{u,w}$. The sequent $\varphi(u,v)\vee A_{u,v}\Rightarrow(TC_{x,y}\varphi)(u,v)$ is also provable in TC_G, hence, $\phi(u,v),\phi(v,w)\Rightarrow\phi(u,w)$ is provable using a cut. Now we can construct the following derivation:

$$\frac{\varphi(x,y)\Rightarrow\varphi(x,y)\vee\exists z\left((TC_{x,y}\varphi)(x,z)\wedge\varphi(z,y)\right)\quad Trans_{x,y}[\phi]}{(TC_{x,y}\varphi)(s,t)\Rightarrow\varphi\left\{\tfrac{s}{x},\tfrac{t}{y}\right\},\exists z\left((TC_{x,y}\varphi)(s,z)\wedge\varphi(z,t)\right)}\quad(18)$$

The proof of the second sequent in (16) is similar. To see that the sequents in (17) are provable, notice that both $\varphi\left\{\tfrac{s}{x},\tfrac{t}{y}\right\}\vee\exists z\left((TC_{x,y}\varphi)(s,z)\wedge\varphi(z,t)\right)\Rightarrow\varphi\left\{\tfrac{t}{y}\right\}$ and $\varphi\left\{\tfrac{s}{x},\tfrac{t}{y}\right\}\vee\exists w\left(\varphi(s,z)\wedge(TC_{x,y}\varphi)(z,t)\right)\Rightarrow\varphi\left\{\tfrac{s}{x}\right\}$ are provable in TC_G. From this, using (16) and cuts, we obtain the desired proofs.   □

Proof of Proposition 5:

Lemma 1. *The following holds:*

- $\left(\varphi\left\{\tfrac{s}{x},\tfrac{t}{y}\right\}\right)^{*}=\varphi^{*}\left\{\tfrac{s}{x},\tfrac{t}{y}\right\}$ *and* $\left(\varphi\left\{\tfrac{s}{x},\tfrac{t}{y}\right\}\right)'=\varphi'\left\{\tfrac{s}{x},\tfrac{t}{y}\right\}$.
- $(\neg\varphi)^{*}=\neg\varphi^{*}$ *and* $(\neg\varphi)'=\neg\varphi'$.
- $(\varphi\circ\psi)^{*}=\varphi^{*}\circ\psi^{*}$ *and* $(\varphi\circ\psi)'=\varphi'\circ\psi'$, *where* $\circ\in\{\wedge,\vee,\rightarrow\}$.
- $(Qx\varphi)^{*}=Qx\varphi^{*}$ *and* $(Qx\varphi)'=Qx\varphi'$, *where* $Q\in\{\forall,\exists\}$.

The proofs of (1) and (2) are carried out by induction, we state here only the cases concerning the TC and RTC operators.

- Rule (13): By standard $\mathcal{LK}_{=}$ rules derive from $\Rightarrow\varphi^{*}\left\{\tfrac{s}{x},\tfrac{t}{y}\right\}$ and the axiom $\Rightarrow(RTC_{x,y}\varphi^{*})(t,t)$ the sequent $\Rightarrow\exists z\left(\varphi^{*}\left\{\tfrac{s}{x},\tfrac{z}{y}\right\}\wedge(RTC_{x,y}\varphi^{*})(z,t)\right)$.
- Rule (14): Rule (6) entails the existence of a proof in RTC_G of the sequent $\exists z\left(\varphi^{*}\left\{\tfrac{r}{x},\tfrac{z}{y}\right\}\wedge(RTC_{x,y}\varphi^{*})(z,t)\right)\Rightarrow(RTC_{x,y}\varphi^{*})(r,t)$. A cut on the hypothesis $\Rightarrow\exists z\left(\varphi^{*}\left\{\tfrac{r}{x},\tfrac{z}{y}\right\}\wedge(RTC_{x,y}\varphi^{*})(z,t)\right)$ results in a proof of the sequent $\Rightarrow(RTC_{x,y}\varphi^{*})(r,t)$. Applying Rule (4) on $\Rightarrow(RTC_{x,y}\varphi^{*})(r,t)$ and $(RTC_{x,y}\varphi^{*})(z,r)\Rightarrow(RTC_{x,y}\varphi^{*})(z,r)$ results in a proof of the sequent $(RTC_{x,y}\varphi^{*})(z,r)\Rightarrow(RTC_{x,y}\varphi^{*})(z,t)$. By standard $\mathcal{LK}_{=}$ rules derive $\exists z\left(\varphi^{*}\left\{\tfrac{s}{x},\tfrac{z}{y}\right\}\wedge(RTC_{x,y}\varphi^{*})(z,r)\right)\Rightarrow\exists z\left(\varphi^{*}\left\{\tfrac{s}{x},\tfrac{z}{y}\right\}\wedge(RTC_{x,y}\varphi^{*})(z,t)\right)$. The desired sequent is obtained by one more cut on the hypothesis $\Rightarrow\exists z\left(\varphi^{*}\left\{\tfrac{s}{x},\tfrac{z}{y}\right\}\wedge(RTC_{x,y}\varphi^{*})(z,r)\right)$.
- Rule (18): From $Trans_{x,y}[\phi^{*}]$ deduce $\phi^{*}(s,x),\phi^{*}(x,y)\Rightarrow\phi^{*}(s,y)$. Using a cut on $\varphi^{*}(x,y)\Rightarrow\phi^{*}(x,y)$ we get $\phi^{*}(s,x),\varphi^{*}(x,y)\Rightarrow\phi^{*}(s,y)$. Applying Rule (5) results in $\phi^{*}(s,z),(RTC_{x,y}\varphi^{*})(z,t)\Rightarrow\phi^{*}(s,t)$. Using a cut on $\varphi^{*}(s,z)\Rightarrow\phi^{*}(s,z)$ we get $\varphi^{*}(s,z),(RTC_{x,y}\varphi^{*})(z,t)\Rightarrow\phi^{*}(s,t)$, from which $\exists z\left(\varphi^{*}(s,z)\wedge(RTC_{x,y}\varphi^{*})(z,t)\right)\Rightarrow\phi^{*}(s,t)$ is easily derivable.

- Axiom (2): The translation of the axiom is $\Rightarrow (TC_{x,y}\varphi')(s,s) \vee s = s$, which is easily derivable from the equality axioms.
- Rule (3): Using Rule and introduction of $\vee$ on the right we can deduce $\Rightarrow (TC_{x,y}\varphi')(s,t) \vee s = t$ from $\Rightarrow \varphi'\left\{\frac{s}{x},\frac{t}{y}\right\}$.
- Rule (4): It is easy to see that $\Rightarrow (TC_{x,y}\varphi')(s,t), s = t$ can be proven from $\Rightarrow (TC_{x,y}\varphi')(s,r), s = r$ and $\Rightarrow (TC_{x,y}\varphi')(r,t), r = t$ using Rule (14) and equality rules.
- Rule (5): Applying Rule (15), which is derivable in TC_G, to the sequent $\psi'(x), \varphi'(x,y) \Rightarrow \psi'\left\{\frac{y}{x}\right\}$ results in the sequent $\psi'\left\{\frac{s}{x}\right\}, (TC_{x,y}\varphi')(s,t) \Rightarrow \psi'\left\{\frac{t}{x}\right\}$. Then, a cut on the provable sequent $\psi'\left\{\frac{s}{x}\right\}, s = t \Rightarrow \psi'\left\{\frac{t}{x}\right\}$ entails a proof of $\psi'\left\{\frac{s}{x}\right\}, (TC_{x,y}\varphi')(s,t) \vee s = t \Rightarrow \psi'\left\{\frac{t}{x}\right\}$. □

Proof of Proposition 6:

If φ does not contain the TC or RTC operator, then $(\varphi')^*$ and $(\varphi^*)'$ are syntactically equal to φ, hence provably equivalent to it.

For (1) assume that $\varphi := (RTC_{x,y}A)(s,t)$. By the induction hypothesis $(A')^* \Rightarrow A$ is provable in RTC_G, thus $\left(RTC_{x,y}(A')^*\right)(s,t) \Rightarrow (RTC_{x,y}A)(s,t)$ is provable by (10). It is easy to check that $\exists z\left((A')^*\left\{\frac{s}{x},\frac{z}{y}\right\} \wedge RTC_{x,y}(A')^*(z,t)\right) \vee s = t \Rightarrow \left(RTC_{x,y}(A')^*\right)(s,t)$ is provable in RTC_G (using (6) and (2)). Then, $\exists z\left((A')^*\left\{\frac{s}{x},\frac{z}{y}\right\} \wedge RTC_{x,y}(A')^*(z,t)\right) \vee s = t \Rightarrow (RTC_{x,y}A)(s,t)$ is provable by a cut on the last two sequents. For the converse, denote by ψ the sequent $\exists z\left((A')^*\left\{\frac{u}{x},\frac{z}{y}\right\} \wedge RTC_{x,y}(A')^*(z,w)\right) \vee s = t$ (notice that $(\varphi')^*$ is $\psi\left\{\frac{s}{u},\frac{t}{w}\right\}$). It is easy to see that $\psi\left\{\frac{s}{u},\frac{x}{w}\right\}, (A')^* \Rightarrow \psi\left\{\frac{s}{u},\frac{y}{w}\right\}$ is provable in RTC_G. An application of Rule (5) results in $\psi\left\{\frac{s}{u},\frac{s}{w}\right\}, (RTC_{x,y}(A')^*)(s,t) \Rightarrow \psi\left\{\frac{s}{u},\frac{t}{w}\right\}$. The sequent $\Rightarrow \psi\left\{\frac{s}{u},\frac{s}{w}\right\}$ is provable using the equality axiom, thus, a cut entails a proof of $\left(RTC_{x,y}(A')^*\right)(s,t) \Rightarrow (\varphi')^*$. By the induction hypothesis $A \Rightarrow (A')^*$ is provable in RTC_G, so $(RTC_{x,y}A)(s,t) \Rightarrow \left(RTC_{x,y}(A')^*\right)(s,t)$ is also provable in RTC_G by (10), and by one cut we obtain $(RTC_{x,y}A)(s,t) \Rightarrow (\varphi')^*$.

For (2) assume that $\varphi := (TC_{x,y}A)(s,t)$. It is easy to check that the sequent $\exists z\left((A^*)'\left\{\frac{s}{x},\frac{z}{y}\right\} \wedge \left(TC_{x,y}(A^*)'(z,t) \vee z = t\right)\right) \Rightarrow \left(TC_{x,y}(A^*)'\right)(s,t)$ is provable in TC_G. By the induction hypothesis we have that $\vdash_{TC_G} (A^*)' \Rightarrow A$, so by the TC-counterpart of (10) the sequent $\left(TC_{x,y}(A^*)'\right)(s,t) \Rightarrow (TC_{x,y}A)(s,t)$ is also provable in TC_G. Now, applying a cut results in a proof of the sequent $\exists z\left((A^*)'\left\{\frac{s}{x},\frac{z}{y}\right\} \wedge \left(TC_{x,y}(A^*)'(z,t) \vee z = t\right)\right) \Rightarrow (TC_{x,y}A)(s,t)$. For the converse, notice that the derivability of (16) in TC_G entails the provability of $\left(TC_{x,y}(A^*)'\right)(s,t) \Rightarrow (A^*)'\left\{\frac{s}{x},\frac{t}{y}\right\} \vee \exists z\left((A^*)'\left\{\frac{s}{x},\frac{z}{y}\right\} \wedge \left(TC_{x,y}(A^*)'\right)(z,t)\right)$. Clearly, the sequent $(A^*)'\left\{\frac{s}{x},\frac{t}{y}\right\} \Rightarrow \exists z\left((A^*)'\left\{\frac{s}{x},\frac{z}{y}\right\} \wedge z = t\right)$ is provable, and again, using the induction hypothesis on A together with the TC-counterpart of (10) we get that $(TC_{x,y}A)(s,t) \Rightarrow \left(TC_{x,y}(A^*)'\right)(s,t)$ is provable in TC_G. Applying cuts results in a proof of the sequent $(TC_{x,y}A)(s,t) \Rightarrow (\varphi^*)'$. □

A Calculus of Anyons

Alessandra Di Pierro and Federica Panarotto

Dipartimento di Informatica, Università di Verona,
Strada le Grazie, 15 – 37134 Verona, Italy

Abstract. Recent developments in theoretical physics have highlighted interesting topological features of some two-dimensional particles, so-called anyons, that can be used to realise robust quantum computation. In this paper we show how an anyon system can be defined as a calculus of *quantum functions*, i.e. linear transformations on the space of all possible physical configurations of a set of anyons. A computation in this calculus represents the braiding of anyons and the final term of a computation corresponds to the outcome of a measurement of the anyons final fusion state, i.e. in general a probability distribution on the set of all possible outcomes. We show that this calculus describes a universal anyonic quantum computer provided that the space of terms satisfies some topological properties.

1 Introduction

The theory of computation has greatly evolved since its introduction that can be dated back to the pioneering works by Church, Gödel, Turing, Kleene, von Neumann and Shannon. Recent research has highlighted the role of the physical laws governing computation and has discovered the world of *Quantum Computation* (see e.g. [4,15]). The quantum theory of computation is a generalisation of the classical theory in the same way as quantum physics is a generalisation of classical physics. Quantum computers could outperform their classical counterparts if they were realisable [19,12]. Unfortunately, the problem of implementing quantum computation that is resilient to errors is enormously difficult, given that it has to obey the laws of quantum mechanics that are easily broken by the environment interaction: local errors, thermic noise and decoherence are the the main obstacles that need to be overcome. *Topological Quantum Computation* (TQC) was introduced in [14] (see also [7,8]) as an alternative paradigm for quantum computation which offers the possibility to face these problems. It exploits the topological properties of *anyons*, i.e. physical (quasi-)particles which are insensitive to local perturbations. They exhibit richer exchange statistics than the particles used in quantum circuits (fermions or bosons): when two anyons are exchanged then the system's wave function is altered, while taking an electron (or boson) around another electron (or boson) does not produce any modification. This implies that anyons can have a richer set of statistical behaviours, possibly different from fermions or bosons. This discovery opened up a new and promising branch in the research in quantum computation.

By focusing in particular on computability theory and formal languages, a formalism similar to the Turing Machine was introduced in [4]. This formalism, called *Quantum Turing Machine* (QTM), provides a model for quantum computation and a base for the

U. Kohlenbach et al. (Eds.): WoLLIC 2014, LNCS 8652, pp. 152–165, 2014.
© Springer-Verlag Berlin Heidelberg 2014

definition of quantum complexity classes generalising the standard ones. In particular, in [2] it is used to define the quantum analog of the class of effectively computable functions on a classical computer, namely the class BQP of functions computable with bounded error, given quantum resources, in polynomial time. The universality of the QTM model was defined in [4] with respect to a 'physical' version of the classical Church-Turing thesis. This states that *Every finitely realisable physical system can be perfectly simulated by a universal model computing machine operating by finite means.*

Like in the classical case, the expressivity of any other quantum computational formalism is asserted by comparison with the QTM: it is universal if it can simulate the universal QTM, and in general we could assert that all 'reasonable' computational models which add the resources of quantum mechanics (or quantum field theory) to classical computation yield classes that can (efficiently) simulate each other, so that there is one quantum theory of computation [8]. Among the various models that have been introduced so far, the quantum circuits [6,5] was shown (quantum) Turing-equivalent in [16] and the quantum λ-calculus [3,18] in [3]. Computational universality has been shown also for other quantum models, whose definition is strongly tied to the properties of quantum mechanics and therefore with no classical analogues(measurement-based quantum computation, adiabatic etc.). The model of TQC has also no direct counterpart in the classical computability theory. It is additionally more distant from the other quantum models which all share the idea of computing with elementary particles (qubits) realisable from fermions or bosons. Moreover, its properties hint to some potentialities that are new compared to qubit-based quantum computation.

The equivalence between TQC and the quantum circuits model is established in [9,10] by showing that for any quantum circuit it is always possible to construct a modular functor (*braiding*) that approximate the circuit up to a given threshold $\delta > 0$ in polynomial time. In this paper we introduce a formalisation of TQC that we call *anyonic calculus*. An anyonic calculus is essentially a rewriting system consisting of a single transformation rule (variable substitution) and a single function definition scheme, just like Church's λ-calculus. However, differently from the latter, our calculus represents an *anyonic computer*, that is a quantum system of *anyons* where computation occurs by braiding a fixed number of anyons among them for some fixed time [21].

We investigate the question of computational universality for TQC, namely the existence of an anyonic quantum computer [21] that is able to simulate any program on any other anyonic quantum computer, and we show that the anyonic calculus satisfies this property.

2 The Anyonic Calculus

Computations with anyons is typically expressed in the languages of physics, abstract algebra or category theory. However, the typical approach to expressing computation in computer science is by means of programming languages and computational models.

For classical computation, the λ-calculus [1] can be seen as the smallest universal programming language: it consists of a single transformation rule (variable substitution) and a single function definition scheme. It was introduced in the 1930s by Alonzo Church as a way of formalising the concept of effective computability. The calculus is *universal* in the sense that any computable function can be expressed and evaluated using this formalism. It is thus equivalent to Turing machines.

In this section we present a formal calculus for TQC which makes this new paradigm more amenable to investigations in the realm of theoretical computer science (such as the study of computability and complexity issues), and provides a more suitable base for the design of quantum programming languages.

We define a calculus of anyons in analogy with the classical λ-calculus, i.e. as a calculus of terms with operations on them corresponding to the lambda-abstraction and function application. We therefore call this calculus the *anyonic λ-calculus*, $\mathcal{A}\lambda$.

For a background on TQC we refer to Appendix A.

2.1 Anyon Trees

In order to correctly re-produce the anyons behaviour in our calculus we need to have a closer look at the Hilbert space of anyons, or *fusion space*, i.e. the space of states that corresponds to the fusion process (see Appendix A). In order to manipulate anyons we must consider the process dual to fusion, or splitting, that applies the same rules but in reverse order. The splitting process creates trees of anyons that we can define as follows:

Definition 1. *An* anyon tree *is a tree where every internal node has two children labelled by the two anyons resulting from a splitting rule applied to the anyon labelling the node.*

We obtain different trees depending on

- the shape of the tree resulting from the choice of the anyon to which we apply the fusion rules;
- the type of the anyons in the tree, resulting from the fusion rule chosen at every step of the construction of the tree.

We call the anyons at the leaves of the tree *leaf anyons*, the anyon at the root *root anyon* and the remaining anyons *internal anyons*. The pentagon and hexagon equations (see Appendix A) relate the different trees obtained by fusing the same set of anyons to the same result. The first are related to the associativity of the fusion rules and the second to the clockwise exchange of two anyons a and b.

In physical terms a system of anyons consists of a closed oriented surface Σ with anyons of types $a_1, \ldots, a_m$ located at distinct points $p_1, \ldots, p_m$, so as to form the desired configuration. In our setting, we look at these anyons as the leaves of a tree constructed according to the splitting rules of a given model, that is as the final configuration of ground states at the end of the splitting process. This is the configuration to which unitary transformations can now be applied in order to perform the desired elaboration of the information encoded in the anyonic system. These final trees are the physical counterparts of the terms of the anyonic calculus we are going to introduce.

There will be as many trees as the number of splitting rules for each internal anyon. These trees representing the same global charge, the root, are orthogonal to each other as they have different internal nodes. We use the orthogonal trees with the same shape as a base of our computational space and we consider a particular shape as the one defining the *standard form* of anyon trees. This is obtained when the tree is constructed by selecting at each level always the leftmost anyon as the splitting element.

In the following we will denote anyons by their types[1], and we will refer to them by $a, b, \ldots$ or $1, 2, \ldots, m$, if $\mathcal{T} = \{a, b, \ldots\}$ or one enumeration of them $\mathcal{T} = \{a_1, a_2, \ldots, a_m\}$ is used to indicate the set, respectively.

Definition 2 (Splitting space). *Given an anyon model that specifies the set $\mathcal{T}$ of anyon types and the set of fusion rules, the splitting space V_{split} is the set of all the anyon trees obtained by fixing the number m and types of the leaf anyons and the root anyon.*

$$\mathcal{M}_m = V_{\text{split}} := V^{a_0}_{a_1 \otimes a_2 \otimes a_3 \cdots \otimes a_m}, \text{ with } a_i \in \mathcal{T}.$$

We can now introduce the notion of *anyonic term* and *anyonic calculus*. We call this calculus $\mathcal{A}\lambda_{\mathcal{T}}$ for its analogy to the classical lambda calculus. The index $\mathcal{T}$ indicates the parametrical definition of the terms and the computational rules of the calculus with respect to the types and the fusion rules of the anyonic system where the computation physically takes place.

2.2 The Language of Terms

Intuitively, an anyonic term represents an element in (a subspace of) the Hilbert space of fusion trees, $\mathcal{M}$, associated to an anyon system.

In order to formally define an anyonic term we assume a fixed set of elementary particles corresponding to the set of anyons that we want to operate on to perform a given computation.

Syntax An anyonic term T in the Hilbert space $\mathcal{M}_m$ associated to an anyon system of m anyons is inductively defined by the following syntax:

$$
\begin{array}{llr}
T ::= & v & \text{variable} \\
      & | \quad \lambda v.T & \text{function abstraction} \\
      & | \quad T_1 T_2 & \text{function application} \\
      & | \quad d_1 T_1 \oplus \ldots \oplus d_n T_n & \text{superposition}
\end{array}
$$

A variable v is any fusion tree in $\mathcal{M}_k$ with $k \leq m$, where $\mathcal{M}_k$ is a subspace of $\mathcal{M}_m$.

Function abstraction is any anyonic term T with some un-specified sub-tree represented by the variable v. The fusion space of v is called the *scope* of the function T and v is said to be bound in $\lambda v.T$. Variables in a term T that are not bound are called free, and an anyonic term with no free variables is called a closed term. We assume here the standard definitions of free variables and variable renaming that can be find in any treatment of the classical λ- calculus (see e.g. [1]).

Function application represents the basic operation of the calculus. It denotes the data T_1 considered as an algorithm applied to the data T_2 considered as input. Operationally,

---

[1] By 'type' we mean here the kind of physical charge of the particle and *not* the type of the term in the sense of type theory.

it captures the notion of *reduction*, i.e. the process of computation; semantically, it describes the output of this process. This will be discussed in Section 3 where we will introduce an operational semantics for this calculus.

Superposition[2] of terms can be seen as a choice among the terms T_i all belonging to the Hilbert subspace $\mathcal{M}_m$. The coefficients d_i are complex numbers whose physical meaning is related to the dimensionality corresponding to the various fusion processes.

Since terms are vectors in a Hilbert space, the λ-abstraction can be seen as both a vector and a linear operators; thus we can define

$$\lambda v.(d_1 T_1 \oplus \ldots \oplus d_n T_n) = d_1 \lambda v.T_1 \oplus \ldots \oplus d_n \lambda v.T_n.$$

Examples Consider an anyon system of four anyons of types $\mathcal{T} = \{a, b, \ldots\}$. Then the following are anyonic terms in standard form (numerical subscripts indicates here a position in the tree):

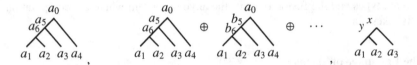

The last term is a variable v representing any tree in the subspace $\mathcal{M}_3 \subseteq \mathcal{M}_4$ (x and y are place-holders for any type $a, b, \ldots$, which is consistent with the fusion rules of the anyon system).

An important notion for defining the meaning of function application is the notion of *substitution*.

Definition 3. *Given an abstraction $\lambda v.T$ and a superposition of terms $T' = d_1 T'_1 \oplus d_2 T'_2 \oplus \cdots \oplus d_k T'_k$, the substitution of T' in T is defined by:*

$$T[T'/v] = d_1 T[T'_1/v] \oplus d_2 T[T'_2/v] \oplus \cdots \oplus d_k T[T'_k/v].$$

Note that T can be a superposition too; in this case the linearity of the abstraction operators applies, as explained before.

In a similar way as in classical λ-calculus, we can define an equivalence relation on anyonic terms by means of a β-conversion rule.

Definition 4. *Let T and T' be two anyonic terms. Then we define*

$$(\lambda v.T)T' = T[T'/v],$$

where we assume that no variable occurring freely in T' becomes bound after its substitution into T.

---

[2] Physicists denote the superposition with a single tree without specifying any internal anyon; this is to stress that the anyon system is unique for all the trees.

3   Operational Semantics

As in the pure λ-calculus, the only means to *compute* in the anyonic calculus is the application of functions to their arguments. This is classically defined by the β-reduction rule [1]. We introduce here a similar computational rule, although, as it will be clear later, the resulting operational semantics is very different from the classical one.

Computational Space   As mentioned before the pentagon and hexagon equations (se Appendix A) induce a tree equivalence that we can define as follows.

Definition 5. $T_1 \sim T_2$ *iff the internal anyons of* T_1 *and* T_2 *have identical types.*

We take as a representative of each equivalence class the tree in standard form in that class. Thus we can define the *computational space* of our calculus as the quotient space $S = V_{\text{split}|\sim}$, and consider only trees in standard form in the definition of the computational process.

Example 1.   The following trees are equivalent

The internal anyons are produced by the same splitting rules and have the same types. The term on the right hand side is in standard form and is the representative of this equivalence class.

Physically a computation on a system of non-Abelian anyons is performed in three basic steps:

1. Create anyons and arrange them in some initial configuration
2. Braid anyons
3. Detect anyonic charge by measuring the final configuration.

In our $\mathcal{A}\lambda$ calculus we suppose that the initial term T encode the creation process; thus T is the fusion state constructed by creating pairs of anyons from the vacuum and successively splitting them until the anyon configuration is obtained that represent the desired input. This term must then be rewritten into a new term by applying a *reduction* step and so on until we reach a term encoding the desired result. Clearly, if we want to model anyons evolution, we need to define the elementary step $T \rightarrow T'$ as a unitary transformation on the leaves of the term T representing the pairwise anyon braiding. This operator can be implemented by means of the F and R matrices of the given anyon system (see Appendix A). Intuitively, the transformation F encodes the associativity of the fusion rules by identifying the fusion spaces $V_{abc}^d = V_{(ab)c}^d = V_{a(bc)}^d$ resulting from the fusion process $abc \rightarrow d$. As the two descriptions $V_{(ab)c}^d$ and $V_{a(bc)}^d$ of the space V_{abc}^d correspond to different orthogonal bases (trees have different shapes), F effectively corresponds to a base change and define the probability of a fusion result. R performs the clockwise exchange of two particles they have a direct fusion channel (they are 'contiguous' leaves in our tree representation). The statistical evolution of the state when this exchange occurs in two spatial dimensions results in a phase factor which is used to calculate the final statistics. When the two particles have no direct fusion channel then F allows us to transform the basis states so as to make them contiguous. Thus the braiding operator $B_{a,b}$ is in general a composition of R and F operators.

The notion of derivation we next introduce captures this stepwise evolution of an anyons configuration.

Definition 6 (Reduction relation). *Given an anyon system with types* $\mathcal{T}$, *let* $T, T' \in \mathcal{A}\lambda_{\mathcal{T}}$. *We define the reduction relation* $\Rightarrow$ *as the symmetric, reflexive and transitive closure of the one-step relation* $\rightarrow_{ij}$ *defined by*

$$T \rightarrow_{ij} T' \ \text{iff} \ T' = B_{ij}T,$$

i.e. the anyon tree corresponding to T' *is obtained by exchanging* a_i *and* a_j *in the anyon tree for* T.

We call a sequence σ *of elementary steps* B_{ij} *a derivation in* $\mathcal{A}\lambda_{\mathcal{T}}$. *If* $T \Rightarrow T'$ *holds then we say that there exists a derivation in* $\mathcal{A}\lambda_{\mathcal{T}}$ *for* T *and we call* T' *the result of the derivation.*

Note that the symmetry of the transition relation $\Rightarrow$ derives from the unitarity of the transformation defining $\rightarrow_{ij}$.

We now define the core of our calculus, that is function application. Given an abstraction term $\lambda v.T$ and a term T', the application of $\lambda v.T$ to T' is obtained in three steps: (1) 'binding' v to T', (2) 'evaluate' T after the instantiation, (3) 'read' the result. In the classical lambda-calculus this is called the β-reduction, and is the base of its operational semantics (in any of the different forms it is defined) [1]. In order to define an operational semantics for our anyonic calculus we therefore only need to formally define what 'binding', 'evaluate' and 'read' means when we work with anyonic terms.

Binding means that we instantiate the subtree v in T by substituting in the abstraction term v with T', i.e. the value of its argument. It represents the physical process of performing in T all the splittings occurring in T'_i in the appropriate sequence.

Function application is realised by applying a braiding, i.e. a sequence of computational steps $\rightarrow_{ij}$, to the instantiated tree $T[T'/v]$. Intuitively this braiding represents the procedure T and realises the computation or term evaluation.

Definition 7. *Let* $\lambda v.T \in \mathcal{A}\lambda_{\mathcal{T}}$ *be an abstraction term and let* $B = T \Rightarrow T_1$ *be a derivation in* $\mathcal{A}\lambda_{\mathcal{T}}$. *Then we define* $(\lambda v.T)T'$ *to be the term* $B(T[T'/v])$.

It is easy to see that this definition of function application is correct, i.e. the term produced by this rule is indeed an anyon tree and therefore a $\mathcal{A}\lambda_{\mathcal{T}}$-term.

Proposition 1. *Let* $\lambda v.T \in \mathcal{A}\lambda_{\mathcal{T}}$. *Then for any* $T' \in \mathcal{A}\lambda_{\mathcal{T}}$ *we have that* $(\lambda v.T)T' \in \mathcal{A}\lambda_{\mathcal{T}}$.

The last step of an anyonic computation consists of reading the result, which in quantum computation essentially means measuring the final state. We can obtain this by applying the same fusion rules used for constructing the initial term (input) in the reverse order and then extract the result produced by the phase introduced by braiding. We illustrate this process in Section 5.

A feature of anyonic quantum computation that distinguishes the $\mathcal{A}\lambda_T$ from classical the λ-calculus is that braiding is an approximation process whose only termination criterion is given by the level of precision one aims to achieve for a given computation. Thus in $\mathcal{A}\lambda_T$, classical notions such as confluence or normalisation play a different role. Anyonic computation does not compute normal forms in the sense of closed terms to which no more reduction steps can be applied, and is *deterministic* (as it is reversible). Nevertheless, we can still impose a necessary condition on termination for the derivation in the $\mathcal{A}\lambda_T$-calculus, namely that a derivation must always produce a closed term as a result. This condition would allow us to guarantee that all functions have been calculated by the anyonic computation.

4 Universality

A well-known result from classical computability theory is the existence of a universal Turing machine that can calculate any recursive function, decide any recursive language, and accept any recursively enumerable language. According to the Church-Turing thesis, the problems solvable by a universal Turing machine are exactly those problems solvable by an algorithm or an effective method of computation, for any reasonable definition of those terms. For these reasons, a universal Turing machine serves as a standard against which to compare computational systems, and a system that can simulate a universal Turing machine is called Turing complete. All known Turing complete systems for classical computation have been shown to be Turing equivalent, i.e. they all compute precisely the same class of functions as do Turing machines. It is well-known that the lambda-calculus is among these systems.

Although, the Quantum Turing Machine provides a formalism similar to the classical Turing Machine for quantum computability, the question of what a universal QTM could be in the sense specified above is not yet completely solved. The extended Church-Turing thesis proposed by Deutsch and mentioned in Section 1 allows us to assert when a quantum model is universal in the sense that it can compute the same class of functions as the Quantum Turing Machine. However, we are not able to say what happens at Planck scale energies: at that level there may well be observables that are not even computable functions in the sense of Turing.

Here we address the restricted notion of computational universality, i.e. the analog of classical Turing-completeness. In quantum computation this is established in terms of approximate realisations of a computable function, thus a *Universal Quantum Computer* is a machine that is able to realise, at any accuracy, all the unitary operators in the space SU(N), i.e. the space of all special unitary matrices $N \times N$, representing all the possible quantum circuits.

According to this notion, is the $\mathcal{A}\lambda$-calculus introduced above universal? The following proposition gives a condition for universality in terms of the property of the anyon system that the calculus represent.

Proposition 2. *Given an anyonic system with types $\mathcal{T}$, the anyonic calculus $\mathcal{A}\lambda_{\mathcal{T}}$ is universal iff the set of all elementary operators $\{B_{ij}\}_{a_i,a_j \in \mathcal{T}}$ generates a group that is dense in $SU(N)$.*

Proof. This is a consequence of the result in [11] showing that if F and R span a dense set of unitaries acting on the qubits, then the corresponding anyonic model supports universal quantum computation implemented just by braiding anyons.

5   An Example

We describe an example that shows how anyonic computation can be expressed in our calculus. We work with the Fibonacci anyon model [20]. In this model the set of anyons types is $\mathcal{T} = \{1, \tau\}$ and there is only one fusion rule with two possible results, namely $\tau \otimes \tau = 1 \oplus \tau$ (τ is the only non-Abelian type). In [13] it is shown how to encode quantum gates in the Fibonacci anyon model. In this encoding a qubit is simulated by means of three τ anyons and a gate by a sequence of rotations on them. We consider the simulation of the Controlled-Phase operation (CP operator), that works on a register of two qubits: if the first (the controller) is set then a change of phase is applied to the second (the target).

The implementation of the CP operator with anyons is given by means of the following two sequences of rotations, calculated in [13],

$$\text{Controller} = B_2 \cdot B_3^4 \cdot B_2^2 \cdot B_3^{-4} \cdot B_2^2 \cdot B_3^2 \cdot B_2^2 \cdot B_3^6 \cdot B_2^2 \cdot B_3^{-2} \cdot B_2^2 \cdot B_3^{-2}$$
$$\cdot B_2^4 \cdot B_3^{-4} \cdot B_2^4 \cdot B_3^{-2} \cdot B_2^{-2} \cdot B_3$$
$$P = B_4^{-2} \cdot B_3^2 \cdot B_4^4 \cdot B_3^4 \cdot B_2^2 \cdot B_4^4 \cdot B_4^4 \cdot B_3^2 \cdot B_4^{-2} \cdot B_3^2 \cdot B_4^4 \cdot B_4^4 \cdot B_2^2 \cdot B_4^4 \cdot B_3^4.$$

The B_i operators are rotations and the whole sequence CP is the braid that encodes an approximation of the Controlled-Phase operator. Each B_i is represented in the model as a matrix of complex numbers, which works on the splitting space $V^1_{\tau \otimes \tau \otimes \tau \otimes \tau \otimes \tau \otimes \tau}$. This space contains trees with six anyons of type τ for the leaves and an anyon of charge 1 as the root. The exponent in the B_i's indicates the power of the matrix, i.e. the number of times it has to be applied. Since the base of $V^1_{\tau \otimes \tau \otimes \tau \otimes \tau \otimes \tau \otimes \tau}$ is composed by five vectors, the B_i's are 5x5 matrices. The sequence of rotations in the anyon model corresponds to the product of all the generators occurring in the braid; thus by multiplying the B_i matrices as expressed in CP we obtain the following matrix

$$CP := \begin{bmatrix} e^{i\alpha} & 0 & 0 & 0 & 0 \\ 0 & e^{i\alpha} & 0 & 0 & 0 \\ 0 & 0 & 1 & 0 & 0 \\ 0 & 0 & 0 & e^{-i\alpha} & 0 \\ 0 & 0 & 0 & 0 & 1 \end{bmatrix}$$

The anyonic terms of our calculus for this computation are terms belonging to the space of anyon trees with six τ anyon leaves and root anyon 1. Consider the following variable term x and function abstraction term $\lambda x.T$.

The second term represents the Controlled-Phase operation applied to a register of two qubits of which the first one (the control qubit) is a variable. Depending on the

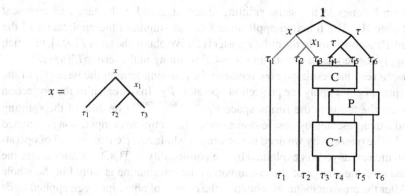

$x =$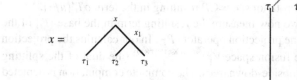

value we will give to the variable x, the change of phase operation is applied or not. The value of the variable x can be taken to simulate the qubit state 1, 0 or any superposition.

Consider the term u:

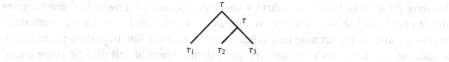

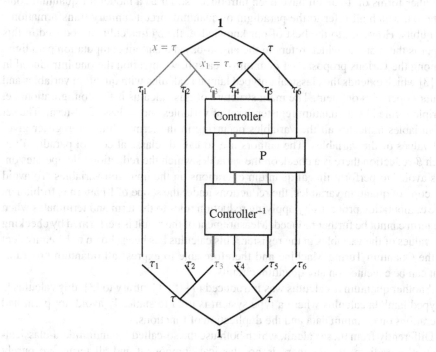

Fig. 1. Simulation of the CP operator

This term belongs to the same splitting space of x and ca be taken to represent the qubit state $|1\rangle$. The function application $(\lambda x.T)u$ simulates the application of the Controlled-Phase operator to the qubit register $|11\rangle$. We obtain the term $T[u/x]$, to which we now apply the sequence of rotations, $\sigma = CP$ resulting in the term $\alpha T'[u/x]$.

In order to read the result we now measure the resulting term in the base $\{T_i\}$ of the splitting space, i.e. we apply the projection operators P_{T_i}. In our calculus the projection operators are the terms in the fusion space $V_1^{((\tau\otimes\tau)\otimes\tau)\otimes(\tau\otimes(\tau\otimes\tau))}$ (the dual of the splitting space) and are represented by upside-down trees. The complete computation is depicted in Figure 1. The reason why we need to construct a whole reverse tree instead of operating just on the leaf anyons is explained by the non-locality of TQC: we cannot fuse the anyons and read the charge of the last anyon as the information is stored in the whole system. After the measurement, we obtain α, the change of phase has been applied since the controller has value 1.

6   Conclusions and Related Work

We have presented a formal calculus for topological quantum computation that captures the essential mechanisms of programming and can therefore be used for the specification of quantum programming languages features, quantum language design and implementation, and the study of quantum type systems. These are all directions we intend to pursue as future work.

Other forms of λ-calculi have been introduced so far as a model for quantum computation, which all refer to the paradigm of quantum circuit (unitary transformations) on qubits. However, to the best of our knowledge, the $\mathcal{A}\lambda$-calculus introduced in this paper is the first one which refers to the anyon-based quantum computation paradigm. Among the various proposals of quantum λ-calculi, we mention the one introduced in by [3] which extends the classical untyped lambda calculus with quantum variables and unitary operators on them. The main structure of this calculus is the configuration, i.e. a triple formed by a quantum register, a set of variables and a classical λ-term. The set of variables indicates all the variables occurring in the term, while the register gives the values of the variables. The authors use so-called 'classical control paradigm': at each β-reduction there is a check on the variables which the reduction rule operates on; this avoids to perform illegal quantum operations on the quantum variables. To avoid the copy of quantum variables, the reductions under the scope of !-pattern is forbidden. The computation proceeds by applying reduction rules to the term and terminates when the term cannot be further reduced. Measurement of the result is performed by checking the values of the variables in the register. This calculus has been shown to be equivalent to the Quantum Turing Machine and therefore able to express all quantum programs that can be executed on any quantum circuits.

Another quantum λ-calculus was introduced in [18]. Contrary to [3], this calculus is a typed lambda calculus where a type system is used to statically avoid not permitted operations on quantum data and the duplication of functions.

Differently from these calculi, which both use the so-called 'quantum-data-classical-control' paradigm, in $\mathcal{A}\lambda$ there is no classical component and all terms are purely quantum.

References

1. Barendregt, H.P.: The Lambda Calculus, revised edn. Studies in Logic and the Foundations of Mathematics, vol. 103. North-Holland, Amsterdam (1991)
2. Bernstein, E., Vazirani, U.V.: Quantum complexity theory. SIAM J. Comput. 26(5), 1411–1473 (1997)
3. dal Lago, U., Masini, A., Zorzi, M.: On a measurement-free quantum lambda calculus with classical control. Mathematical. Structures in Comp. Sci. 19, 297–335 (2009)
4. Deutsch, D.: Quantum Theory, the Church-Turing Principle and the Universal Quantum Computer. Proceedings of the Royal Society of London. A. Mathematical and Physical Sciences 400(1818), 97–117 (1985)
5. Deutsch, D.: Quantum Computational Networks. Proceedings of the Royal Society of London. A. Mathematical and Physical Sciences 425, 73–90 (1989)
6. Feynman, R.P.: Simulating physics with computers. International Journal of Theoretical Physics 21(6-7), 467–488 (1982)
7. Freedman, M.H., Kitaev, A., Larsen, M.J., Wang, Z.: Topological Quantum Computation. Physical Review Letters 40(12), 120402 (2001), http://arxiv.org/abs/quant-ph/0101025
8. Freedman, M.H., Kitaev, A., Larsen, M.J., Wang, Z.: Topological Quantum Computation. Bull. Amer. Math. Soc. 40, 31–38 (2003)
9. Freedman, M.H., Kitaev, A., Wang, Z.: Simulation of topological filed theories by quantum computers. Commun. Math. Phys. 227(3), 587–603 (2002)
10. Freedman, M.H., Larsen, M., Wang, Z.: A modular functor which is universal for quantum computation. Commun. Math. Phys. 227(3), 605–622 (2002)
11. Freedman, M.H., Wang, Z.: The Two-Eigenvalue Pproblem and density of Jones Representation of Braid Groups. Commun. Math. Phys. 228, 177–199 (2002)
12. Grover, L.K.: Quantum mechanics helps in searching for a needle in a haystack. Physical Review Letters 79(2), 325 (1997)
13. Hormozi, L., Zikos, G., Bonesteel, N.E., Simon, S.H.: Topological quantum compiling. Physical Review B 75(16), 165310 (2007), http://prb.aps.org/abstract/PRB/v75/i16/e165310
14. Kitaev, A.: Fault-tolerant quantum computation by anyons. Annals of Physics 303(1), 27 (1997), http://arxiv.org/abs/quant-ph/9707021
15. Nielsen, M., Chuang, I.: Quantum Computation and Quantum Information. Cambridge University Press, Cambridge (2000)
16. Nishimura, H., Ozawa, M.: Computational complexity of uniform quantum circuit families and quantum turing machines. Theoretical Computer Science 276(1), 147–181 (2002)
17. Pachos, J.K.: Introduction to Topological Quantum Computation. Cambridge University Press (2012)
18. Selinger, P., Valiron, B.: A lambda calculus for quantum computation with classical control. Mathematical Structures in Computer Science 16, 527–552 (2006)
19. Shor, P.W.: Polynomial-time algorithms for prime factorization and discrete logarithms on a quantum computer. SIAM Journal on Computing 26(5), 1484–1509 (1997)
20. Trebst, S., Troyer, M., Wang, Z., Ludwig, A.W.W.: A short introduction to fibonacci anyon models. Progress of Theoretical Physics 176, 384–407 (2008)
21. Wang, Z.: Topological Quantum Computation. American Mathematical Soc. (2010)

A   Physics Background

Anyons are quasiparticles that live in a two-dimensional space. Their behaviour is described by exchanging one particle with another. This exchange rotates the system's quantum state. They were discovered at the end of the 1970's when Leinaas and Myrheim observed that these particles did not obey either Bose-Einstein statistics or Fermi-Dirac statistics and therefore could not be identified neither with bosons nor with fermions, the only two physical particles known up to then.

From two anyons we can generate a new one by applying the *fusion rules* $a \otimes b = N_{ab}^c c$, where the number $N_{ab}^c (\in \mathbb{N})$ indicates the different ways of fusing a and b into c. These rules give the charge of a composed particle in terms of its constituents, and determine the particular anyonic model. We can use them in the opposite direction in order to split c into a and b and obtain two anyons from one. In this case we refer to the rules as *splitting rules*.

An anyon type a for which $\sum_c N_{ab}^c > 1$ is called *non-Abelian*. In other words, a non-Abelian anyon is one for which the fusion with another anyon may result in anyons of more than one type. This property is essential for computation because it implies a quantum dimension greater than 1 and therefore allows us to construct non-trivial computational spaces, i.e. spaces of dimension $n \geq 1$ of ground states where to store and elaborate information.

Considering the dual splitting process, a non-Abelian anyon can therefore have more than one splitting rule that applies to it, e.g. $a \otimes b = c$ and $e \otimes b = c$. Given an anyon of type c we can split it into two new anyons a, b and obtain a tree, then we can apply another rule and obtain a tree with leaf anyons c, d, b and root c. We can also split c into e, b and, supposing that there exists a fusion rule that split e into c, d, we can again split this state. The two resulting trees have leaf anyons and root anyon of same type, and differ only for the internal anyon a, e. Since we have obtained a, e splitting c, this two trees are orthogonal. Fixing the root and a fixed set of leaf anyon types, the set of orthogonal trees forms a base of the splitting space. Applying all the possible fusion rules that generate the leaf anyons we obtain also other (non orthogonal) trees which have different shapes and contain only copies of information. This is because the total charge is conserved by locally exchanging two anyons. Thus, trees with different shapes which are obtained by applying the same splitting rules are 'equivalent' from a computational viewpoint, since they have the same information content.

The dimension $D(m)$ of a splitting space H_m is calculated from the fusion rules by the formula

$$\sum_{x \in \text{Types}} N_{a_1 a_2}^{x_{m+1}} N_{x_{m+1} x_3}^{x_{m+2}} \cdots N_{x_{2m-1} a_m}^{a_0},$$

where $N_{a_j a_{j+1}}^{x_{m+1+j}}$ is the coefficient of the fusion rule that fuses the anyons $a_j a_{j+1}$ in the anyon x_{m+1+j}.

Example 2 (Splitting space dimension). Consider the Fibonacci anyon model [17]. In this model the only types are $\{1, \tau\}$. The splitting space dimensions are calculated by the formula $(\tau \otimes \tau) \otimes \tau = 2 \cdot \tau + 1 \cdot 1$. We therefore obtain the following spaces:

– S_3^τ of dimension 2 composed of the following two trees t_1 and t_2

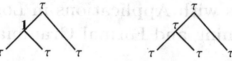

– S_3^1 of dimension 1 composed of the tree

Consider now the Ising anyon model [17]. This model has types $\{1, \sigma, \psi\}$. The splitting spaces are calculated by the formula $(\sigma \otimes \sigma) \otimes \sigma = 0 \cdot \sigma + 2 \cdot \psi + 2 \cdot \psi$. In this model we can construct the spaces:

– S_4^σ of dimension 0,
– S_4^ψ of dimension 2 composed of the trees

– S_4^1 of dimension 2 composed of the trees

The different trees obtained by fusing the same set of anyons to the same result are related by a set of equations, called the *pentagon identities*, expressing the associativity of the fusion rules via the operator F:

$$(F_{abz}^c)_o^l (F_{oae}^c)_b^z = \sum_{o'} (F_{bae}^l)_e^c (F_{ao'e}^l)_b^l (F_{aba}^b)_o^{o'} .$$

To use an anyon system for performing computations essentially means to look at the exchanges of the anyons of the system as a process evolving in time, i.e. looking at the system as a 2+1 dimensional space. This corresponds to braiding the threads (worldlines) starting from each anyon of the system. The braiding process causes non-trivial unitary rotations of the fusion space that are governed by so-called *hexagon* equations which relates the fusion of three anyons by a sequence of fusion rearrangements and braiding operations:

$$\sum_b (F_{231}^4)_b^c R_4^{1b} (F_{123}^4)_a^b = R_c^{13} (F_{213}^4)_a^c R_a^{12},$$

where R_c^{ab} is the phase given by the braiding operation between anyons a and b with fusion outcome c. We refer to [17,21] for a more complete treatment of TQC.

Theorems on Pre-fixed Points of Non-Monotonic Functions with Applications in Logic Programming and Formal Grammars*

Zoltán Ésik[1] and Panos Rondogiannis[2]

[1] Department of Computer Science, University of Szeged, Hungary
ze@inf.u-szeged.hu
[2] Department of Informatics and Telecommunications, University of Athens, Greece
prondo@di.uoa.gr

Abstract. We derive two novel theorems regarding pre-fixed points of non-monotonic functions and demonstrate that they have immediate applications in logic programming and formal grammars. In particular, the results of the paper lead to a natural generalization of the model intersection theorem for definite logic programs, to the more general class of normal logic programs. Moreover, the obtained results also offer the first (to our knowledge) model intersection result for Boolean grammars.

1 Introduction

The area of non-monotonicity has proven to be a quite fruitful one for Computer Science. On the practical side, non-monotonicity has found numerous applications in artificial intelligence, databases, programming languages and formal language theory. On the more theoretical side, the study of non-monotonicity has triggered the extension of classical fixed-point theory in order to apply to the case of non-monotonic functions [3,4].

This paper further contributes to the development of a novel non-monotonic fixed point theory. Our starting point is a recently introduced formal framework [5,6] for studying the properties of a broad class of non-monotonic functions over specially structured complete lattices. The key objective of this new framework is to obtain novel fixed point results regarding functions that are not necessarily monotonic. More specifically, the starting point of the framework developed in [5,6], is a complete lattice $(L, \leq)$ equipped with a family of pre-orderings, that give rise to another ordering relation $\sqsubseteq$. It was proved in [5,6]

* This research is being supported by the Greek General Secretariat for Research and Technology, the National Development Agency of Hungary, and the European Commission (European Regional Development Fund) under a Greek-Hungarian intergovernmental programme of Scientific and Technological collaboration. Project title: "Extensions and Applications of Fixed Point Theory for Non-Monotonic Formalisms". It is also supported by grant no. ANN 110883 from the National Foundation of Hungary for Scientific Research.

U. Kohlenbach et al. (Eds.): WoLLIC 2014, LNCS 8652, pp. 166–180, 2014.
© Springer-Verlag Berlin Heidelberg 2014

that if the preorderings over L obey certain simple and natural axioms, then the structure $(L, \sqsubseteq)$ is also a complete lattice. Moreover, it was demonstrated that a large class of functions $f : L \to L$ which may not be monotonic with respect to $\sqsubseteq$, possess a $\sqsubseteq$-least fixed point. Finally, it was shown that the new fixed point theorem generalizes both the Knaster-Tarski and the Kleene fixed point theorems (when f is monotonic or continuous respectively).

In this paper we extend the study carried out in [5,6] and we obtain two novel theorems regarding pre-fixed points of functions $f : L \to L$. The first theorem states that if L satisfies certain simple axioms and if X is a set of pre-fixed points of f with respect to $\leq$, then the greatest lower bound of X with respect to the relation $\sqsubseteq$ is also a pre-fixed point of f with respect to $\leq$. The second theorem concerns the case where L satisfies fewer axioms. In this case it is demonstrated that if X is a set of pre-fixed points of f with respect to $\sqsubseteq$ then the greatest lower bound of X with respect to $\sqsubseteq$ is also a pre-fixed point of f with respect to $\sqsubseteq$. Dual versions of both theorems regarding post-fixed points, are also obtained.

The above results have direct implications in the study of two areas of research where non-monotonicity plays an important role, namely in logic programming with negation [8,1] and in Boolean grammars [9]. It is known that under the infinite-valued semantics of logic programs with negation [10], the models of a program correspond *exactly* to the set of pre-fixed points with respect to the order relation $\leq$ of the immediate consequence operator of the program. Using the theorems obtained in the present paper, we get a straightforward proof of a *model intersection theorem* for logic programming with negation (a less direct proof of this theorem has recently also been obtained in [2]). In particular, we demonstrate that if X is an arbitrary set of models (in the sense of [10]) of a given program, then the greatest lower bound of X with respect to the relation $\sqsubseteq$, is also a model of the program. In a similar way, we obtain a model intersection theorem for Boolean grammars, which (to our knowledge) is the first such result for this class of grammars.

Model intersection theorems are of foundational importance for recursive formalisms. When they hold, they demonstrate that for every instance of the recursive formalism (eg. program, grammar, etc) the set of its models has a unique minimum element, which can be taken as the intended meaning of the instance. In other words, a model intersection theorem provides a pure, logical characterization of the semantics of the formalism under consideration.

2   Mathematical Preliminaries

In this section we provide some basic material from [5,6] that will be needed throughout the paper.

We consider complete lattices $(L, \leq)$; the least and greatest element of L will be denoted by $\bot$ and $\top$ respectively and the least upper bound and the greatest lower bound operation will be denoted by $\bigvee$ and $\bigwedge$ respectively. The results of this paper apply to every such complete lattice $(L, \leq)$ that has a special, refined structure. More specifically, we require that there also exist certain preorderings defined on L, which offer to the lattice this more refined structure. We start by the formal definitions and then give an intuitive description of the structure of elements of L. Let $\kappa > 0$ be a fixed ordinal; the results of the paper hold for every possible value of κ. More formally:

Definition 1. *Let $(L, \leq)$ be a complete lattice, let $\kappa > 0$ be a fixed ordinal and assume that for each ordinal $\alpha < \kappa$ there exists a preordering $\sqsubseteq_\alpha$ on L. We write $x =_\alpha y$ iff $x \sqsubseteq_\alpha y$ and $y \sqsubseteq_\alpha x$. We write $x \sqsubset_\alpha y$ iff $x \sqsubseteq_\alpha y$ but $x =_\alpha y$ does not hold. We write $x \sqsubset y$ iff $x \sqsubset_\alpha y$ for some $\alpha < \kappa$. We define $x \sqsubseteq y$ iff $x \sqsubset y$ or $x = y$. Let $x \in L$ and $\alpha < \kappa$. We define $(x]_\alpha = \{y : \forall \beta < \alpha\ x =_\beta y\}$ and $[x]_\alpha = \{y \in L : x =_\alpha y\}$.*

Intuitively, one can think of elements of L as consisting of "smaller components", one for every $\alpha < \kappa$. Roughly speaking, the relations $\leq$ and $\sqsubseteq$ correspond to two different ways of comparing elements of L: the relation $\leq$ corresponds to a "pointwise" comparison of elements of L; the relation $\sqsubseteq$ corresponds to a "lexicographic" comparison, in which we start comparing the 0-th level components, then the first level components, and so on, until we reach a decision. Therefore, when we write $x \leq y$, we mean that every component of x is less than or equal to the corresponding component of y (at least when Axiom 6 holds, see below). When we write $x =_\alpha y$ we mean that x and y are equal for all components up to level α. The relation $x \sqsubseteq_\alpha y$ means that x is equal to y for all $\beta < \alpha$ and it is either equal or smaller than y at level α. Finally, $x \sqsubseteq y$ means that either $x = y$ or there exists some α such that x and y are equal in all components less than α and x is genuinely smaller than y in the α-th component. The axioms that will be given shortly express these intuitions (and certain additional properties of our ordering relations).

A key property that will be used throughout the paper is that if the preordering relations defined above satisfy certain simple axioms, then $(L, \sqsubseteq)$ is also a complete lattice; one can then derive certain novel fixed point results regarding a class of (potentially) non-monotonic functions over this lattice. The first five of the axioms given below, were introduced in [5,6]; the last one has been added for the purposes of this paper. The following definition is needed before formally presenting the axioms:

Definition 2. *Let $x \in L$, $\alpha < \kappa$, and $X \subseteq (x]_\alpha$. Assume there exists $y \in (x]_\alpha$ such that:*

- *$X \sqsubseteq_\alpha y$, and*
- *for all $z \in (x]_\alpha$, if $X \sqsubseteq_\alpha z$ then $y \sqsubseteq_\alpha z$ and $y \leq z$.*

Then, y is unique and will be denoted by $\bigsqcup_\alpha X$. In the special case where $X = \{x\}$ we will denote $\bigsqcup_\alpha X$ by $x|_\alpha$.

The element y specified by the above definition represents an element of L that is an upper bound of the elements of X with respect to the $\sqsubseteq_\alpha$ relation; since it is also $\leq$ than any possible other such upper bound, we can view it as a special upper bound of the elements of X (and that's why it is denoted by $\bigsqcup_\alpha X$). For every X, if such an element y exists, then it is clearly unique. When $X = \{x\}$ then the element $x|_\alpha = \bigsqcup_\alpha \{x\}$ intuitively represents the α-th component of x. We can now formally state the axioms:

1. **Axiom 1:** For all ordinals $\alpha < \beta < \kappa$, $\sqsubseteq_\beta$ is included in $=_\alpha$.
2. **Axiom 2:** $\bigcap_{\alpha<\kappa} =_\alpha$ is the equality relation on L.
3. **Axiom 3:** For each $x \in L$, for every ordinal $\alpha < \kappa$, and for any $X \subseteq (x]_\alpha$, $\bigsqcup_\alpha X$ exists.
4. **Axiom 4:** For every non-empty $X \subseteq L$ and ordinal $\alpha < \kappa$, if $y =_\alpha x$ for all $x \in X$, then $y =_\alpha (\bigvee X)$.
5. **Axiom 5:** For all $x, y \in L$ and $\alpha < \kappa$, if $x \leq y$ and $x =_\beta y$ for all $\beta < \alpha$, then $x \sqsubseteq_\alpha y$.
6. **Axiom 6:** For all $x, y \in L$, if $x \leq y$ then $x|_\alpha \leq y|_\alpha$ for all $\alpha < \kappa$.

As it has been demonstrated in [6], there exist many natural structures that satisfy Axioms 1-5 above. In subsequent sections we will exhibit two well-known structures that satisfy all the above axioms.

For all $x, y \in L$ and $\alpha < \kappa$, let us define $x \geq y$ iff $y \leq x$ and $x \sqsupseteq_\alpha y$ iff $y \sqsubseteq_\alpha x$. The *dual of an axiom* is obtained by replacing each occurrence of $\leq$ by $\geq$, each occurrence of $\sqsubseteq_\alpha$ by $\sqsupseteq_\alpha$, each occurrence of $\bigvee$ by $\bigwedge$, each occurrence of $\bigsqcup_\alpha$ by $\bigsqcap_\alpha$ and finally each occurrence of $|_\alpha$ by $|^\alpha$. We say that an axiom is *self-dual* if every L that satisfies it, also satisfies the dual of the axiom. It is clear that Axioms 1, 2 and 5 are self-dual.

We will call L *a strong model* if it satisfies Axioms 1–6. L will be called *a dual strong model* if it satisfies the duals of these axioms. A *symmetric strong model* is both a strong model and a dual strong model.

We now repeat two results that have been established in [6] and that will be used in our subsequent exposition.

Lemma 1 (Lemma 3.11 of [6]). *Suppose that L is a model satisfying Axioms 1–5. Then for all $x \in L$, $\alpha < \kappa$ and $y, z \in (x]_\alpha$, $y|_\alpha \leq z|_\alpha$ iff $y|_\alpha \sqsubseteq_\alpha z|_\alpha$ iff $y \sqsubseteq_\alpha z$.*

In the rest of the paper we write $X|_\alpha$ to denote the set $\{x|_\alpha : x \in X\}$.

Lemma 2 (Proposition 5.7 of [6]). *Suppose that L is a model satisfying Axioms 1–5. Then for all $x \in L$, $\alpha < \kappa$ and $X \subseteq (x]_\alpha$, $\bigvee X|_\alpha = \bigsqcup_\alpha X|_\alpha$.*

We will also use the easy fact that when $X \subseteq L$ and Axioms 1–4 hold, then $\bigsqcup_\alpha X|_\alpha = \bigsqcup_\alpha X$, and that by [6][Lemma 3.14], for all $x, y \in L$ it holds $x = y$ iff $x|_\alpha = y|_\alpha$ for all $\alpha < \kappa$. In fact, $x = \bigvee_{\alpha<\kappa} x|_\alpha$ holds for all $x \in L$ as shown in [6][Lemma 3.16]. Thus, as it has been demonstrated in [6][Lemma 3.17], if L satisfies Axioms 1–4 then for all $x, y \in L$, if for all $\alpha < \kappa$ it holds $x|_\alpha \leq y|_\alpha$, then $x \leq y$. Therefore, if L satisfies Axioms 1–4 and additionally Axiom 6, then for all $x, y \in L$, we have $x \leq y$ iff $x|_\alpha \leq y|_\alpha$ for all $\alpha < \kappa$.

The following theorem has been established in [6]:

Theorem 1 (Theorem 4.1 of [6]). *Assume that L satisfies Axioms 1-4. Then, $(L, \sqsubseteq)$ is a complete lattice.*

It can be shown (see [6]) that the least and greatest elements of L with respect to $\sqsubseteq$ are also $\bot$ and $\top$ (ie., they coincide with the least and greatest elements of L with respect to $\leq$). We will denote by $\sqcap$ and $\sqcup$ the corresponding glb and lub operations of the lattice $(L, \sqsubseteq)$ implied by the above lemma. It was shown in [6] that we can construct $\sqcap$ and $\sqcup$ in terms of $\sqcap_\alpha$ and $\sqcup_\alpha$ respectively. Given $X \subseteq L$, we now demonstrate how we can construct $\sqcup X$. The construction of $\sqcap X$ is symmetric. Our exposition follows [6][Section 4]. We will make use of the following definition:

Definition 3. *A sequence $(x_\alpha)_{\alpha<\kappa}$ of elements of L is called* compatible *if each x_α is the $\leq$-least element of $[x_\alpha]_\alpha$ and if $x_\alpha =_\alpha x_\beta$ for all $\alpha < \beta$.*

As shown in [6][Lemma 3.20 and Corollary 3.21], a sequence $(x_\alpha)_{\alpha<\kappa}$ is compatible iff there exists (a necessarily unique) $x \in L$ with $x|_\alpha = x_\alpha$ for all $\alpha < \kappa$. In fact, this unique element is $x = \bigvee_{\alpha<\kappa} x|_\alpha$. Moreover, if a sequence $(x_\alpha)_{\alpha<\kappa}$ is compatible, then $x_\alpha =_\alpha x_\beta$ and $x_\alpha \leq x_\beta$ holds for all $\alpha \leq \beta < \kappa$. For each ordinal $\alpha < \kappa$ we define the sets X_α, which are then used in order to obtain $\sqcup X$. For each α, let $X_\alpha = \{y \in X : \forall \beta < \alpha \ y =_\beta x\} = \{y \in X : \forall \beta < \alpha \ y|_\beta = x|_\beta\}$. Notice that $X_0 = X$. It has been demonstrated in [6] that for all $\alpha < \kappa$, if $X_\alpha \neq \emptyset$ then $x|_\alpha = \sqcup_\alpha X_\alpha$; moreover, if $X_\alpha = \emptyset$ then $x|_\alpha = \bigvee_{\beta<\alpha} x|_\beta$ and $x|_\alpha$ is a $\sqsubseteq_\alpha$-least and the $\leq$-least element of $(x|_\alpha)_\alpha$. Finally, we define $x_\infty = \bigvee_{\alpha<\kappa} x|_\alpha$. It can be shown (see [6][Theorem 4.1]) that $x_\infty = \sqcup X$. Moreover, it is easy to prove that by construction it holds $x|_\alpha = x|_\beta$ and $x|_\beta \leq x|_\alpha$ for all $\beta < \alpha$ and that the sequence $(x|_\beta)_{\beta<\kappa}$ is compatible.

We will be considering functions that are generally non-monotonic with respect to $\sqsubseteq$ (and also non-monotonic with respect to $\leq$) but exhibit a restricted form of monotonicity with respect to the relations $\sqsubseteq_\alpha$.

Definition 4. *Let $\alpha < \kappa$. A function $f : L \to L$ is called α-monotonic if for all $x, y \in L$, if $x \sqsubseteq_\alpha y$ then $f(x) \sqsubseteq_\alpha f(y)$.*

The functions that will be the objects of our study will be α-monotonic for all $\alpha < \kappa$. However, it should be noted that even if a function is α-monotonic for all $\alpha < \kappa$, then it need not be necessarily monotonic with respect to the relation $\sqsubseteq$ (for a counterexample, see [10, Example 5.7, pages 453–454]). Therefore, the standard tools of classical fixed point theory do not suffice in order to establish properties of such functions, and stronger tools need to be devised.

3   Pre-fixed Point and Post-fixed Point Theorems

In this section we develop two novel theorems regarding post-fixed points of functions $f : L \to L$ that are α-monotonic for all $\alpha < \kappa$. Using duality, we obtain respective results regarding pre-fixed points of such functions. Recall that:

Definition 5. *Let L be a complete lattice, let $\leq$ be an ordering relation on L and let $f : L \to L$. An element $x \in L$ will be called a* post-fixed point *of f with respect to $\leq$ (respectively* pre-fixed point*) if $f(x) \geq x$ (respectively $f(x) \leq x$).*

The first of the theorems concerns the case where L is a strong model. In this case it is demonstrated that if X is a set of post-fixed points of f with respect to $\leq$ then $\bigsqcup X$ is also a post-fixed point of f with respect to $\leq$. The dual of this theorem states that if X is a set of pre-fixed points of f with respect to $\leq$ then $\bigsqcap X$ is also a pre-fixed point of f with respect to $\leq$. As it will be demonstrated in subsequent sections, this last result will be used to obtain general model intersection results for normal logic programs and for Boolean grammars. This is due to the known fact that the set of models of a logic program coincides with the sets of pre-fixed points of the immediate consequence operator for these programs, which is an α-monotonic function for all countable ordinals α. A similar result will also be demonstrated for Boolean grammars.

The second of the theorems concerns the case where L satisfies only Axioms 1-4. In this case it is demonstrated that if X is a set of post-fixed points of f with respect to $\sqsubseteq$ then $\bigsqcup X$ is also a post-fixed point of f with respect to $\sqsubseteq$. The dual of this theorem also immediately holds.

Our first theorem concerns post-fixed points of functions $f : L \to L$, where L is a strong model and f is α-monotonic for all $\alpha < \kappa$.

Theorem 2. *Suppose that L is a strong model. Suppose that $f : L \to L$ is α-monotonic for each $\alpha < \kappa$. Let $X \subseteq L$ be a set of post-fixed points of f with respect to $\leq$. Then $\bigsqcup X$ is also a post-fixed point of f with respect to $\leq$.*

Proof. Our claim is clear when X is empty, since in that case $\bigsqcup X = \bot$. But our argument below works for all X, including the empty set.

Let $x = \bigsqcup X$. We want to show that $x|_\alpha \leq f(x)|_\alpha$ for all $\alpha < \kappa$, since it then follows that $x = \bigvee_{\alpha < \kappa} x|_\alpha \leq \bigvee_{\alpha < \kappa} f(x)|_\alpha = f(x)$.

We argue by induction on α.

Recall now from the background material given in Section 2 that for each α, $X_\alpha = \{y \in X : \forall \beta < \alpha \; y =_\beta x\} = \{y \in X : \forall \beta < \alpha \; y|_\beta = x|_\beta\}$. Also recall that for all $\alpha < \kappa$, if $X_\alpha \neq \emptyset$ then $x|_\alpha = \bigsqcup_\alpha X_\alpha = \bigsqcup_\alpha \{y|_\alpha : y \in X_\alpha\} = \bigvee \{y|_\alpha : y \in X_\alpha\}$, by Lemma 2 and since $y|_\beta = x|_\beta$ for all $y \in X_\alpha$ and $\beta < \alpha$, and if $X_\alpha = \emptyset$ then $x|_\alpha = \bigvee_{\beta < \alpha} x|_\beta$. We distinguish cases based on whether $X_\alpha = \emptyset$ or not.

Suppose that $X_\alpha \neq \emptyset$. Now $y \sqsubseteq_\alpha x =_\alpha x|_\alpha$ for all $y \in X_\alpha$, and since f is α-monotonic, $f(y) \sqsubseteq_\alpha f(x)$, or equivalently, $f(y)|_\alpha \sqsubseteq_\alpha f(x)|_\alpha$ for all $y \in X_\alpha$. Since $f(y) \sqsubseteq_\alpha f(x)$ for all $y \in X_\alpha$, also $f(y) =_\beta f(x)$ and $f(y)|_\beta = f(x)|_\beta$ for all $y \in X_\alpha$ and $\beta < \alpha$. It follows by Lemma 1 that $f(y)|_\alpha \leq f(x)|_\alpha$ for all $y \in X_\alpha$. Now also $y \leq f(y)$ for all $y \in X_\alpha$, so that $y|_\alpha \leq f(y)|_\alpha$ for all $y \in X_\alpha$ by Axiom 6. We conclude that $y|_\alpha \leq f(x)|_\alpha$ for all $y \in X_\alpha$. Since $x|_\alpha = \bigvee \{y|_\alpha : y \in X_\alpha\}$ and $y|_\alpha \leq f(x)|_\alpha$ for all $y \in X_\alpha$, we conclude that $x|_\alpha \leq f(x)|_\alpha$.

Suppose that $X_\alpha = \emptyset$. Then $x|_\alpha = \bigvee_{\beta < \alpha} x|_\beta \leq \bigvee_{\beta < \alpha} f(x)|_\beta \leq f(x)|_\alpha$, using the induction hypothesis. $\qquad\square$

By duality, we get the following corollary regarding the greatest lower bound of a set of pre-fixed points of f.

Corollary 1. *Suppose that L is a dual strong model. Suppose that $f : L \to L$ is α-monotonic for each $\alpha < \kappa$. Let $X \subseteq L$ be a set of pre-fixed points of f with respect to $\leq$. Then $\bigsqcap X$ is also a pre-fixed point of f with respect to $\leq$.*

Corollary 2. *Suppose that L is a symmetric strong model. Suppose that f : $L \to L$ is α-monotonic for each $\alpha < \kappa$. Let X be a set of pre-fixed points and Y a set of post-fixed points of f with respect to $\leq$. Then $\bigcap X$ is a pre-fixed point and $\bigsqcup Y$ is a post-fixed point of f with respect to $\leq$.*

As it turns out, if we assume that L only satisfies Axioms 1-4, then we can still get pre-fixed and post-fixed point results, but this time with respect to the $\sqsubseteq$ relation. We will make use of the following lemma:

Lemma 3. *Suppose that L satisfies Axioms 1-4 and $f : L \to L$ is α-monotonic for all $\alpha < \kappa$. Suppose that X is a set of post-fixed points of f with respect to the relation $\sqsubseteq$. Let $x = \bigsqcup X$. Then for all $\alpha < \kappa$, either there exists $\beta < \alpha$ such that $x|_\beta \sqsubset_\beta f(x|_\beta)$, or $x|_\beta =_\beta f(x|_\beta)$ for all $\beta \leq \alpha$.*

Proof. Our claim is clear when X is empty since in that case $x = \bot$ and $\bot$ is the least element of L w.r.t. the partial order $\sqsubseteq$. So below we assume that X is not empty. Recall now from the background material given in Section 2 that for each $\alpha < \kappa$, $X_\alpha = \{y \in X : \forall \beta < \alpha \; y =_\beta x\} = \{y \in X : \forall \beta < \alpha \; y|_\beta = x|_\beta\}$, so that $X_0 = X$. Recall also that for all $\alpha < \kappa$, if $X_\alpha \neq \emptyset$ then $x|_\alpha = \bigsqcup_\alpha X_\alpha$, and if $X_\alpha = \emptyset$ then $x|_\alpha = \bigvee_{\beta < \alpha} x|_\beta$ and $x|_\alpha$ is a $\sqsubseteq_\alpha$-least and the $\leq$-least element of $(x|_\alpha]_\alpha$.

Suppose that the claim does not hold and let α_0 denote the least ordinal for which it fails. Then $x|_\beta =_\beta f(x|_\beta)$ for all $\beta < \alpha_0$ and $x|_{\alpha_0} \not\sqsubseteq_{\alpha_0} f(x|_{\alpha_0})$. Since the sequence $(x|_\alpha)_{\alpha < \kappa}$ is compatible, we have that $x|_\beta =_\beta x|_{\alpha_0}$ and thus also $f(x|_\beta) =_\beta f(x|_{\alpha_0})$ for all $\beta < \alpha_0$. We conclude that $x|_\beta =_\beta x|_{\alpha_0} =_\beta f(x|_{\alpha_0})$ for all $\beta < \alpha_0$.

Case 1: $X_{\alpha_0} = \emptyset$. Then $x|_{\alpha_0}$ is a $\sqsubseteq_{\alpha_0}$-least element of $(x|_{\alpha_0}]_{\alpha_0}$. Since $x|_\beta =_\beta x|_{\alpha_0} =_\beta f(x|_{\alpha_0})$ for all $\beta < \alpha_0$, $f(x|_{\alpha_0}) \in (x|_{\alpha_0}]_{\alpha_0}$. It follows that $x|_{\alpha_0} \sqsubseteq_{\alpha_0} f(x|_{\alpha_0})$, a contradiction.

Case 2: $X_{\alpha_0} \neq \emptyset$. Then $x|_{\alpha_0} = \bigsqcup_{\alpha_0} X_{\alpha_0}$. Recall that $X_{\alpha_0} = \{z \in X : \forall \beta < \alpha_0 \; z =_\beta x|_\beta\}$. But we know that $x|_\beta =_\beta f(x|_\beta)$ for all $\beta < \alpha_0$, and by our assumption on f, also $f(x|_\beta) =_\beta f(z)$ for all $z \in X_{\alpha_0}$ and $\beta < \alpha_0$. So we conclude that $z =_\beta f(z)$ holds for all $z \in X_{\alpha_0}$ and $\beta < \alpha_0$. Since also $z \sqsubseteq f(z)$ for all $z \in X_{\alpha_0}$, we obtain that $z \sqsubseteq_{\alpha_0} f(z)$ for all $z \in X_{\alpha_0}$.

But for all $z \in X_{\alpha_0}$, $z \sqsubseteq_{\alpha_0} x|_{\alpha_0}$ and thus $z \sqsubseteq_{\alpha_0} f(z) \sqsubseteq_{\alpha_0} f(x|_{\alpha_0})$. Thus, $f(x|_{\alpha_0})$ is an $\sqsubseteq_{\alpha_0}$-upper bound of X_{α_0}, and by the definition of $x|_{\alpha_0}$, we have $x|_{\alpha_0} \sqsubseteq_{\alpha_0} f(x|_{\alpha_0})$, a contradiction again. □

Now, the following theorem follows easily by the previous lemma:

Theorem 3. *Suppose that L satisfies Axioms 1-4 and $f : L \to L$ is α-monotonic for all $\alpha < \kappa$. Let $X \subseteq L$ be a set of post-fixed points of f with respect to $\sqsubseteq$. Then $\bigsqcup X$ is a post-fixed point of f with respect to $\sqsubseteq$.*

Proof. Let $x = \bigsqcup X$. If $x =_\alpha f(x)$ for all $\alpha < \kappa$, then x is a fixed point by Axiom 2. Otherwise, by Lemma 3, there is some $\alpha < \kappa$ with $x =_\alpha x|_\alpha \sqsubset_\alpha f(x|_\alpha) =_\alpha f(x)$ and then $x \sqsubset_\alpha f(x)$ and $x \sqsubseteq f(x)$. □

By duality, we get the following corollary regarding the greatest lower bound of a set of pre-fixed points of f (recall that the duals of Axioms 1 and 2 are the Axioms 1 and 2 themselves).

Corollary 3. *Suppose that L satisfies the duals of Axioms 1–4. Let $f : L \to L$ be α-monotonic for all $\alpha < \kappa$. Let $X \subseteq L$ be a set of pre-fixed points of f with respect to $\sqsubseteq$. Then $\sqcap X$ is also a pre-fixed point of f with respect to $\sqsubseteq$.*

Corollary 4. *Suppose that L is a symmetric strong model. Let $f : L \to L$ be α-monotonic for all $\alpha < \kappa$. Suppose that $X \subseteq L$ is a set of pre-fixed points of f and Y is a set of post-fixed points of f with respect to $\sqsubseteq$. Then $\sqcap X$ is a pre-fixed point and $\sqcup Y$ is a post fixed-point of f with respect to $\sqsubseteq$.*

4   An Application to Non-Monotonic Logic Programming

In this section we demonstrate that the results of the previous section can be used in order to obtain a novel result regarding logic programs with negation. In particular, we obtain a model intersection theorem for these programs, which generalizes the classical model intersection theorem [8] for definite (ie., negation-less) logic programs.

In the rest of this section, we adopt a common assumption in the area of logic programming and study programs that are propositional and have a countable number of rules. Formally:

Definition 6. *A normal program clause is a clause whose body is a conjunction of propositional literals, where a literal is either a propositional atom or the negation of an atom. A normal logic program is a countable set of normal program clauses.*

In order to apply the results of the previous section to the semantics of normal logic programs, it suffices to show that the set of interpretations of such programs is a symmetric strong model. As it turns out, the infinite-valued approach for logic programs [10], obeys this requirement. We start by presenting the main ideas of this approach by following [10].

The basic idea behind the infinite-valued approach is to express the semantics of normal logic programs using a refined multiple-valued logic which contains one truth value F_α and one T_α for each countable ordinal α, with an element 0 in the middle. The ordering of the truth values is as follows:

$$F_0 < F_1 < \cdots < F_\alpha < \cdots < 0 < \cdots < T_\alpha < \cdots < T_1 < T_0$$

Intuitively, F_0 and T_0 are the classical *False* and *True* values and 0 is the *undefined* value. The intuition behind the new values is that they express different levels of truth and falsity.

In the following we denote by V the set consisting of the above truth values, ie., $V = \{F_\alpha : \alpha < \Omega\} \cup \{T_\alpha : \alpha < \Omega\} \cup \{0\}$, where Ω is the first uncountable ordinal. We have the following definition:

Definition 7. *The order of a truth value is defined as follows:* $order(T_\alpha) = \alpha$, $order(F_\alpha) = \alpha$ *and* $order(0) = \Omega$.

Let Z be the set of propositional atoms that appear in program P. Interpretations are defined as follows:

Definition 8. *An (infinite-valued) interpretation I of a program P is a function from the set Z of propositional atoms of P to V.*

Interpretations can be extended to apply to literals and to conjunctions of literals. The most interesting case is the handling of negation: in order to negate a non-zero truth value, we take the symmetric truth value (with respect to zero) and then move one further step towards zero. The intuition here is that cycles through negation may lead to the zero truth value (eg. as in the case of the program $r \leftarrow \sim r$). The interested reader can consult [10] for a more detailed discussion on infinite-valued interpretations and their properties.

Definition 9. *Let $I \in V^Z$ be an interpretation of a program P. Then, I can be extended as follows:*

– *For every negative atom $\sim p$ appearing in P:*

$$I(\sim p) = \begin{cases} T_{\alpha+1} & \text{if } I(p) = F_\alpha \\ F_{\alpha+1} & \text{if } I(p) = T_\alpha \\ 0 & \text{if } I(p) = 0 \end{cases}$$

– *For every conjunction of literals $l_1, \ldots, l_n$ appearing as the body of a clause in P:*

$$I(l_1, \ldots, l_n) = min\{I(l_1), \ldots, I(l_n)\}$$

The notion of satisfiability of a clause can be defined in a direct manner:

Definition 10. *Let P be a program and $I \in V^Z$ an interpretation of P. Then, I satisfies a clause $p \leftarrow l_1, \ldots, l_n$ of P if $I(p) \geq I(l_1, \ldots, l_n)$. Moreover, I is a model of P if I satisfies all clauses of P.*

Given an interpretation of P, we adopt a specific notation for the set of propositional symbols of the program that are assigned a specific truth value and for the subset of the interpretation that corresponds to a particular order:

Definition 11. *Let P be a program, $I \in V^Z$ an interpretation of P and $v \in V$. Then $I \parallel v = \{p \in Z \mid I(p) = v\}$.*

We now proceed to define the relations $\leq$, $\sqsubseteq_\alpha$ for all $\alpha < \Omega$. Notice that since V contains values indexed by countable ordinals, it suffices to define relations $\sqsubseteq_\alpha$ for all $\alpha < \Omega^1$. We do not need to define the $\sqsubset$ and $\sqsubseteq$ relations, because their definition is immediate from the $\sqsubseteq_\alpha$ relations (see Section 2).

---

[1] In other words, the ordinal κ introduced in the previous sections, is taken to be equal to Ω for the particular application. At the end of this section we further discuss the consequences of choosing an alternative value of κ.

The definitions of the following relations are driven by the following goal. Assume we are given a set X of infinite-valued models of a program. We would like to find an interpretation that is the greatest lower bound of the set X and that is also a model of the program. This interpretation must have the smallest possible set of T_0 values and the largest possible set of F_0 values (as dictated by the elements of X), and among the models having exactly these features, it must have the smallest possible set of T_1 values and the largest possible set of F_1 values, and so forth, for every T_α, F_α and every countable ordinal α; all atoms that do not receive a value in this way, are assigned the value 0. Therefore, our relations must be defined in such a way that they allow this kind of "lexicographic comparison".

Definition 12. *Let $I, J \in V^Z$ be interpretations of a given program P. We write $I \leq J$ if for all $p \in Z$, $I(p) \leq J(p)$.*

Definition 13. *Let $I, J \in V^Z$ be interpretations of a given program P and $\alpha < \Omega$. We write $I =_\alpha J$, if for all $\beta \leq \alpha$, $I \parallel T_\beta = J \parallel T_\beta$ and $I \parallel F_\beta = J \parallel F_\beta$.*

Definition 14. *Let $I, J \in V^Z$ be interpretations of a given program P and $\alpha < \Omega$. We write $I \sqsubset_\alpha J$, if for all $\beta < \alpha$, $I =_\beta J$ and either $I \parallel T_\alpha \subset J \parallel T_\alpha$ and $I \parallel F_\alpha \supseteq J \parallel F_\alpha$, or $I \parallel T_\alpha \subseteq J \parallel T_\alpha$ and $I \parallel F_\alpha \supset J \parallel F_\alpha$. We write $I \sqsubseteq_\alpha J$ if $I =_\alpha J$ or $I \sqsubset_\alpha J$.*

We have the following lemma:

Lemma 4. *The set V^Z of infinite-valued interpretations $\mathcal{I}_P$ of a normal logic program P is a symmetric strong model.*

Proof. It has been been demonstrated in [6] that V^Z (the set of infinite-valued interpretations of P) satisfies Axioms 1-5. We first demonstrate that it also satisfies Axiom 6, ie., that if $I, J \in V^Z$ with $I \leq J$, then $I|_\alpha \leq J|_\alpha$, for all $\alpha < \Omega$.

Suppose that $I \leq J$. We want to show that $I|_\alpha(z) \leq J|_\alpha(z)$ for all $z \in Z$. If $I|_\alpha(z) = I(z)$ and $J|_\alpha(z) = J(z)$, our claim is clear. Suppose that $I|_\alpha(z) \neq I(z)$. Then $I|_\alpha(z) = F_{\alpha+1}$ and $order(I(z)) > \alpha$. Thus, since $I(z) \leq J(z)$, either $order(J(z)) > \alpha$ or $J(z) \geq T_\alpha$. In the first case $J|_\alpha(z) = F_{\alpha+1}$, and in the second case $J|_\alpha(z) = J(z) \geq T_\alpha$. In either case, $I|_\alpha(z) \leq J|_\alpha(z)$.

It remains to deal with the case $I|_\alpha(z) = I(z)$ but $J|_\alpha(z) \neq J(z)$. In that case $F_{\alpha+1} < J(z) < T_\alpha$ and $I(z) \leq F_{\alpha+1}$. Thus $J|_\alpha(z) = F_{\alpha+1}$ and $I|_\alpha(z) \leq J|_\alpha(z)$ again.

In conclusion, V^Z is a strong model. Consider the bijection $\varphi : V \to V$ defined by $\varphi(F_\alpha) = T_\alpha$, $\varphi(T_\alpha) = F_\alpha$, for all $\alpha < \Omega$, and $\varphi(0) = 0$. We extend φ to a bijection $V^Z \to V^Z$ in the expected way so that $\varphi(I)(z) = \varphi(I(z))$ for all $I \in V^Z$ and $z \in Z$. It is clear that for all interpretations $I, J \in V^Z$ and for all $\alpha < \kappa$, $I \leq J$ iff $\varphi(J) \leq \varphi(I)$ and $\varphi(I) \sqsubseteq_\alpha \varphi(J)$ iff $\varphi(J) \sqsubseteq_\alpha \varphi(I)$. It follows that $I \sqsubseteq J$ iff $\varphi(J) \sqsubseteq \varphi(I)$. Thus, φ is a "dual isomorphism". Since V^Z is a strong model and φ is a dual isomorphism, V^Z is a dual strong model and thus a symmetric strong model. $\square$

Consider now the *immediate consequence operator* for normal logic programs defined in [10]:

$$T_P(I)(p) = \bigvee \{I(l_1, \ldots, l_n) \mid (p \leftarrow l_1, \ldots, l_n) \in P\}$$

It can easily be shown (see [10] as-well-as [6]) that T_P is α-monotonic for each $\alpha < \Omega$. Moreover, it is easy to check (see again [10]) that the set of models of a program P coincides with the set of pre-fixed points of T_P with respect to the ordering $\leq$. Then, the following theorem follows directly from Corollary 1:

Theorem 4. *Let P be a normal logic program and let X be a set of infinite-valued models of P. Then, $T_P(\bigsqcap X) \leq \bigsqcap X$, ie., $\bigsqcap X$ is also an infinite-valued model of P.*

As a direct consequence of Theorem 4, the greatest lower bound of the set of all infinite-valued models of a program P, is the unique minimum (with respect to $\sqsubseteq$) infinite-valued model of P. Notice that this is an *existence* result and it does not provide any direct way for actually *constructing* the minimum model; a constructive way is given in [10]. However, in certain cases we can identify the minimum model through the definition of $\bigsqcap$, as the following example illustrates.

Example 1. Consider the following program P:

$$p \leftarrow \sim q$$
$$r \leftarrow \sim r$$

Let X be the set of all infinite-valued models of P. We calculate $\bigsqcap X$, ie., the minimum infinite-valued model of P. We will make use of the construction of $\bigsqcap X$ in terms of its "approximations" $\bigsqcap_\alpha X_\alpha$, as explained in Section 2.

By the definition of model (Definition 10) and from the syntax of the program, we get that every model of P is of the form $\{(p, v_p), (q, v_q), (r, v_r)\}$ where $0 \leq v_r \leq T_0$ (the values below 0 do not satisfy the second clause), $F_0 \leq v_q \leq T_0$ and $v_p \geq \sim v_q$. Notice that there exists at least one model with $v_q = F_0$ (for example, $\{(p, T_0), (q, F_0), (r, T_0)\}$). Therefore, by the definition of $\bigsqcap_0$ we get that $(\bigsqcap_0 X)(q) = F_0$ and therefore $(\bigsqcap X)(q) = F_0$. Notice now that there exists at least one model in X where p is equal to T_1 (for example, $\{(p, T_1), (q, F_0), (r, T_0)\}$). By the definition of $\bigsqcap_0$ we get that $(\bigsqcap_0 X)(p) = T_1$. Notice now that in every model it must be $v_p \geq \sim v_q$ and since $(\bigsqcap X)(q) = F_0$, we get that $(\bigsqcap X)(p) = T_1$. Consider now the sets X_α used in the construction of $\bigsqcap X$ (see Section 2). It is easy to see that for all $\alpha \geq 0$ and $0 \leq v \leq T_\alpha$, X_α has models with $v_r = v$. This implies that $(\bigsqcap X)(r) = 0$. In conclusion, $\bigsqcap X = \{(p, T_1), (q, F_0), (r, 0)\}$. $\quad\square$

Theorem 4 generalizes the well-known *model intersection theorem* for logic programs without negation (see for example [8]). To our knowledge, this is the first completely general model intersection theorem for normal logic programs. A weaker result of this form was reported in [10, Theorem 8.6], in which X was the set of *all* the models of a program (and not an arbitrary subset of them).

We close this section by discussing the fact that for the case of logic programs, we actually chose κ to be equal to Ω and the set of truth values to be indexed by ordinals $\alpha < \Omega$. It is natural to wonder what would be the effect of choosing an alternative value of κ and correspondingly, a different set V_κ of truth values that would be indexed by ordinals $\alpha < \kappa$. Due to lack of space, our explanations below are given at an intuitive level. First notice that due to the semantics of the $\sim$ operator, κ must always be a limit ordinal. This is because if $\kappa = \beta + 1$ for some β, then the value $\sim T_\beta$ will not belong to V_κ. Assume therefore that κ is a limit ordinal. Now, intuitively, if we choose a small κ then we "lose precision" while if we choose a very large κ, then we "don't gain any extra precision". More formally, if we choose a small κ, then in the minimum model of a program certain variables would be forced to have the value 0 (due to the lack of truth values of larger order). On the other hand, if we choose $\kappa > \Omega$, then the extra truth values will not contribute anything to the minimum model (which will only use values of order less than Ω); this is due to the fact that, by definition, our programs consist of a countable number of rules, and therefore a countable number of truth values suffices in order to assign to them the correct meaning.

5   An Application to Non-Monotonic Formal Grammars

In this section we demonstrate that we can derive a model intersection theorem for Boolean grammars [9], a relatively recent extension of context-free grammars with conjunction and complementation. The semantics of Boolean grammars was developed in [7] using a three-valued logic. In the rest of this section we demonstrate that it can also be defined using an infinite-valued logic similar to that of the previous section.

Definition 15 ([9]). *A Boolean grammar is a quadruple $G = (\Sigma, N, P, S)$, where Σ and N are disjoint finite non-empty sets of terminal and non-terminal symbols respectively, P is a finite set of rules, each of the form*

$$A \to u_1 \& \cdots \& u_m \& \neg v_1 \& \cdots \& \neg v_n \qquad (m + n \geq 1, u_i, v_j \in (\Sigma \cup N)^*),$$

and $S \in N$ is the start symbol of the grammar. We will call the non-terminal A the head *of the rule, the u_i's* positive conjuncts *and the $\neg v_j$'s* negative ones.

As it was demonstrated in [7], in order to construct the three-valued semantics of a Boolean grammar, it suffices to iterate an appropriate operator for ω times (and not Ω times as is the case for logic programs with negation). Therefore, the set of truth values now becomes: $V = \{F_\alpha : \alpha < \omega\} \cup \{T_\alpha : \alpha < \omega\} \cup \{0\}$ with the obvious ordering:

$$F_0 < F_1 < \cdots < 0 < \cdots < T_1 < T_0$$

We can now define the notion of an infinite-valued language:

Definition 16. *Let Σ be a finite non-empty set of symbols. Then, an infinite-valued language over Σ is a function from Σ^* to the set V.*

The classical notion of concatenation also needs to be generalized:

Definition 17. *Let Σ be a finite non-empty set of symbols and let $L_1, \ldots, L_n$ be infinite-valued languages over Σ. We define the* infinite-valued concatenation *of the languages $L_1, \ldots, L_n$ to be the language L such that:*

$$L(w) = \max_{\substack{(w_1, \ldots, w_n): \\ w = w_1 \cdots w_n}} \left(\min_{1 \leq i \leq n} L_i(w_i) \right)$$

for every $w \in \Sigma^*$.

Similarly to logic programs, we can define infinite-valued interpretations for Boolean grammars:

Definition 18. *An* infinite-valued *interpretation I of a Boolean grammar $G = (\Sigma, N, P, S)$ is a function $I : (N \times \Sigma^*) \to V$.*

Notice that an interpretation may also be viewed as an assignment of an infinite-valued language to each nonterminal.

An interpretation I can be recursively extended to apply to expressions that appear in the right-hand sides of Boolean grammar rules:

Definition 19. *Let $G = (\Sigma, N, P, S)$ be a Boolean grammar and let I be an interpretation of G. Then, I can be extended as follows:*

- *For every $w \in \Sigma^*$, it is $I(\epsilon, w) = T_0$ if $w = \epsilon$, and $I(\epsilon, w) = F_0$ otherwise.*
- *Let $a \in \Sigma$. Then, for every $w \in \Sigma^*$, it is $I(a, w) = T_0$ if $w = a$, and $I(a, w) = F_0$ otherwise.*
- *Let $u = u_1 \cdots u_n$, $n \geq 2$, $u_i \in \Sigma \cup N$. Then, for every $w \in \Sigma^*$, it is*

$$I(u, w) = \max_{\substack{(w_1, \ldots, w_n): \\ w = w_1 \cdots w_n}} \left(\min_{1 \leq i \leq n} I(u_i, w_i) \right)$$

- *Let $u \in (\Sigma \cup N)^*$. Then, for every $w \in \Sigma^*$, it is:*

$$I(\neg u, w) = \begin{cases} T_{\alpha+1} & \text{if } I(u, w) = F_\alpha \\ F_{\alpha+1} & \text{if } I(u, w) = T_\alpha \\ 0 & \text{if } I(u, w) = 0 \end{cases}$$

- *Let $l_1, \ldots, l_n$ be conjuncts. Then, for every $w \in \Sigma^*$, it is $I(l_1 \& \cdots \& l_n, w) = \min\{I(l_1, w), \ldots, I(l_n, w)\}$.*

We are now in a position to define the notion of a model of a Boolean grammar:

Definition 20. *Let $G = (\Sigma, N, P, S)$ be a Boolean grammar and I an interpretation of G. Then, I is a model of G if for every rule $A \to l_1 \& \cdots \& l_n$ in P and for every $w \in \Sigma^*$, it is $I(A, w) \geq I(l_1 \& \cdots \& l_n, w)$.*

Definition 21. *Let $G = (\Sigma, N, P, S)$ be a Boolean grammar, I an interpretation of G and $v \in V$. We define $I \parallel v = \{(A, w) \in N \times \Sigma^* \mid I(A, w) = v\}$.*

The definitions of $\leq$, $=_\alpha$ and $\sqsubseteq_\alpha$ are then identical to those given in Definitions 12, 13 and 14. The following lemma can then be established in an entirely analogous way as the corresponding one for logic programs (Lemma 4):

Lemma 5. *The set of infinite-valued interpretations $\mathcal{I}_G$ of a Boolean grammar G is a symmetric strong model.*

The immediate consequence operator for Boolean grammars can be defined analogously to that for logic programs:

$$T_G(I)(A, w) = \bigvee \{I(l_1 \& \cdots \& l_n, w) \mid (A \leftarrow l_1 \& \cdots \& l_n) \in P\}$$

By Corollary 7.8 of [6] it follows that T_G is α-monotonic for all $\alpha < \omega$. Moreover, it is straightforward to check that an interpretation I is a model of a grammar G iff $T_G(I) \leq I$. Based on the above, the following theorem follows directly using Corollary 1:

Theorem 5. *Let G be a Boolean grammar and let X be a set of infinite-valued models of G. Then, $T_G(\bigsqcap X) \leq \bigsqcap X$, ie., $\bigsqcap X$ is also an infinite-valued model of G.*

Example 2. Consider the boolean grammar G given by the following rules:

$$S \to \neg A$$
$$A \to \epsilon$$
$$A \to aAb$$

Let $U = \{a^n b^n \mid n \geq 0\}$. Moreover, let X be the set of all infinite-valued models of G. We calculate $\bigsqcap X$, ie., the minimum infinite-valued model of G. By the definition of model of a grammar (Definition 20) and from the syntax of the rules of the grammar, we get that every model M of G has the property that for all $w \in U$, $M(A, w) = T_0$. Notice now that there exists a model such that for every $w \notin U$, $M(A, w) = F_0$. Therefore, by the definition of $\bigsqcap_0$ we have $(\bigsqcap_0 X)(A, w) = F_0$ for all $w \notin U$ and consequently $(\bigsqcap X)(A, w) = F_0$ for all $w \notin U$. Consider now the set X_1 used in the construction of $\bigsqcap X$ (see Section 2). Notice that there exists a model $M_1 \in X_1$ such that $M_1(S, w) = F_1$, for all $w \in U$. Therefore, by the definition of $\bigsqcap_1$, we have $(\bigsqcap_1 X_1)(S, w) = F_1$ and therefore $(\bigsqcap X)(S, w) = F_1$, for all $w \in U$. Similar arguments give $(\bigsqcap X)(S, w) = T_1$, for all $w \notin U$. In conclusion, we have:

$$(\bigsqcap X)(S, w) = \begin{cases} F_1 \text{ if } w \in U \\ T_1 \text{ if } w \notin U \end{cases} \qquad (\bigsqcap X)(A, w) = \begin{cases} T_0 \text{ if } w \in U \\ F_0 \text{ if } w \notin U \end{cases}$$

which is the minimum infinite-valued model for the given grammar. $\square$

Theorem 5 provides a new, infinite-valued characterization of the semantics of Boolean grammars. Moreover, it is the first (to our knowledge) model intersection theorem given for this class of formal grammars.

6   Conclusions

We have derived two novel pre-fixed point theorems for non-monotonic functions and have demonstrated that they have direct applications in the areas of logic programming and formal language theory. In this way we have made a further step towards the development of an abstract fixed point theory for non-monotonicity, which we believe will have other interesting applications in the future. We are currently investigating such alternative possibilities.

References

1. Krzysztof, R.: Apt. Logic programming. In: Handbook of Theoretical Computer Science. Formal Models and Semantics, vol. B, pp. 493–574 (1990)
2. Charalambidis, A., Ésik, Z., Rondogiannis, P.: Minimum model semantics for extensional higher-order logic programming with negation. In: International Conference on Logic Programming (to appear, 2014)
3. Denecker, M., Marek, V.W., Truszczynski, M.: Approximations, stable operators, well-founded fixpoints and applications in nonmonotonic reasoning. In: Minker, J. (ed.) Logic-Based Artificial Intelligence, pp. 127–144. Kluwer Academic Publishers (2000)
4. Denecker, M., Marek, V.W., Truszczynski, M.: Ultimate approximation and its application in nonmonotonic knowledge representation systems. Inf. Comput. 192(1), 84–121 (2004)
5. Ésik, Z., Rondogiannis, P.: A fixed point theorem for non-monotonic functions. In: Proceedings of 13th Panhellenic Logic Symposium, Athens, Greece (2013)
6. Ésik, Z., Rondogiannis, P.: A fixed point theorem for non-monotonic functions. CoRR, abs/1402.0299 (2014)
7. Kountouriotis, V., Nomikos, C., Rondogiannis, P.: Well-founded semantics for boolean grammars. Information and Computation 207(9), 945–967 (2009)
8. Lloyd, J.W.: Foundations of Logic Programming. Springer (1987)
9. Okhotin, A.: Boolean grammars. Information and Computation 194(1), 19–48 (2004)
10. Rondogiannis, P., Wadge, W.W.: Minimum model semantics for logic programs with negation-as-failure. ACM Transactions on Computational Logic 6(2), 441–467 (2005)

Common Knowledge Semantics
of Armstrong's Axioms

Zachary Heckle and Pavel Naumov

Department of Mathematics and Computer Science
McDaniel College, Westminster, Maryland 21157, USA
{zbh001,pnaumov}@mcdaniel.edu

Abstract. Armstrong's axioms were originally proposed to describe functional dependency between sets of attributes in relational databases. The database semantics of these axioms can be easily rephrased in terms of distributed knowledge in multi-agent systems. The paper proposes alternative semantics of the same axioms in terms of common knowledge. The main technical result of this work is soundness and completeness of Armstrong's axioms with respect to the proposed semantics. An important implication of this result is an unexpected duality between notions of distributed and common knowledge.

1 Introduction

1.1 Armstrong's Axioms

For any two variables a and b, we say that a functionally determines b if for each possible value of a there is a unique value of b. We denote this by $a \rhd b$. For example, the length of a side of an equilateral triangle functionally determines the area of the triangle: $length \rhd area$.

Similarly, one can define functional dependency between two *sets* of variables. For example, two legs of a right triangle uniquely determine its hypotenuse and area:

$$leg_1, leg_2 \rhd hypotenuse, area.$$

The functional dependency relation has been first studied in the context of database theory, where functional dependency is defined between two sets of attributes. Armstrong [1] proposed the following axiomatization of this relation:

1. *Reflexivity*: $A \rhd B$, if $A \supseteq B$,
2. *Augmentation*: $A \rhd B \to A, C \rhd B, C$,
3. *Transitivity*: $A \rhd B \to (B \rhd C \to A \rhd C)$,

where here and everywhere below A, B denotes the union of sets A and B. He proved soundness and completeness of this logical system with respect to a database semantics. The above axioms became known in database literature as Armstrong's axioms [2, p. 81]. Beeri, Fagin, and Howard [3] suggested a variation of Armstrong's axioms that describes properties of multi-valued dependence.

U. Kohlenbach et al. (Eds.): WoLLIC 2014, LNCS 8652, pp. 181–194, 2014.
© Springer-Verlag Berlin Heidelberg 2014

Armstrong's axioms also describe properties of functional dependency in settings different from database theory. More and Naumov [4] investigated functional dependency between secrets shared over a network with a fixed topology. They presented a sound and complete axiomatization of this type of functional dependency consisting of Armstrong's axioms and one additional Gateway axiom that captures properties specific to the topology of the network. Harjes and Naumov [5] considered functional dependency between strategies of players in a Nash equilibrium of a strategic game. They gave a sound a complete axiomatization of this relation for games with a fixed dependency graph of a pay-off function. Their axiomatization also consists of Armstrong's axioms and one additional Contiguity axiom that captures properties specific to the topology of the graph. In another work, they axiomatized functional dependency between single strategies in Nash equilibria of cellular games [6]. Instead of considering secrets shared over a network with a fixed topology, one can consider a fixed group of symmetries of such a network. The complete axiomatization of functional dependency in such a setting [7] also consists of Armstrong's axioms and two additional axioms specific to the group of symmetries.

A logical system that simultaneously describes properties of functional dependency between single variables and properties of the nondeducibility relation [8] has been proposed earlier [9]. A different type of dependency in strategic games has been studied by Naumov and Nicholls [10]. They called it *rationally* functional dependence. The axioms of rationally functional dependence are significantly different from Armstrong's axioms discussed in this paper.

1.2   Distributed Knowledge

In the original Armstrong setting, in the functional dependency predicate $A \triangleright B$, sets A and B are sets of database attributes. Let us assume now that each of the attributes in a database is known to a specific distinct agent and only to this agent. Furthermore, suppose that each of these agents knows nothing else but the value of the corresponding attribute. Under these assumptions, we can informally identify attributes with the agents that know them. Thus, relation $A \triangleright B$ can now be viewed as a relation between sets of agents. Two sets of agents are in this relation if agents in set A collectively know all what is known to each agent in set B. In other words, everything distributively known to agents in set B is also distributively known to agents in set A. To paraphrase it once again: distributed knowledge of a set of agents B is a subset of distributed knowledge of a set of agents A:

$$DK(A) \supseteq DK(B), \tag{1}$$

where $DK(A)$ informally represents distributed knowledge of set of agents A. Later in this paper, we formally specify the meaning of statement (1) and claim soundness and completeness of Armstrong's axioms with respect to semantics of distributed knowledge. The proofs are given in the appendix. These proofs are, essentially, translations of Armstrong's [1] arguments from database language to Kripke semantics language. The main focus of this paper is on *common* knowledge.

1.3 Common Knowledge

Let $CK(A)$, informally, denote all common knowledge [11] of set of agents A. By analogy with relation (1), one can consider relation

$$CK(A) \supseteq CK(B)$$

between sets of agents A and B. This relation does not satisfy Armstrong's axioms since the Reflexivity axiom does not hold (common knowledge of a subgroup, generally speaking, is not a common knowledge of a group). However, it turns out that relation

$$CK(A) \subseteq CK(B) \tag{2}$$

does satisfy Armstrong's axioms. Furthermore, the main technical result of this paper is the completeness theorem for Armstrong's axioms with respect to common knowledge semantics informally specified by relation (2).

The significant implication of this result is the duality between distributed knowledge and common knowledge captured by relations (1) and (2). Properties of both of them are described by Armstrong's axioms.

In the rest of the paper we first further discuss and formally define relation (1) in terms of epistemic Kripke frames. We then show how this definition can be modified to formally specify relation (2). We conclude the paper with the proof of soundness and completeness of Armstrong's axioms with respect to common knowledge semantics. In the appendix we show soundness and completeness of Armstrong's axioms with respect to distributed knowledge semantics.

2   Distributed Knowledge Semantics

Definition 1. *A Kripke frame is a triple* $(W, \mathcal{A}, \{\sim_a\}_{a \in \mathcal{A}})$, *where*

1. W *is a nonempty set of "epistemic worlds",*
2. $\mathcal{A}$ *is a set of "agents",*
3. $\sim_a$ *is an ("indistinguishability") equivalence relation on set W for each $a \in \mathcal{A}$.*

For any two epistemic worlds $u, v \in W$ and any set of agents $A \subseteq \mathcal{A}$, we write $u \sim_A v$ if $u \sim_a v$ for each $a \in A$.

According to the standard Kripke semantics of distributed knowledge [12, p. 24], $u \Vdash \Box_B p$ means that $v \Vdash p$ for each $v \in W$ such that $u \sim_B v$. If we want statement $u \Vdash \Box_B p \rightarrow \Box_A p$ to be true no matter how propositional variable p is evaluated over the given Kripke frame, we need to require that

$$\{v \in W \mid u \sim_A v\} \subseteq \{v \in W \mid u \sim_B v\}. \tag{3}$$

Epistemic logic usually studies the validity of formulas in a particular epistemic world. In Armstrong's database semantics, however, statement $A \triangleright B$ means that the values of the attributes in set B are functionally determined by the values of attributes in set A for *all possible values* of the attributes in set A. Thus, under the corresponding distributed knowledge semantics, statement $A \rhd B$ should mean that (3) is true for each $u \in W$. In other words, informal statement (1) could be formally specified as

$$\forall u, v \in W (u \sim_A v \to u \sim_B v). \tag{4}$$

We use this specification in Definition 3.

Definition 2. *For any set of "agents" $\mathcal{A}$, by $\Phi(\mathcal{A})$ we mean the minimal set of formulas such that*

1. $\perp \in \Phi(\mathcal{A})$,
2. $A \rhd B \in \Phi(\mathcal{A})$ *for all finite subsets $A, B \subseteq \mathcal{A}$,*
3. *if $\varphi, \psi \in \Phi(\mathcal{A})$, then $\varphi \to \psi \in \Phi(\mathcal{A})$.*

Definition 3. *For any Kripke frame $K = (W, \mathcal{A}, \{\sim_a\}_{a \in \mathcal{A}})$ and any $\varphi \in \Phi(\mathcal{A})$, we define relation $K \vDash \varphi$ recursively:*

1. $K \nvDash \perp$,
2. $K \vDash A \rhd B$ *iff for each $u, v \in W$, if $u \sim_A v$, then $u \sim_B v$,*
3. $K \vDash \varphi \to \psi$ *iff $K \nvDash \varphi$ or $K \vDash \psi$.*

Theorem 1 (Armstrong [1]). *$K \vDash \varphi$ for each Kripke frame K whose set of agents contains all agents from formula φ if and only if formula φ is provable from Armstrong's axioms and propositional tautologies using the Modus Ponens inference rule.*

This theorem has been originally proven by Armstrong for database semantics, but, as we show in the appendix, his proof could be easily adopted to Kripke frames.

3   Common Knowledge Semantics

As usual, common knowledge of p between a group of agents A means that each agent knows p, each agent knows that each agent knows p, and so on ad infinitum. In epistemic logic notations [13,12], in epistemic world u there is a common knowledge of p between a group of agents A if for each sequence $a_1, \ldots, a_m$ of elements of A, possibly with repetitions,

$$u \Vdash \square_{a_1} \square_{a_2} \ldots \square_{a_m} p.$$

Thus, in a given epistemic world u there is a common knowledge of p by the group of agents A if $v_m \Vdash p$ for each sequence $a_1, \ldots, a_m$ of elements of A and each sequence of worlds $v_0, v_1, \ldots, v_m \in W$ such that

$$u = v_0 \sim_{a_1} v_1 \sim_{a_2} v_2 \sim_{a_3} \cdots \sim_{a_m} v_m.$$

If we want common knowledge of p by group A in epistemic world u to imply common knowledge of p by group B in u no matter how propositional variable p is evaluated over the given Kripke frame, we need to require that

$$\{w_k \in W \mid u = w_0 \sim_{b_1} w_1 \sim_{b_2} w_2 \sim_{b_3} \cdots \sim_{b_n} w_n \text{ and } b_1, \ldots, b_n \in B\} \subseteq$$
$$\{v_n \in W \mid u = v_0 \sim_{a_1} v_1 \sim_{a_2} v_2 \sim_{a_3} \cdots \sim_{a_m} v_m \text{ and } a_1, \ldots, a_m \in A\}.$$

Hence, for common knowledge by group A to imply common knowledge by group B in *each* epistemic world u we need to require that for each $x, y \in W$, if there exist $n \geq 0$, $w_0, \ldots, w_n \in W$, and $b_1, \ldots, b_n \in B$ such that

$$x = w_0 \sim_{b_1} w_1 \sim_{b_2} w_2 \sim_{b_3} \cdots \sim_{b_n} w_n = y,$$

then there must exist $m \geq 0$, $v_0, \ldots, v_m \in W$, and $a_1, \ldots, a_m \in A$ such that

$$x = v_0 \sim_{a_1} v_1 \sim_{a_2} v_2 \sim_{a_3} \cdots \sim_{a_m} v_m = y.$$

In the definition below, we take this requirement as the formalization of (2).

Definition 4. *Let $\vDash$ be the relation between a Kripke frame $K = (W, \mathcal{A}, \{\sim_a\}_{a \in \mathcal{A}})$ and a propositional formula in $\Phi(\mathcal{A})$ such that:*

1. *$K \nvDash \bot$,*
2. *$K \vDash A \triangleright B$ iff for each $x, y \in W$, if there exist $n \geq 0$, $w_0, \ldots, w_n \in W$, and $b_1, \ldots, b_n \in B$ such that*

$$x = w_0 \sim_{b_1} w_1 \sim_{b_2} w_2 \sim_{b_3} \cdots \sim_{b_n} w_n = y,$$

then there exist $m \geq 0$, $v_0, \ldots, v_m \in W$, $a_1, \ldots, a_m \in A$ such that

$$x = v_0 \sim_{a_1} v_1 \sim_{a_2} v_2 \sim_{a_3} \cdots \sim_{a_m} v_m = y,$$

3. *$K \vDash \varphi \rightarrow \psi$ iff $K \nvDash \varphi$ or $K \vDash \psi$.*

4   Axioms

For any given set of agents $\mathcal{A}$, our logical system consists of all propositional tautologies in language $\Phi(\mathcal{A})$, the Modus Ponens inference rule, and Armstrong's axioms:

1. *Reflexivity:* $A \triangleright B$, if $A \supseteq B$,
2. *Transitivity:* $A \triangleright B \rightarrow (B \triangleright C \rightarrow A \triangleright C)$,
3. *Augmentation:* $A \triangleright B \rightarrow A, C \triangleright B, C$,

where, as we have mentioned earlier, A, B stands for the union of sets A and B. We write $X \vdash \varphi$ if statement φ is provable in our logical system using additional set of axioms X. We abbreviate $\varnothing \vdash \varphi$ as $\vdash \varphi$.

5   Example

The soundness of Armstrong's axioms with respect to common knowledge semantics will be shown in the next section. Note that soundness of the Reflexivity and Transitivity axioms is intuitively clear, but soundness of the Augmentation axiom is, perhaps, unexpected. Below we illustrate our logical system by stating and proving from Armstrong's axioms an even less intuitively clear property of common knowledge:

Theorem 2. $\vdash A \triangleright B \to (C \triangleright D \to A, C \triangleright B, D)$.

Proof. Suppose that $A \triangleright B$ and $C \triangleright D$. Thus, by the Augmentation axiom, $A, C \triangleright B, C$ and $B, C \triangleright B, D$. Therefore, by the Transitivity axiom, $A, C \triangleright B, D$.   □

6   Soundness

In this section we prove soundness of our logical system with respect to common knowledge semantics. Soundness of propositional tautologies and the Modus Ponens inference rule is straightforward. We prove soundness of each of Armstrong's axioms as a separate lemma.

Lemma 1. $K \vDash A \triangleright B$ *for each* $K = (W, \mathcal{A}, \{\sim_a\}_{a \in \mathcal{A}})$ *and each* $B \subseteq A \subseteq \mathcal{A}$.

Proof. Consider any $x, y \in W$. Let there exist $n \geq 0$, $w_0, \ldots, w_n \in W$, and $b_1, \ldots,$
$b_n \in B$ such that

$$x = w_0 \sim_{b_1} w_1 \sim_{b_2} w_2 \sim_{b_3} \cdots \sim_{b_n} w_n = y.$$

Note that $b_1, \ldots, b_n \in A$ because $B \subseteq A$.   □

Lemma 2. *For each* $K = (W, \mathcal{A}, \{\sim_a\}_{a \in \mathcal{A}})$ *and each* $A, B, C \subseteq \mathcal{A}$, *if* $K \vDash A \triangleright B$ *and* $K \vDash B \triangleright C$, *then* $K \vDash A \triangleright C$.

Proof. Consider any $x, y \in W$. Let there exist $n \geq 0$, $w_0, \ldots, w_n \in W$, and $c_1, \ldots,$
$c_n \in C$ such that

$$x = w_0 \sim_{c_1} w_1 \sim_{c_2} w_2 \sim_{c_3} \cdots \sim_{c_n} w_n = y.$$

Thus, by assumption $K \vDash B \triangleright C$, there exist $m \geq 0$, $v_0, \ldots, v_m \in W$, and $b_1, \ldots, b_m \in C$ such that

$$x = v_0 \sim_{b_1} v_1 \sim_{b_2} v_2 \sim_{b_3} \cdots \sim_{b_m} v_m = y.$$

Hence, by assumption $K \vDash A \triangleright B$, there exist $k \geq 0$, $u_0, \ldots, u_k \in W$, and $a_1, \ldots, a_k \in A$ such that

$$x = u_0 \sim_{a_1} u_1 \sim_{a_2} u_2 \sim_{a_3} \cdots \sim_{a_k} u_k = y.$$

□

Lemma 3. *For each* $K = (W, \mathcal{A}, \{\sim_a\}_{a \in \mathcal{A}})$ *and each* $A, B, C \subseteq \mathcal{A}$, *if* $K \vDash A \rhd B$, *then* $K \vDash A, C \rhd B, C$.

Proof. Consider any $x, y \in W$. Let there exist $n \geq 0$, $w_0, \ldots, w_n \in W$, and $e_1, \ldots,$
$e_n \in B \cup C$ such that

$$x = w_0 \sim_{e_1} w_1 \sim_{e_2} w_2 \sim_{e_3} \cdots \sim_{e_{n-1}} w_{n-1} \sim_{e_n} w_n = y.$$

We will show that there exist $m \geq 0$, $v_0, \ldots, v_m \in W$, and $f_1, \ldots, f_m \in A \cup C$ such that

$$x = v_0 \sim_{f_1} v_1 \sim_{f_2} v_2 \sim_{f_3} \cdots \sim_{f_m} v_m = y$$

by induction on n. If $n = 0$, then $x = y$ and $m = 0$.

Let $n > 0$. By the induction hypothesis, there exist $k \geq 0$, $u_0, \ldots, u_k \in W$, and $g_1, \ldots, g_k \in A \cup C$ such that

$$x = u_0 \sim_{g_1} u_1 \sim_{g_2} u_2 \sim_{g_3} \cdots \sim_{g_k} u_k = w_{n-1}.$$

Case I: $e_n \in C$. Then,

$$x = u_0 \sim_{g_1} u_1 \sim_{g_2} \cdots \sim_{g_k} u_k = w_{n-1} \sim_{e_n} w_n = y$$

and $g_1, g_2, \ldots, g_k, e_n \in A \cup C$.

Case II: $e_n \in B$. By assumption $K \vDash A \rhd B$, there exist $\ell \geq 0$, $t_0, t_1, \ldots, t_\ell \in W$, and $h_0, h_1, \ldots, h_\ell \in A$ such that

$$w_{k-1} = t_0 \sim_{h_1} t_1 \sim_{h_2} \cdots \sim_{h_\ell} t_\ell = w_n.$$

Therefore,

$$x = u_0 \sim_{g_1} u_1 \sim_{g_2} u_2 \sim_{g_3} \cdots \sim_{g_k} u_k = w_{n-1} = t_0 \sim_{h_1} t_1 \sim_{h_2} \cdots \sim_{h_\ell} t_\ell = w_n = y,$$

where $g_1, g_2, \ldots, g_k, h_1, h_2, \ldots, h_\ell \in A \cup C$. $\qquad\square$

7   Two-World Kripke Frames

In this section we define a simple two-world Kripke frame. Later, multiple instances of such frames will be combined together to prove completeness of Armstrong's axioms with respect to common knowledge semantics.

Definition 5. *For any set of agents* $\mathcal{A}$ *and any subset* $D \subseteq \mathcal{A}$, *let* $K(\mathcal{A}, D)$ *be the Kripke frame* $(W, \mathcal{A}, \{\sim_a\}_{a \in \mathcal{A}})$ *such that*

1. W *is the two-element set* $\{w_0, w_1\}$,
2. $w_0 \sim_a w_1$ *if and only if* $a \notin D$.

Informally, D is the set of all "distinguishers" who can distinguish world w_0 from world w_1.

Lemma 4. *For any set of agents $\mathcal{A}$ and any subset $D \subseteq \mathcal{A}$, $K(\mathcal{A}, D) \vDash A \triangleright B$ if and only if at least one of the following conditions is satisfied:*

1. $A \nsubseteq D$,
2. $B \subseteq D$.

Proof. ($\Rightarrow$) Suppose $K \vDash A \triangleright B$ as well as $A \subseteq D$ and $B \nsubseteq D$. Since $B \nsubseteq D$, there exists $b_0 \in B$ such that $b_0 \notin D$. Thus, $w_0 \sim_{b_0} w_1$, by Definition 5. Hence, by the assumption $K \vDash A \triangleright B$, there exist $n \geq 0$, $v_0, \ldots, v_n \in W$, and $a_1, \ldots, a_n \in A$ such that

$$w_0 = v_0 \sim_{a_1} v_1 \sim_{a_2} \cdots \sim_{a_{n-1}} v_{n-1} \sim_{a_n} v_n = w_1,$$

which is a contradiction to $A \subseteq D$ and Definition 5.

($\Leftarrow$) First, assume that $A \nsubseteq D$. Thus, there exists $a_0 \in A$ such that $a_0 \notin D$. Hence, by Definition 5, $x \sim_{a_0} y$ for each $x, y \in W$. Thus, $K(\mathcal{A}, D) \vDash A \triangleright B$.

Next, suppose $B \subseteq D$. To prove $K(\mathcal{A}, D) \vDash A \triangleright B$, consider any $x, y \in W$. Let $n \geq 0$, $v_0, \ldots, v_n \in W$, and $b_1, \ldots, b_n \in B$ be such that

$$x = v_0 \sim_{b_1} v_1 \sim_{b_2} \cdots \sim_{b_{n-1}} v_{n-1} \sim_{b_n} v_n = y.$$

Thus, $x = y$ by the assumption $B \subseteq D$ and Definition 5. $\square$

8   Product of Kripke Frames

In this section we define a composition operation on Kripke frames and prove a fundamental property of this operation. Later we use this operation to combine several different two-world frames, defined in the previous section, into a single Kripke frame needed to prove completeness of Armstrong's axioms with respect to common knowledge semantics.

Definition 6. *For any set of agents $\mathcal{A}$ and any family of Kripke frames $\{K^i\}_{i=0}^n = \{(W^i, \mathcal{A}, \{\sim_a^i\}_{a \in \mathcal{A}})\}_{i=0}^n$ we define the product $\prod_{i=0}^n K^i$ to be the Kripke frame $K = (W, \mathcal{A}, \{\sim_a\}_{a \in \mathcal{A}})$, where*

1. *W is the Cartesian product $\prod_{i=0}^n W^i$ of the sets of epistemic words of individual frames,*
2. *for any $\langle u_i \rangle_{i \leq n}, \langle v_i \rangle_{i \leq n} \in W$, let $\langle u_i \rangle_{i \leq n} \sim_a \langle v_i \rangle_{i \leq n}$ if $u_i \sim_a^i v_i$ for each $i \leq n$.*

Theorem 3. *Let $\mathcal{A}$ be any set of agents. If A and B are any two finite subsets of $\mathcal{A}$ and $\{K^i\}_{i=1}^n$ is any family of Kripke frames with set of agents $\mathcal{A}$, then $\prod_{i=1}^n K^i \Vdash A \triangleright B$ if and only if $K^i \Vdash A \triangleright B$ for each $i \leq n$.*

Proof. Let $\{K^i\}_{i=1}^n = \{(W^i, \mathcal{A}, \{\sim_a^i\}_{a \in \mathcal{A}})\}_{i=1}^n$.

($\Rightarrow$) Assume $i_0 \leq n$ and $x, y \in W^{i_0}$ are such that there exist $k \geq 0$, $v_0, \ldots, v_k \in W$, and $b_1, \ldots, b_k \in B$ such that

$$x = v_0 \sim_{b_1} v_1 \sim_{b_2} v_2 \sim_{b_3} \cdots \sim_{b_k} v_k = y.$$

We will show that there exist $m \geq 0$, $u_0, \ldots, u_m \in W$, and $a_1, \ldots, a_m \in A$ such that

$$x = u_0 \sim_{a_1} u_1 \sim_{a_2} u_2 \sim_{a_3} \cdots \sim_{a_m} u_m = y.$$

Indeed, due to Definition 1, for each $i \leq n$ there is at least one epistemic world $w^i \in W^i$. Then,

$$\langle w_1, \ldots, w_{i_0-1}, v_0, w_{i_0+1}, \ldots, w_n \rangle \sim_{b_1} \langle w_1, \ldots, w_{i_0-1}, v_1, w_{i_0+1}, \ldots, w_n \rangle \sim_{b_2}$$
$$\langle w_1, \ldots, w_{i_0-1}, v_2, w_{i_0+1}, \ldots, w_n \rangle \sim_{b_3} \cdots \sim_{b_k} \langle w_1, \ldots, w_{i_0-1}, v_k, w_{i_0+1}, \ldots, w_n \rangle$$

due to Definition 6 and reflexivity of relations $\{\sim_{b_i}\}_{i=1}^k$. Hence, by the assumption of the theorem, there exist $m \geq 0$, $\langle z_0^1, z_0^2, \ldots, z_0^n \rangle$, $\langle z_1^1, z_1^2, \ldots, z_1^n \rangle$, $\ldots, \langle z_m^1, z_m^2, \ldots, z_m^n \rangle$ in $\prod_{i=1}^n W^i$, and $a_1, \ldots, a_m$ in A such that

$$\langle w_1, \ldots, w_{i_0-1}, v_0, w_{i_0+1}, \ldots, w_n \rangle = \langle z_0^1, z_0^2, \ldots, z_0^n \rangle \sim_{a_1} \langle z_1^1, z_1^2, \ldots, z_1^n \rangle \sim_{a_2}$$
$$\cdots \sim_{a_m} \langle z_m^1, z_m^2, \ldots, z_m^n \rangle = \langle w_1, \ldots, w_{i_0-1}, v_k, w_{i_0+1}, \ldots, w_n \rangle.$$

Therefore, by Definition 6,

$$x = v_0 = z_0^{i_0} \sim_{a_1} z_1^{i_0} \sim_{a_2} z_2^{i_0} \sim_{a_3} \cdots \sim_{a_m} z_m^{i_0} = v_k = y.$$

($\Leftarrow$) Consider any $X, Y \in \prod_{i=1}^n W^i$. Suppose there exist $m \geq 0$, $b_1, \ldots, b_m \in B$, and $\langle w_0^1, \ldots, w_0^n \rangle$, $\langle w_1^1, \ldots, w_1^n \rangle$, $\ldots$, $\langle w_m^1, \ldots, w_m^n \rangle$ in $\prod_{i=1}^n W^i$ such that

$$X = \langle w_0^1, \ldots, w_0^n \rangle \sim_{b_1} \langle w_1^1, \ldots, w_1^n \rangle \sim_{b_2}$$
$$\langle w_2^1, \ldots, w_2^n \rangle \sim_{b_3} \cdots \sim_{b_m} \langle w_m^1, \ldots, w_m^n \rangle = Y.$$

Thus, by Definition 6, for each $i \leq n$,

$$w_0^i \sim_{b_1} w_1^i \sim_{b_2} \cdots \sim_{b_m} w_m^i.$$

Hence, by the assumption of the theorem, for each $i \leq n$ there exist $k^i \geq 0$, $u_0^i, \ldots, u_{k^i}^i \in W^i$, and $a_1^i, a_2^i, \ldots, a_{k^i}^i \in A$ such that

$$w_0^i = u_0^i \sim_{a_1^i} u_1^i \sim_{a_2^i} u_2^i \sim_{a_3^i} \cdots \sim_{a_{k^i}^i} u_{k^i}^i = w_m^i.$$

Therefore, by Definition 6,

$$X = \langle w_0^1, w_0^2, w_0^3, \ldots, w_0^{n-1}, w_0^n \rangle = \langle u_0^1, w_0^2, w_0^3, \ldots, w_0^{n-1}, w_0^n \rangle \sim_{a_1^1}$$
$$\langle u_1^1, w_0^2, w_0^3, \ldots, w_0^{n-1}, w_0^n \rangle \sim_{a_2^1} \langle u_2^1, w_0^2, w_0^3, \ldots, w_0^{n-1}, w_0^n \rangle \sim_{a_3^1} \cdots \sim_{a_{k^1}^1}$$
$$\langle u_{k^1}^1, w_0^2, w_0^3, \ldots, w_0^{n-1}, w_0^n \rangle = \langle w_m^1, w_0^2, w_0^3, \ldots, w_0^{n-1}, w_0^n \rangle =$$
$$\langle w_m^1, u_0^2, w_0^3, \ldots, w_0^{n-1}, w_0^n \rangle \sim_{a_1^2} \langle w_m^1, u_1^2, w_0^3, \ldots, w_0^{n-1}, w_0^n \rangle \sim_{a_2^2}$$
$$\langle w_m^1, u_2^2, w_0^3, \ldots, w_0^{n-1}, w_0^n \rangle \sim_{a_3^2} \cdots \sim_{a_{k^2}^2} \langle w_m^1, u_{k^2}^2, w_0^3, \ldots, w_0^{n-1}, w_0^n \rangle =$$
$$\langle w_m^1, w_m^2, w_0^3, \ldots, w_0^{n-1}, w_0^n \rangle = \cdots = \langle w_m^1, w_m^2, w_m^3, \ldots, w_m^{n-1}, w_0^n \rangle =$$
$$\langle w_m^1, w_m^2, w_m^3, \ldots, w_m^{n-1}, u_0^n \rangle \sim_{a_1^n} \langle w_m^1, w_m^2, w_m^3, \ldots, w_m^{n-1}, u_1^n \rangle \sim_{a_2^n}$$
$$\langle w_m^1, w_m^2, w_m^3, \ldots, w_m^{n-1}, u_2^n \rangle \sim_{a_3^n} \cdots \sim_{a_{k^n}^n} \langle w_m^1, w_m^2, w_m^3, \ldots, w_m^{n-1}, u_{k^1}^n \rangle =$$
$$\langle w_m^1, w_m^2, w_m^3, \ldots, w_m^{n-1}, w_m^n \rangle = Y.$$

$\square$

9   Star Closure

In this section we introduce a technical notion of A^* closure of a set of agents A and prove basic properties of this notion. The closure is used in the next section to prove the completeness theorem.

Let $\mathcal{A}$ be any finite set of agents and M be any fixed subset of $\Phi(\mathcal{A})$.

Definition 7. *For any subset $A \subseteq \mathcal{A}$, let A^* be the set*

$$\{a \in \mathcal{A} \mid M \vdash A \triangleright a\}.$$

Set A^* is finite due the assumption that set $\mathcal{A}$ is finite.

Lemma 5. $A \subseteq A^*$, *for each $A \subseteq \mathcal{A}$.*

Proof. Let $a \in A$. By the Reflexivity axiom, $\vdash A \triangleright a$. Hence, $a \in A^*$. $\square$

Lemma 6. $M \vdash A \triangleright A^*$, *for each $A \subseteq \mathcal{A}$.*

Proof. Let $A^* = \{a_1, \ldots, a_n\}$. By the definition of A^*, $M \vdash A \triangleright a_i$, for each $i \leq n$. We will prove, by induction on k, that $M \vdash (A \triangleright a_1, \ldots, a_k)$ for each $0 \leq k \leq n$.
Base Case: $M \vdash A \triangleright \varnothing$ by the Reflexivity axiom.
Induction Step: Assume that $M \vdash (A \triangleright a_1, \ldots, a_k)$. By the Augmentation axiom,

$$M \vdash A, a_{k+1} \triangleright a_1, \ldots, a_k, a_{k+1}. \tag{5}$$

Recall that $M \vdash A \triangleright a_{k+1}$. Again by the Augmentation axiom, $M \vdash (A \triangleright A, a_{k+1})$. Hence, $M \vdash (A \triangleright a_1, \ldots, a_k, a_{k+1})$, by (5) and the Transitivity axiom. $\square$

10   Completeness

We are now ready to prove completeness of Armstrong's axioms with respect to common knowledge semantics.

Theorem 4. *If $K \vDash \varphi$ for each Kripke frame K whose set of agents contains all agents in formula φ, then $\vdash \varphi$.*

Proof. Suppose $\nvdash \varphi$. Let $\mathcal{A}$ be the finite set of all agents mentioned in formula φ and M be a maximal consistent subset of $\Phi(\mathcal{A})$ containing formula $\neg\varphi$.

Definition 8. *Let Kripke frame K be $\prod_{A \subseteq \mathcal{A}} K(\mathcal{A}, A^*)$.*

Lemma 7. $M \vdash B \triangleright C$ *if and only if $K \vDash B \triangleright C$, for all subsets B and C of $\mathcal{A}$.*

Proof. ($\Rightarrow$) First, suppose that $M \vdash B \triangleright C$ and $\prod_{A \subseteq \mathcal{A}} K(\mathcal{A}, A^*) \nvDash B \triangleright C$. Thus, by Theorem 3, there exists $A_0 \subseteq \mathcal{A}$ such that $K(\mathcal{A}, A_0^*) \nvDash B \triangleright C$. Hence, by Lemma 4, $B \subseteq A_0^*$ and

$$C \nsubseteq A_0^*. \tag{6}$$

Then, by the Reflexivity axiom, $\vdash A_0^* \triangleright B$. By assumption $M \vdash B \triangleright C$ and the Transitivity axiom, $M \vdash A_0^* \triangleright C$. By Lemma 6, $M \vdash A_0 \triangleright A_0^*$. Thus, by the

Transitivity axiom, $X \vdash A_0 \triangleright C$. By the Reflexivity axiom, $\vdash C \triangleright c$ for all $c \in C$. Hence, by the Transitivity axiom, $M \vdash A_0 \triangleright c$ for all $c \in C$. Then, $c \in A_0^*$ for all $c \in C$. Thus, $C \subseteq A_0^*$, which is a contradiction to (6).

($\Leftarrow$) Next, suppose $\prod_{A \subseteq \mathcal{A}} K(\mathcal{A}, A^*) \vDash B \triangleright C$. Then, $K(\mathcal{A}, B^*) \vDash B \triangleright C$. Hence, by Lemma 4, either $B \nsubseteq B^*$ or $C \subseteq B^*$. The former is not possible due to Lemma 5. Thus, $C \subseteq B^*$. Hence, by the Reflexivity axiom, $\vdash B^* \triangleright C$. Note that $M \vdash B \triangleright B^*$, by Lemma 6. Therefore, by the Transitivity axiom, $M \vdash B \triangleright C$. □

Lemma 8. $\psi \in M$ if and only if $K \vDash \psi$ for each $\psi \in \Phi(\mathcal{A})$.

Proof. Induction on the structural complexity of formula ψ. The base case follows from Lemma 7. The induction step follows from Definition 4 as well as maximally and consistency of set M in the standard way.   □

Note that $K \nvDash \psi$ due to assumption $\neg \varphi \in M$, Lemma 8, and consistency of set M. This concludes the proof of Theorem 4.   □

11   Conclusion

In this paper we proposed common knowledge semantics for Armstrong's axioms and proved corresponding soundness and completeness theorems. This result shows that relations (1) and (2) have the same logical properties and, thus, demonstrates a certain duality between common knowledge and distributed knowledge.

A possible extension of our work could be developing a logical system that deals with relations (1) and (2) at the same time. Another possible extension of this work is to consider common knowledge on hypergraphs in the same way that it has been done [4] for distributed knowledge.

A   Appendix: Distributed Knowledge Semantics

In this appendix we prove soundness and completeness of Armstrong's axioms with respect to the distributed knowledge semantics specified by Definition 3. Thus, everywhere in this section $\vDash$ refers to the relation from Definition 3 and not the one from Definition 4. The main result of this section is the completeness proof. In the presentation of this proof we follow the general outline of our completeness proof with respect to common knowledge semantics. In particular, we reuse earlier defined notions of two-world Kripke frames, product of Kripke frames, and star closure.

A.1   Soundness

In this section we prove soundness of our logical system with respect to distributed knowledge semantics. Soundness of propositional tautologies and the Modus Ponens inference rule is straightforward. We prove soundness of each of Armstrong's axioms as a separate lemma.

Lemma 9. $K \vDash A \rhd B$ *for each* $K = (W, \mathcal{A}, \{\sim_a\}_{a \in \mathcal{A}})$ *and each* $B \subseteq A \subseteq \mathcal{A}$.

Proof. Consider any $x, y \in W$. Suppose that $x \sim_A y$. Therefore, $x \sim_B y$ due to assumption $B \subseteq A$. □

Lemma 10. *For each* $K = (W, \mathcal{A}, \{\sim_a\}_{a \in \mathcal{A}})$ *and each* $A, B, C \subseteq \mathcal{A}$, *if* $K \vDash A \rhd B$ *and* $K \vDash B \rhd C$, *then* $K \vDash A \rhd C$.

Proof. Consider any $x, y \in W$. Suppose that $x \sim_A y$. Hence, $x \sim_B y$ by assumption $K \vDash A \rhd B$. Thus, $x \sim_C y$ by assumption $K \vDash B \rhd C$. □

Lemma 11. *For each* $K = (W, \mathcal{A}, \{\sim_a\}_{a \in \mathcal{A}})$ *and each* $A, B, C \subseteq \mathcal{A}$, *if* $K \vDash A \rhd B$, *then* $K \vDash A, C \rhd B, C$.

Proof. Consider any $x, y \in W$. Suppose that $x \sim_{A,C} y$. Thus, $x \sim_A y$ and $x \sim_C y$. Hence, $x \sim_B y$ by assumption $K \vDash A \rhd B$. Therefore, $x \sim_{B,C} y$. □

A.2   Completeness

We start with the distributed knowledge version of Lemma 4.

Lemma 12. *For any set of agents* $\mathcal{A}$ *and any subset* $D \subseteq \mathcal{A}$, $K(\mathcal{A}, D) \vDash A \rhd B$ *if and only if at least one of the following conditions is satisfied:*

1. $A \cap D \neq \varnothing$,
2. $B \cap D = \varnothing$.

Proof. ($\Rightarrow$) Suppose that $A \cap D = \varnothing$ and $B \cap D \neq \varnothing$. The former, by Definition 5, implies that $w_0 \sim_A w_1$, where w_0 and w_1 are the two worlds of Kripke frame $K(\mathcal{A}, D)$. The latter implies that there exists $b_0 \in B \cap D$. Thus, $w_0 \sim_B w_1$ due to assumption $K(\mathcal{A}, D) \vDash A \rhd B$. Hence, $w_0 \sim_{b_0} w_1$ because $b_0 \in B$, which is a contradiction to Definition 5 since $b_0 \in D$.

($\Leftarrow$) First, suppose that $A \cap D \neq \varnothing$. Thus, there exists $d_0 \in A \cap D$. To show that $K(\mathcal{A}, D) \vDash A \rhd B$, consider any $x, y \in \{w_0, w_1\}$ and assume that $x \sim_A y$. We will show that $x \sim_B y$. Indeed, $x \sim_A y$ implies that $x \sim_{d_0} y$ since $d_0 \in A$. Thus, $x = y$ by Definition 5 and assumption $d_0 \in D$. Therefore, $x \sim_B y$ due to reflexivity of relation $\sim_B$.

Second, assume that $B \cap D = \varnothing$. Thus, $x \sim_B y$ for each $x, y \in \{w_0, w_1\}$ due to Definition 5. Therefore, $K(\mathcal{A}, D) \vDash A \rhd B$, by Definition 3. □

Next is the distributed knowledge version of Theorem 3.

Theorem 5. *Let* $\mathcal{A}$ *be any set of agents. If* A *and* B *are any two finite subsets of* $\mathcal{A}$, *and* $\{K^i\}_{i=1}^n$ *is any family of Kripke frames with set of agents* $\mathcal{A}$, *then* $\prod_{i=1}^n K^i \Vdash A \rhd B$ *if and only if* $K^i \Vdash A \rhd B$ *for each* $i \leq n$.

Proof. ($\Rightarrow$) Consider any $i_0 \leq n$ and any $x, y \in W^{i_0}$ such that $x \sim_A y$. We will show that $x \sim_B y$. By Definition 1, for each $i \leq n$ there is at least one epistemic world $w^i \in W^i$. Note that

$$\langle w_1, w_2, \ldots, w_{i_0-1}, x, w_{i_0+1}, \ldots, w_n \rangle \sim_A$$
$$\langle w_1, w_2, \ldots, w_{i_0-1}, y, w_{i_0+1}, \ldots, w_n \rangle$$

due to Definition 6, assumption $x \sim_A y$, and reflexivity of relation $\sim_A$. Hence,

$$\langle w_1, w_2, \ldots, w_{i_0-1}, x, w_{i_0+1}, \ldots, w_n \rangle \sim_B$$
$$\langle w_1, w_2, \ldots, w_{i_0-1}, y, w_{i_0+1}, \ldots, w_n \rangle$$

by assumption $\prod_{i=1}^n K^i \Vdash A \rhd B$. Therefore, $x \sim_B y$ by Definition 6.

($\Leftarrow$) Consider any tuples $\langle x_1, x_2, \ldots, x_n \rangle$ and $\langle y_1, y_2, \ldots, y_n \rangle$ in $\prod_{i=1}^n W^i$ such that

$$\langle x_1, x_2, \ldots, x_n \rangle \sim_A \langle y_1, y_2, \ldots, y_n \rangle.$$

By Definition 6, $x_i \sim_A y_i$ for each $i \leq n$. Hence, $x_i \sim_B y_i$ for each $i \leq n$, due to the assumption of the theorem. Therefore, $\langle x_1, x_2, \ldots, x_n \rangle \sim_B \langle y_1, y_2, \ldots, y_n \rangle$.
$\square$

We are now ready to prove completeness of Armstrong's axioms with respect to distributed knowledge semantics. This result has been earlier claimed as a part of Theorem 1.

Theorem 6. *If $K \vDash \varphi$ for each Kripke frame K whose set of agents contains all agents in formula φ, then $\vdash \varphi$.*

Proof. Suppose $\nvdash \varphi$. Let $\mathcal{A}$ be the finite set of all agents mentioned in formula φ and M be a maximal consistent subset of $\Phi(\mathcal{A})$ containing formula $\neg\varphi$.

Definition 9. *Let Kripke frame K be*

$$\prod_{A \subseteq \mathcal{A}} K(\mathcal{A}, \mathcal{A} \setminus A^*).$$

Lemma 13. *$M \vdash B \rhd C$ if and only if $K \vDash B \rhd C$, for all finite subsets B and C of $\mathcal{A}$.*

Proof. ($\Rightarrow$) Suppose that $M \vdash B \rhd C$ and

$$\prod_{A \subseteq \mathcal{A}} K(\mathcal{A}, \mathcal{A} \setminus B^*) \nvDash B \rhd C.$$

Thus, by Theorem 5, there exists $A_0 \subseteq \mathcal{A}$ such that $K(\mathcal{A}, \mathcal{A} \setminus A_0^*) \nvDash B \rhd C$. Hence, by Lemma 12,

$$B \cap (\mathcal{A} \setminus A_0^*) = \varnothing$$

and

$$C \cap (\mathcal{A} \setminus A_0^*) \neq \varnothing.$$

In other words, $B \subseteq A_0^*$ and

$$C \nsubseteq A_0^*. \tag{7}$$

Then, by the Reflexivity axiom, $\vdash A_0^* \rhd B$. By assumption $M \vdash B \rhd C$ and the Transitivity axiom, $M \vdash A_0^* \rhd C$. By Lemma 6, $M \vdash A_0 \rhd A_0^*$. Thus, by the Transitivity axiom, $M \vdash A_0 \rhd C$. By the Reflexivity axiom, $\vdash C \rhd c$ for all $c \in C$.

Hence, by the Transitivity axiom, $M \vdash A_0 \rhd c$ for all $c \in C$. Then, $c \in A_0^*$ for all $c \in C$. Thus, $C \subseteq A_0^*$, which is a contradiction to (7).

($\Leftarrow$) Suppose $\prod_{A \subseteq \mathcal{A}} K(\mathcal{A}, \mathcal{A} \setminus A^*) \vDash B \rhd C$. Then, $K(\mathcal{A}, \mathcal{A} \setminus B^*) \vDash B \rhd C$, by Theorem 5. Hence, by Lemma 12, either $B \cap (\mathcal{A} \setminus B^*) \neq \varnothing$ or $C \cap (\mathcal{A} \setminus B^*) = \varnothing$. In other words, either $B \nsubseteq B^*$ or $C \subseteq B^*$. The former is not possible due to Lemma 5. Thus, $C \subseteq B^*$. Hence, by the Reflexivity axiom, $\vdash B^* \rhd C$. Note that $M \vdash B \rhd B^*$ by Lemma 6. Therefore, by the Transitivity axiom, $M \vdash B \rhd C$. $\square$

Lemma 14. $\psi \in M$ *if and only if* $K \vDash \psi$ *for each* $\psi \in \Phi(\mathcal{A})$.

Proof. Induction on the structural complexity of formula ψ. The base case follows from Lemma 13. The induction step follows from the maximally and consistency of set M in the standard way. $\square$

Note that $K \nvDash \psi$ due to assumption $\neg\varphi \in M$, Lemma 14, and consistency of set M. This concludes the proof of Theorem 6. $\square$

References

1. Armstrong, W.W.: Dependency structures of data base relationships. In: Proc. IFIP Congress on Information Processing 1974, Stockholm, pp. 580–583. North-Holland, Amsterdam (1974)
2. Garcia-Molina, H., Ullman, J., Widom, J.: Database Systems: The Complete Book, 2nd edn. Prentice-Hall (2009)
3. Beeri, C., Fagin, R., Howard, J.H.: A complete axiomatization for functional and multivalued dependencies in database relations. In: SIGMOD 1977: Proceedings of the 1977 ACM SIGMOD International Conference on Management of Data, pp. 47–61. ACM, New York (1977)
4. More, S.M., Naumov, P.: The functional dependence relation on hypergraphs of secrets. In: Leite, J., Torroni, P., Ågotnes, T., Boella, G., van der Torre, L. (eds.) CLIMA XII 2011. LNCS, vol. 6814, pp. 29–40. Springer, Heidelberg (2011)
5. Harjes, K., Naumov, P.: Functional dependence in strategic games. In: 1st International Workshop on Strategic Reasoning, Rome, Italy. Electronic Proceedings in Theoretical Computer Science 112, pp. 9–15 (March 2013)
6. Harjes, K., Naumov, P.: Cellular games, nash equilibria, and Fibonacci numbers. In: Grossi, D., Roy, O., Huang, H. (eds.) LORI. LNCS, vol. 8196, pp. 149–161. Springer, Heidelberg (2013)
7. Kane, J., Naumov, P.: Symmetry in information flow. Annals of Pure and Applied Logic 165(1), 253–265 (2014)
8. Sutherland, D.: A model of information. In: Proceedings of Ninth National Computer Security Conference, pp. 175–183 (1986)
9. Kelvey, R., Miner More, S., Naumov, P., Sapp, B.: Independence and functional dependence relations on secrets. In: Proceedings of 12th International Conference on the Principles of Knowledge Representation and Reasoning, Toronto, pp. 528–533. AAAI (2010)
10. Naumov, P., Nicholls, B.: Rationally functional dependence. In: 10th Conference on Logic and the Foundations of Game and Decision Theory, LOFT (2012)
11. Lewis, D.: Convention: a philosophical study. Harvard University Press (1969)
12. Fagin, R., Halpern, J.Y., Moses, Y., Vardi, M.Y.: Reasoning about knowledge. MIT Press, Cambridge (1995)
13. Halpern, J.Y., Moses, Y.: Knowledge and common knowledge in a distributed environment. J. ACM 37(3), 549–587 (1990)

On Distributed Stochastic Logics for Mobile Systems

Juliana Küster Filipe Bowles[1] and Petrucio Viana[2]

[1] School of Computer Science, University of St Andrews
Jack Cole Building, North Haugh, St Andrews KY16 9SX, UK
jkfb@st-andrews.ac.uk

[2] Departamento de Análise, Instituto de Matemática e Estatística
Universidade Federal Fluminense, Rua Mario Santos Braga, s/n
Campus do Valonguinho, Niterói, RJ, 24020-140, Brazil
petrucio@cos.ufrj.br

Abstract. We introduce a family of logics for expressing and reasoning about stochastic properties over mobile distributed systems. More specifically, to reason about state properties, action sequences, location-based properties and differentiate between local and global object-based behaviour, we define an $ASMC^+$ extending an action and state-based Markov chain with object and location labels. We introduce a *distributed stochastic logic* DSL interpreted over $ASMC^+$s. For an $ASMC^+$ model M, we define quotient structures equivalent to M which induce sublogics over DSL. The logics include a *global stochastic logic* GSL and *local stochastic logics* LSL for locations in the system. In general, the logics can be used to capture different quantitative dependability properties for distributed mobile systems. We point out that the interpretation of the sublogics over equivalence preserving state space aggregations brings considerable advantages for verification.

1 Introduction

Service-oriented, cloud and mobile computing is changing the way applications are developed, as services and resources become more commonly distributed and accessed over the internet. Consequently there is an increasing need for dedicated programming and specification formalisms that address aspects such as (code and object) mobility, remote execution, dynamic distribution of resources over a network, security, reliability and performance. Ideally such high-level modelling languages allow us to select the best design for a system, and rank alternatives on the basis of particular metrics for non-functional properties of interest. For performance common metrics include response time, throughput and utilisation. If specification formalisms allow the statement of required dependability and performance measures (the later for example are increasingly used for service-level agreements), efficient and reliable computation of these measures should then become common practice for distributed mobile application development.

There are a number of languages and frameworks that aim to address some of the above challenges including ULM [1], KLAIM [2], StoKLAIM [3], and various process algebra-based approaches including stochastic π-calculus [4], PEPA nets [5], and the calculus of mobile ambients [6]. High-level performance languages such as PEPA nets

U. Kohlenbach et al. (Eds.): WoLLIC 2014, LNCS 8652, pp. 195–205, 2014.
© Springer-Verlag Berlin Heidelberg 2014

and StoKLAIM support the modelling of distributed mobile systems with notions of location, context and evaluation environment. As such it is possible to clearly distinguish computation at one of the locations from the movement of (code-containing) objects from one location to another. Mobile objects behave in accordance with their location and what other objects are available for interaction. In such languages actions have a random duration governed by a negative exponential distribution. The resulting operational model is a Continuous-Time Markov Chain (CTMC), one of the most popular models used for performance and dependability evaluation.

We introduce a family of stochastic logics suitable for expressing properties over mobile distributed systems. We use CTMCs with actions and state labels (ASMCs) as defined in [7] adapted to a distributed mobile system context. We extend the model with object and location labels which we call $ASMC^+$. We first introduce DSL, a *distributed stochastic logic* interpreted over $ASMC^+$. To separate global from local property specification, we define as sublogics a *global stochastic logic* GSL and a family of *local stochastic logics* LSL for each location in the system. The sublogics share a semantics at the level of the $ASMC^+$ but can be interpreted directly over quotient structures defined over the original model and reflecting the intended separation. The interpretation of the sublogics over equivalence preserving state space aggregations constitutes a favourable approach for verification and is well known. GSL expresses global properties of the system such as how an object moves between locations, global reachability properties, and how many resources there are at equilibrium in one or more locations of the system. For each location l, LSL_l describes individual properties of location l in the system such as available resources, resource constraints, and local object interaction.

Many (temporal) logics have been proposed to express stochastic behaviour (e.g., [8,7] among many others) and separately system properties related to mobility (e.g., [9]). However, to the best of our knowledge there is only one probabilistic logic for mobility called MOSL and additional variant MOSL+[3]. Stochastic logics such as CSL [8] are not able to capture distribution and mobility properties directly, nor express both state and action sequences. The logic asCSL [7] is an extension of CSL which allows the specification of execution paths of Markov chains with actions and state labels. In asCSL, path formulae are defined as regular expressions over an alphabet combining state formulae and an action set. MOSL is an action-based stochastic extension of state-based logic MOMO [9], whereby MOSL+ is both action and state-based.

The main logic we propose, the distributed stochastic logic DSL, is action and state-based and can be used to describe global/local properties of distributed systems taking into account system structure, locations and object mobility. Similarly to MOSL it describes path formulae using a bounded until operator equipped with action sets. The logics DSL and MOSL+ have a comparable expressiveness, but differ in many crucial ways. Syntactically MOSL+ is very closely linked to the specification language StoK-LAIM [3] it was developed for which severely compromises its usability elsewhere. MOSL operators reappear in SoSL to express properties over MarCaSPiS, a service-oriented calculus developed within the SENSORIA project [10]. Our logic does not make any assumption on the high-level formalism used and can be used in combination with PEPA net specifications or other suitable modelling approaches. An advantage of our approach is that we exploit quotient structures defined over an $ASMC^+$ to represent different abstractions of the distributed mobile system. A high-level or global view of the system is used directly by the global stochastic logic GSL and its semantics. Additional location and object-based views can be constructed for localised behaviour specification. The distributed nature and location awareness of a StoKLAIM specification is lost in the actual verification process as MOSL+ flattens the underlying model and hence loses explicit knowledge of location and structure in the underlying Markov-based model. It is well known that retaining and exploiting the structure of the model can bring considerable advantages for verification and lead to more efficient analysis of large system models by reducing overall required state space.

Paper structure: Section 2 defines an ASMC$^+$ as an extension of ASMCs with object and location labels. In Section 3 we introduce the syntax and semantics of a distributed stochastic logic DSL for specifying properties over distributed mobile systems and show some examples of its expressive power. In Section 4 we explore different quotient structures over an ASMC$^+$ and how these correspond to sublogics of DSL. Section 5 concludes the paper.

2   Extending ASMCs for Mobile Systems

We consider a mobile system as a network of distributed locations, whereby the objects in the system can move between locations and interact with other objects provided they are at the same location when they interact. Objects move by performing a *global* action, and interact with other objects by performing a shared *local* action. Synchronisation on global actions is not allowed. Let $\mathcal{L}$ be a finite set of *locations*, $\mathcal{O}$ be a finite set of *object types*, and $\mathcal{I}_o$ be a finite set of *object instances* of type $o \in \mathcal{O}$. We write $\mathcal{I} = \bigcup_{o \in \mathcal{O}} \mathcal{I}_o$ for the overall set of instances (pairwise disjoint). Let $Act = Act_G \uplus Act_L$ be a finite set of actions consisting of global and local actions respectively where $Act_G \cap Act_L = \emptyset$. In addition $Act_L = \bigcup_{l \in \mathcal{L}} Act_L(l)$ where $Act_L(l)$ refers to the local actions in location l, where we assume here that $Act_L(l_1) \cap Act_L(l_2) = \emptyset$ for any two $l_1 \neq l_2 \in \mathcal{L}$. We also write $Act_L(N)$ for the local actions in region $N \subseteq \mathcal{L}$, and $Act_{L,o}$ for the local actions of object type $o \in \mathcal{O}$.

The model underlying the distributed mobile system can be represented using an action and state-based Markov chain (ASMC) [7] extended here to be defined over a set of object types $\mathcal{O}$, set of locations $\mathcal{L}$, and set of instances $\mathcal{I}_o$ of object type $o \in \mathcal{O}$. We refer to our extension as an ASMC$^+$. We assume $\mathcal{O}$, $\mathcal{L}$, and $\mathcal{I}$ fixed throughout.

Definition 1. *An ASMC$^+$ is a tuple $\mathcal{M} = (S, Act, AP, F, H, R)$ where:*

- *S is a finite set of states with initial state s_0,*
- *Act is a finite set of action labels,*
- *AP is a finite set of atomic propositions,*
- *$F : S \times \mathcal{I}_o \rightarrow 2^{AP}$ is a labelling function which assigns to each state $s \in S$ and object instance $i \in \mathcal{I}_o$ the set $F(s, i)$ of atomic propositions that are assumed to be valid in s for instance i.*
- *$H : S \times \mathcal{I}_o \rightarrow \mathcal{L}$ is a labelling function which assigns to each state $s \in S$ and object instance $i \in \mathcal{I}_o$ a location $H(s, i) \in \mathcal{L}$ where i is located at state s.*
- *$R : S \times Act \times S \rightarrow \mathbb{R}_{\geq 0}$ is a rate matrix.*

In Definition 1, we extend the usual labelling function that assigns to each state $s \in S$ a set of atomic propositions that are (assumed) valid in s. Instead we refer in addition to a specific object instance, and consider two functions F and H, where F associates to each (s, i) pair the set of atomic propositions valid in that state for instance i, and H specifies the location where i is residing at state s. It is possible for object instances to remain in the same location throughout and for such instances H is a constant. For a state s, the current number of instances of type o at a location l, can be inferred as follows. Let $C_o^s(l) \subseteq \mathcal{I}_o$ be the subset of instances of type $o \in \mathcal{O}$ at location l in state s, i.e., $C_o^s(l) = \{c \in \mathcal{I}_o \mid H(s, c) = l\}$. The number of instances of type o in state s at location l is given by $|C_o^s(l)|$. We make use of a function $max : \mathcal{L} \times \mathcal{O} \to \mathbb{N}$ which returns the maximum number of instances of object type o possible at the location at any moment in time. Similarly, we can consider a min function to denote the minimum number of required instances of type o at a location.

A state change occurs when one object instance performs an action either on its own or with another object. According to the rate matrix, if $R(s, a, s') = \lambda > 0$ then there is a transition labelled with action a from state s to state s' whose delay is given by an exponential distribution with rate λ. More precisely the meaning of $R(s, a, s') = \lambda > 0$ is that the probability that transition $s \xrightarrow{a} s'$ is enabled within the next t time units from the current state s is given by $1 - e^{-\lambda \cdot t}$. If more than one state s' can be reached from s with rate greater than zero then there is a race between the outgoing transitions from s.

The underlying continuous-time Markov chain (CTMC) of an ASMC$^+$ is straighthforward and defined as the tuple (S, AP, F, R') that arises from $\mathcal{M}$ by removing the action set and accumulating the rates of all transitions between the same two states in S, i.e., $R'(s, s') = \sum_{a \in Act} R(s, a, s')$ (cf. [7]).

Finite and infinite paths in an ASMC$^+$ are as expected and correspond to finite and infinite words over $(S \times Act \times \mathbb{R}_{>0})$. A finite path σ is written as a sequence of transitions $\sigma = s_0 \xrightarrow{a_0, t_0} s_1 \xrightarrow{a_1, t_1} \dots \xrightarrow{a_{n-1}, t_{n-1}} s_n$ where t_i is the sojourn time in state s_i. We denote $|\sigma|$ the number of transitions in path σ and $\sigma[i] = S_i$, the ith state of σ. Moreover, we denote by $T(\sigma) = \sum_{j=0}^{n-1} t_j$ the execution time of σ, and $\sigma(i, j)$ is the part of the path starting at state s_i and ending at state s_j. $Path_{fin}^{\mathcal{M}}$ denotes the set of all finite paths in $\mathcal{M}$, and $Path_w^{\mathcal{M}}$ denotes the set of infinite paths in $\mathcal{M}$. Moreover, by $Path_{fin}^{\mathcal{M}}(s_i)$ and $Path_w^{\mathcal{M}}(s_i)$ we denote the set of all finite and infinite paths in $\mathcal{M}$ with initial state s_i.

3   The Logic DSL

The (stochastic) properties of interest for mobile distributed systems include quantitative logical statements on their structure, resource distribution, object mobility as well as local behaviour of objects and their interactions at given system locations. This section describes a logic suitable for expressing stochastic properties over mobile distributed systems.

Stochastic logics such as CSL [8], aCSL [11] and asCSL [7] are interpreted over Markov chains, action-labelled Markov chains, and action and state-labelled Markov chains respectively. The underlying model used depends on the intention of the logic. With an action-based logic such as aCSL, for instance, we want to specify properties such as *"there is at least a 90% chance that action a occurs within 2 time units"*. In aCSL, we can only make statements about actions and not about the state the system may be when something occurs, as this logic does not have atomic propositions to describe states. The logic asCSL is action and state-based, and are thus more expressive enabling both kinds of properties to be captured (over actions and over states). The distributed stochastic logic DSL proposed here is also action and state-based.

In our setting, we distinguish between *locations* ($\mathcal{L}$) and other atomic propositions (AP), as well as between *local* (Act_L) and *global* (Act_G) actions. In addition, we want to express properties over objects and their local interactions. In line with our object-based extension of ASMCs we define the grammar DSL as a location and object-based extension of action and state-based CSL. Note that for the moment we do not differentiate between local and global actions and behaviour. We come back to that later on.

We define the distributed stochastic logic DSL over sets $\mathcal{L}$, $\mathcal{O}$, Act and AP, which we may write as $DSL[\mathcal{L}, \mathcal{O}, Act, AP]$ or just highlighting specific sets if the others are clear from context.

Syntax. The symbols of DSL are:

Object types	: $o \in \mathcal{O}$
Instances of an object type	: $i, j \in \mathcal{I}_o$
Locations	: $l \in \mathcal{L}$
Actions	: $a \in Act$
Atomic propositions	: $ap \in AP$
Probability values	: $p \in [0, 1]$
Comparisons	: $\bowtie \in \{<, \leq, >, \geq\}$
Time interval	: $I = [t, t'] \subseteq \mathbb{R}_0^+$
Sets of actions	: $A, B \subseteq Act$

The formulae of DSL are defined by the grammar, where Φ is a state formula, Ψ is a location formula, and φ is a path formula:

$$\Phi ::= \top \mid \Phi \wedge \Phi \mid \neg \Phi \mid \mathcal{S}_{\bowtie p}(\Phi) \mid \mathcal{P}_{\bowtie p}(\varphi) \mid l : \Psi$$
$$\Psi ::= ap \mid i.ap \mid i \leftrightarrow j.a \mid l : o[\,] \mid o[-] \mid o[i]$$
$$\varphi ::= \Phi_*\mathcal{U}_B^I \Phi \mid \Phi_A \mathcal{U}_B^I \Phi \mid \Phi_A \mathcal{U}_*^I \Phi$$

The intended meanings of state formulae are:

$\top$	:	true
$\wedge$	:	conjunction
$\neg$	:	negation
$\mathcal{S}_{\bowtie p}(\Phi)$	:	steady state operator, i.e., the probability of being in a Φ-state in the long run obeys to the bound $\bowtie p$
$\mathcal{P}_{\bowtie p}(\varphi)$	:	probabilistic operator, i.e., the probability measure of all infinite paths satisfying φ obeys to the bound $\bowtie p$
$l : \Psi$	:	location l satisfies property Ψ

More specifically:

$l : ap$: ap holds in location l
$l : i.ap$: in location l object i satisfies ap
$l : i \leftrightarrow j.a$: in location l objects i and j are synchronised on shared action a
$l : o[\,]$: location l has an available space for an object of type o
$l : o[-]$: an arbitrary object of type o is currently at location l
$l : o[i]$: instance i of type o is currently at location l

Finally, the intended meanings of path formulae are:

$\Phi_1 {}_* \mathcal{U}^I_B \Phi_2$: eventually, within time bound I, by performing an action in B, we reach a state that satisfies Φ_2, and until then all states satisfy Φ_1

$\Phi_1 {}_A \mathcal{U}^I_B \Phi_2$: eventually, within time bound I, by performing an action in B, we reach a state that satisfies Φ_2, and until then all states satisfy Φ_1 and only actions in A can occur

$\Phi_1 {}_A \mathcal{U}^I_* \Phi_2$: eventually, within time bound I, we reach a state that satisfies Φ_2 and until then all states satisfy Φ_1 and only actions in A can occur

In addition, we can extend location specific properties over regions $N \subseteq \mathcal{L}$ introducing shorthand notation:

$$\langle N \rangle : \Psi \equiv l_1 : \Psi \vee \cdots \vee l_n : \Psi$$

and

$$[N] : \Psi \equiv l_1 : \Psi \wedge \cdots \wedge l_n : \Psi$$

The intended meanings of these defined operators are easy to grasp: $\langle N \rangle : \Psi$ means that Ψ holds *somewhere* in the region N of the system, and $[N] : \Psi$ means that Ψ holds *everywhere* in the region N of the system.

Semantics over an ASMC$^+$. We define the semantics of DSL by means of a satisfaction relation $\models$ over the states of an ASMC$^+$ $\mathcal{M}$ w.r.t. object types $\mathcal{O}$ and locations $\mathcal{L}$ (cf. Definition 1).

The relation "state formula Φ is satisfied (holds) at a point s of a ASMC$^+$ $\mathcal{M}$", denoted $\mathcal{M}, s \models \Phi$, is defined as follows:

- $\mathcal{M}, s \models \top$
- $\mathcal{M}, s \models \Phi_1 \wedge \Phi_2$ iff $\mathcal{M}, s \models \Phi_1$ and $\mathcal{M}, s \models \Phi_2$
- $\mathcal{M}, s \models \neg \Phi$ iff $\mathcal{M}, s \models \Phi$ does not hold
- $\mathcal{M}, s \models l : ap$ iff $ap \in AP$ and there is an $i \in \mathcal{I}_o$ with $ap \in F(s, i)$ and $H(s, i) = l$
- $\mathcal{M}, s \models l : i.ap$ iff $ap \in F(s, i)$, $i \in \mathcal{I}_o$, $ap \in AP$, and $H(s, i) = l$
- $\mathcal{M}, s \models l : i \leftrightarrow j.a$ iff there is a path in $\mathcal{M}$ containing s such that there is a transition labelled with $a \in Act$ leaving s, a is a shared action between i and j and $H(s, i) = H(s, j) = l$
- $\mathcal{M}, s \models l : o[\,]$ iff there is available space for an instance of type o in l, i.e., $|C^s_o(l)| < max(l, o)$
- $\mathcal{M}, s \models l : o[-]$ iff an instance of type o is currently in l, i.e., there exists an $i \in \mathcal{I}_o$ with $H(s, i) = l$

- $\mathcal{M}, s \models l : o[i]$ iff $H(s, i) = l$ and $i \in \mathcal{I}_o$
- $\mathcal{M}, s \models S_{\bowtie p}(\Phi)$ iff $\Pi^{\mathcal{M}}(s, Sat_{\mathcal{M}}(\Phi)) \bowtie p$
- $\mathcal{M}, s \models \mathcal{P}_{\bowtie p}(\varphi)$ iff $Prob^{\mathcal{M}}(s, \varphi) \bowtie p$

Above, $Sat_{\mathcal{M}}(\Phi) = \{s \in S \mid s \models \Phi\}$, and $\Pi^{\mathcal{M}}(s, S')$ denotes the steady-state probability to be in a state of set S' when starting in s. It is defined in terms of a probability measure $Pr^{\mathcal{M}}$ on the set of infinite paths η with initial state s. Formally, $\Pi^{\mathcal{M}}(s, S') = \lim_{t \to \infty} Pr^{\mathcal{M}}\{\eta \in Path_w^{\mathcal{M}}(s) \mid \eta@t \in S'\}$. In addition, $Prob^{\mathcal{M}}(s, \varphi)$ is the probability measure of all infinite paths starting in s satisfying φ, i.e., $Prob^{\mathcal{M}}(s, \varphi) = Pr^{\mathcal{M}}\{\eta \in Path_w^{\mathcal{M}}(s) \mid \eta \models \varphi\}$.

Finally, the relation "path formula φ is satisfied (holds) at an infinite path η of a ASMC$^+$ $\mathcal{M}$", denoted $\mathcal{M}, \eta \models \varphi$, is defined as follows:

- $\mathcal{M}, \eta \models \Phi_1 \,_* \mathcal{U}_B^I \, \Phi_2$ iff there is a finite prefix σ of η, with initial state s_n and end state s_j, such that for $n \leq p \leq j - 1$, $s_p \models \Phi_1$ and $s_j \models \Phi_2$, and s_j was reached by the occurrence of an action in B
- $\mathcal{M}, \eta \models \Phi_1 \,_A \mathcal{U}_B^I \, \Phi_2$ iff there is a finite prefix σ of η, with initial state s_n and end state s_j, such that for $n \leq p \leq j - 1$, $s_p \models \Phi_1$ and $s_j \models \Phi_2$, and only actions in A occur up to s_{j-1} and s_j was reached by the occurrence of an action in B
- $\mathcal{M}, \eta \models \Phi_1 \,_A \mathcal{U}_*^I \, \Phi_2$ holds iff there is a finite prefix σ of η, with initial state s_n and end state s_j, such that for $n \leq p \leq j - 1$, $\mathcal{M}, s_p \models \Phi_1$ and $\mathcal{M}, s_j \models \Phi_2$, and only actions in A occur up to s_{j-1}

A few examples of DSL properties we may want to express are described next. In location l, object i_1 interacts with i_2 on action a is written

$$l : i_1 \leftrightarrow i_2.a \tag{1}$$

There is an (arbitrary) object of type o_1 in region $N = \{l_1, l_2\}$ of the model is given by

$$\langle N \rangle : o_1[-] \tag{2}$$

An example of a DSL path formula states that if object i_1 is at location l and satisfies $state_1$ then it will eventually do action a in less than 5 time units and satisfy $state_2$ is written

$$(l : i_1.state_1) \,_* \mathcal{U}_{\{a\}}^{\leq 5} (l : i_1.state_2) \tag{3}$$

We can capture as a path formula that if the system is currently in a state such that l_1 has a space for an object of type o_1 and l_2 an occupied space for an object of type o_1 then a *move* can eventually occur and l_1 will have an occupied space for an object of type o_1 is written

$$(l1 : o_1[] \wedge l_2 : o_1[-]) \,_* \mathcal{U}_{\{move\}} (l_1 : o_1[-]) \tag{4}$$

In equilibrium, the accumulated probability of both spaces of type o_1 and o_2 in l being full at the same time is at least 0.8 is written

$$S_{\geq 0.8}(l : o_1[-] \wedge l : o_2[-]) \tag{5}$$

Finally, the next formula states that the probability of object i_1 leaving location l without performing action a is at most 0.5. Let $A = Act \setminus \{a\}$ and $R = \mathcal{L} \setminus \{l\}$, the formula is given by

$$\mathcal{P}_{\leq 0.5}((l : i_1)_A \mathcal{U}_* (\langle R \rangle : o_1[i_1])) \tag{6}$$

4   The Global and Local Sublogics

From the previously defined syntax of DSL$[\mathcal{L}, \mathcal{O}, Act, AP]$, we now want to separate what captures *global behaviour* from *local behaviour and interactions* at locations in the distributed mobile system. This gives us two sublogics of DSL, namely a *global logic* GSL and a family of *local logics* over the set of locations $\mathcal{L}$ given by $\{\text{LSL}_l\}_{l \in \mathcal{L}}$. We assume that $\mathcal{L}, \mathcal{O}, Act_G$ and Act_L are nonempty sets.

The global logic GSL is a restriction of DSL to global actions Act_G, i.e., GSL = DSL$[Act_G]$. This means that we cannot have properties of the kind $l : i \leftrightarrow j.a$ in GSL since a is necessarily a local action (recall that global actions cannot be shared). Any formula in DSL which only makes use of global actions is a valid GSL formula.

The local logic LSL$_l$ for location l is a restriction of DSL to a set of locations $\mathcal{L} = \{l\}$ and local actions in $Act_L(l)$, i.e, LSL$_l$ = DSL$[\{l\}, Act_L(l)]$. This means that we can only refer to properties of l and not other locations in the system, and the behaviour of objects within the location. Any formula in DSL which only makes use of local actions to location l and properties over objects at l is a valid LSL$_l$ formula.

Proposition 1 GSL $\subseteq$ DSL *and* LSL$_l$ $\subseteq$ DSL

Proof. This follows directly from the definition of GSL and LSL. Since $Act_G \subseteq Act$, we have GSL = DSL$[Act_G]$ = DSL$[\mathcal{L}, \mathcal{O}, Act_G, AP]$ $\subseteq$ DSL$[\mathcal{L}, \mathcal{O}, Act, AP]$. Similarly for LSL. □

Given our restrictions above, it is clear that from the formulae given in the previous section: formulae (2), (4), (5) and (6) are GSL formulae, and formulae (1), (3) and (5) are LSL$_l$ formulae. In particular, note that a formula may be both a local and a global formula, as is the case with (5). However, this is only possible if the state formula does not contain the occurrence of any action, as the logics use a disjoint set of actions. A further comment relates to formula (6) which may be a global formula only if global actions are used. Note that as a DSL formula, formula (6) can be more general as the only restriction we made on the set A was that it excluded the local action a. If A includes a mixture of local and global actions, this formula is not a GSL formula. This shows that DSL is more expressive than GSL and LSL.

Since the sublogics are only syntactic restrictions on the sets over which DSL is defined, their semantics is the same as for DSL when defined over the states of the ASMC$^+$ $\mathcal{M}$. The semantics of the sublogics of DSL can, however, also be defined directly over particular quotient structures defined over $\mathcal{M}$. We describe how to obtain the quotient associated to GSL first.

Definition 2. *Let* $\mathcal{M} = (S, Act, AP, F, H, R)$ *be an ASMC$^+$, let* Γ *be a set of* GSL *formulae, and let* $\equiv_\Gamma$ *be an equivalence relation on S defined via* $s_1 \equiv_\Gamma s_2$ *iff for all* $\varphi \in \Gamma$, $\mathcal{M}, s_1 \models \varphi$ *iff* $\mathcal{M}, s_2 \models \varphi$. *Use* $\lfloor s \rfloor$ *to denote* $\{s' \mid s \equiv_\Gamma s'\}$. *Then the global quotient structure of $\mathcal{M}$ by* $\equiv_\Gamma$ *is the structure* $\mathcal{M}/_{\equiv_\Gamma} = (S', Act_G, AP, F', H', R')$ *where* $S' = \{\lfloor s \rfloor \mid s \in S\}$, $F'(\lfloor s \rfloor, i) = \cap_{s' \in \lfloor s \rfloor} F(s', i)$, $H'(\lfloor s \rfloor, i) = H(s, i)$, *and* $R'(\lfloor s \rfloor, a, \lfloor s' \rfloor) = R(s, a, s')$.

We need to establish that the obtained distributed quotient agrees with $\mathcal{M}$ in the equivalence of Γ.

Proposition 2 *Let $\mathcal{M}$ be an ASMC$^+$ and $\varphi \in \Gamma \subseteq$ GSL.*

$$\mathcal{M}, s \models \varphi \text{ iff } \mathcal{M}/_{\equiv_\Gamma}, \lfloor s \rfloor \models \varphi$$

Proof. The proof can be done using structural induction on the syntax of state, location and path properties. Most arguments from [12] for CSL, and [7] for asCSL apply for state and path formula. We give a brief idea of how to address location-based properties.

We prove it by contradiction for $\varphi = l : i.ap$. Assume in one direction that $\mathcal{M}/_{\equiv_\Gamma}$, $\lfloor s \rfloor \models \varphi$ holds, but that $\mathcal{M}, s \not\models \varphi$. The former means that $ap \in F'(\lfloor s \rfloor, i) = \cap_{s' \in \lfloor s \rfloor} F(s', i)$ and $H'(\lfloor s \rfloor, i) = l$ consequently since $s \in \lfloor s \rfloor$ necessarily, then we must have $ap \in F(s, i)$ and $H(s, i) = l$ by definition of H'. It then follows that $\mathcal{M}, s \models l : i.ap$ which contradicts our assumption. The reverse case can also be proved by contradiction. We proceed like this for all cases of location formulae allowed in GSL. Notice that as mentioned before $l : i \leftrightarrow j.a$ is not a global formula and does not belong to Γ. Consider $\varphi = l : o[i]$. Assume in one direction that $\mathcal{M}/_{\equiv_\Gamma}, \lfloor s \rfloor \models \varphi$ holds, but that $\mathcal{M}, s \not\models \varphi$. The former means that we have $H'(\lfloor s \rfloor, i) = l$, and by definition of H' we have $H'(\lfloor s \rfloor, i) = H(s, i)$ so we must have $\mathcal{M}, s \models \varphi$ which contradicts our assumption. $\square$

The above result establishes that $\mathcal{M}$ and $\mathcal{M}/_{\equiv_\Gamma}$ are bisimilar wrt GSL formulae in Γ. The fundamental advantage resulting from this is that we can reduce verification of global properties to a quotient of the original model which corresponds to an equivalence preserving state space aggregation. In other words, if a global property holds in the *global quotient structure* of a model $\mathcal{M}$ then it holds in the original model $\mathcal{M}$.

In exactly the same way, we can obtain quotient structures over $\mathcal{M}$ for subsets of local formulae in LSL$_l$ for some location $l \in \mathcal{L}$.

Definition 3. *Let $\mathcal{M} = (S, Act, AP, F, H, R)$ be an ASMC$^+$, let Δ be a set of LSL$_l$ formulae, and let $\equiv_\Delta$ be an equivalence relation on S defined via $s_1 \equiv_\Delta s_2$ iff for all $\varphi \in \Delta$, $\mathcal{M}, s_1 \models \varphi$ iff $\mathcal{M}, s_2 \models \varphi$. Use $\lfloor s \rfloor_l$ to denote $\{s' \mid s \equiv_\Delta s'\}$. Then the local quotient structure for l of $\mathcal{M}$ by $\equiv_\Delta$ is the structure $\mathcal{M}/_{\equiv_\Delta} = (S', Act_L(l), AP, F', H', R')$ where $S' = \{\lfloor s \rfloor_l \mid s \in S\}$, $F'(\lfloor s \rfloor_l, i) = \cup_{s' \in \lfloor s \rfloor_l} F(s', i)$, $H'(\lfloor s \rfloor_l, i) = H(s, i) = l$, and $R'(\lfloor s \rfloor_l, a, \lfloor s' \rfloor_l) = R(s, a, s')$.*

In the above definition, the local quotient structure for l is such that all equivalence classes over S contain (at most) state s with an object located at l.

A similar proposition can be formulated stating that the obtained local quotient for l agrees with $\mathcal{M}$ in the equivalence of Δ.

5   Conclusion

We presented a family of stochastic logics for the performance analysis of distributed mobile systems. We defined ASMC$^+$s as an object and location-based extension of action and state-based Markov chains, and introduced a new distributed stochastic logic DSL interpreted over an ASMC$^+$ $\mathcal{M}$. The logic extends known stochastic logics with the ability to express location and resource specific properties as well as both state and action-based properties. We then separated *global* and *local properties* and defined two sublogics of DSL, namely a *global stochastic logic* GSL and a collection of *local stochastic logics* LSL_l over a location $l \in \mathcal{L}$.

Our formulae capture properties of mobile distributed systems and can be used as a reasoning framework for any high-level model representation which have an underlying CTMC model, such as PEPA nets, StoKLAIM and various stochastic Petri net models. In general, the logic can be used to capture quantitative dependability and performance properties for distributed mobile systems, and contributes to a more efficient computation of dependability and performance measures of distributed mobile systems.

A further contribution in this paper is an initial attempt towards a method to achieve state space aggregation in an $ASMC^+$ model $\mathcal{M}$ induced by sets of formulae and the quotient structure that they define on $\mathcal{M}$. In particular, we defined how to obtain quotient structures for GSL and LSL_l (for a location $l \in \mathcal{L}$). By exploring the structure of $\mathcal{M}$ and using that as an interpretation of the sublogics we obtain a more efficient approach for property verification. We need to extend this research further, explore concrete examples and identify how we can exploit the sets of formulae further to obtain useful state space aggregations with respect to the kinds of properties we are interested in verifying. In addition, we want to explore the formal relation between common notions of strong and weak bisimulation equivalences and simulation pre-orders (see for example [13,12,14]) or non-bisimulation-based Markovian behavioural equivalences (see [15]) in our setting.

Further considerations on verification and model checking are outside the scope of the present paper. However, it can be noted that a general strategy for model checking DSL over an $ASMC^+$ $\mathcal{M}$ and its sublogics and associated models derived from $\mathcal{M}$ follows the standard approach for verification of stochastic logics. For a given DSL (GSL or LSL_l) formula Φ, the model checking algorithm recursively computes the sets of states satisfying the subformulas of Φ, and then obtains from them the sets of states that satisfy the overall formula. Usually in the context of stochastic logics the difficulties are in the steady-state properties and the probabilistic quantifier. Since our DSL logic extends features from CSL [8] and aCSL [11], algorithms defined in that context can be exploited. Indeed, we could consider using an existing state-based stochastic model-checker such as MRMC[16] or PRISM[17], wrapping them on a DSL model-checking algorithm.

Acknowledgements. We thank Joost-Pieter Katoen and the anonymous reviewers for valuable comments on an earlier version of this paper.

References

1. Boudol, G.: ULM: A core programming model for global computing. In: Schmidt, D. (ed.) ESOP 2004. LNCS, vol. 2986, pp. 234–248. Springer, Heidelberg (2004)
2. Bettini, L., De Nicola, R., Loreti, M.: Formalizing properties of mobile agent systems. In: Arbab, F., Talcott, C. (eds.) COORDINATION 2002. LNCS, vol. 2315, pp. 72–87. Springer, Heidelberg (2002)
3. de Nicola, R., Katoen, J.P., Latella, D., Loreti, M., Massink, M.: Model checking mobile stochastic logic. TCS 382, 42–70 (2007)

4. Priami, C.: Stochastic π calculus. Comput. J. 38, 578–589 (1995)
5. Gilmore, S., Hillston, J., Kloul, L., Ribaudo, M.: PEPA nets: a structured performance modelling formalism. Performance Evaluation 54, 79–104 (2003)
6. Cardelli, L., Gordon, A.: Mobile ambients. TCS 240, 177–213 (2000)
7. Baier, C., Cloth, L., Haverkort, B.R., Kuntz, M., Siegle, M.: Model checking markov chains with actions and state labels. IEEE Transactions on Software Engineering 33, 209–224 (2007)
8. Aziz, A., Sandwal, K., Singhal, V., Brayton, R.: Model checking continuous time Markov chains. ACM Transactions on Computational Logic 1, 162–170 (2000)
9. De Nicola, R., Loreti, M.: MOMO: A modal logic for reasoning about mobility. In: de Boer, F.S., Bonsangue, M.M., Graf, S., de Roever, W.-P. (eds.) FMCO 2004. LNCS, vol. 3657, pp. 95–119. Springer, Heidelberg (2005)
10. De Nicola, R., Latella, D., Loreti, M., Massink, M.: SOSL: A service-oriented stochastic logic. In: Wirsing, M., Hölzl, M. (eds.) SENSORIA Project. LNCS, vol. 6582, pp. 447–466. Springer, Heidelberg (2011)
11. Hermanns, H., Katoen, J.-P., Meyer-Kayser, J., Siegle, M.: Towards model checking stochastic process algebra. In: Grieskamp, W., Santen, T., Stoddart, B. (eds.) IFM 2000. LNCS, vol. 1945, pp. 420–439. Springer, Heidelberg (2000)
12. Desharnais, J., Panangaden, P.: Continuous stochastic logic characterizes bisimulation of continuous-time markov processes. The Journal of Logic and Algebraic Programming 56, 99–115 (2003)
13. Baier, C., Katoen, J.P., Hermanns, H., Wolf, V.: Comparative branching-time semantics for markov chains. Information and Computation 200, 149–214 (2005)
14. Bernardo, M., de Nicola, R., Loreti, M.: Relating strong behavioral equivalences for processes with nondeterminism and probabilities. Theoretical Computer Science (2014) (in print)
15. Bernardo, M.: Non-bisimulation-based markovian behavioral equivalences. Journal of Logic and Algebraic Programming 72, 3–49 (2007)
16. Katoen, J.P., Zapreev, I., Hahn, E.M., Hermanns, H., Jansen, D.N.: The Ins and Outs of The Probabilistic Model Checker MRMC. Performance Evaluation 68, 90–104 (2011)
17. Kwiatkowska, M., Norman, G., Parker, D.: PRISM 4.0: Verification of probabilistic real-time systems. In: Gopalakrishnan, G., Qadeer, S. (eds.) CAV 2011. LNCS, vol. 6806, pp. 585–591. Springer, Heidelberg (2011)

On the Construction of Analytic Sequent Calculi for Sub-classical Logics*

Ori Lahav and Yoni Zohar

School of Computer Science, Tel Aviv University, Israel
orilahav@post.tau.ac.il, yoni.zohar@cs.tau.ac.il

Abstract. We study the question of when a given set of derivable rules in some basic analytic propositional sequent calculus forms itself an analytic calculus. First, a general syntactic criterion for analyticity in the family of pure sequent calculi is presented. Next, given a basic calculus admitting this criterion, we provide a method to construct weaker pure calculi by collecting simple derivable rules of the basic calculus. The obtained calculi are analytic-by-construction. While the criterion and the method are completely syntactic, our proofs are semantic, based on interpretation of sequent calculi via non-deterministic valuation functions. In particular, this method captures calculi for a wide variety of paraconsistent logics, as well as some extensions of Gurevich and Neeman's primal infon logic.

1 Introduction

Proof theory reveals a wide mosaic of possibilities for sub-classical logics. These are logics that are strictly contained (as consequence relations) in classical logic. Thus, by choosing a subset of axioms and derivation rules that are derivable in (some proof system for) classical logic, one easily obtains a (proof system for a) sub-classical logic. Various important and useful non-classical logics can be formalized in this way, with the most prominent example being intuitionistic logic. In general, the resulting logics come at first with no semantics. They might be also unusable for computational purposes, since the new calculi might not be *analytic*: it is often the case that proofs of some formula φ must contain formulas that are not subformulas of φ. This is evident within the framework of Hilbert-style calculi, that are rarely analytic. But, even for Gentzen-type sequent calculi, where the initial proof system for classical logic **LK** is analytic, there is no guarantee that an arbitrary collection of classically derivable sequent rules constitutes an analytic sequent calculus.

In this paper, we focus on a general family of relatively simple sequent calculi for propositional logics, called *pure* sequent calculi (originally studied in [2]), of which (the propositional fragment of) **LK** is the prototype example. Our contribution is twofold. First, we generalize the coherence condition from [4] to provide a decidable sufficient syntactic criterion for analyticity of a given pure

---

* This research was supported by The Israel Science Foundation (grant no. 280-10).

U. Kohlenbach et al. (Eds.): WoLLIC 2014, LNCS 8652, pp. 206–220, 2014.
© Springer-Verlag Berlin Heidelberg 2014

sequent calculus. Here we employ a general concept of analyticity, based on a parametrized notion of a subformula, that shares the attractive features with the usual subformula property. This criterion is useful in many cases, e.g. for proving the analyticity of a sequent calculus for the logic of first-degree entailment [1], and of course, the analyticity of the propositional fragment of **LK**. Second, we show that calculi admitting this criterion can be utilized for constructing other analytic sequent calculi. Taking a basic calculus **B**, we present a method for obtaining other analytic-by-construction sub-calculi of **B**, by collecting derivable rules of **B** that have a certain "safe" form.

The proposed method is general enough to capture a wide variety of known sequent calculi for sub-classical logics. This includes:

- A large family of sequent calculi for propositional paraconsistent logics, originated from philosophical motivations, and obtained by replacing the usual left introduction rule of negation with weaker rules, each of which is derivable in **LK**.
- A sequent calculus for primal logic (without quotations) from [7], as well as some natural extensions of it. This calculus originated from practical computational motivations, aiming to allow efficient proof search. It is obtained by replacing the usual right introduction rule of implication with a weaker rule, and discarding the rule for introducing disjunction on the left hand-side.

Our approach is semantic: We formulate and use a semantic property of sequent calculi that is equivalent to analyticity. The semantics, however, plays a role only in our arguments, while the actual use of the proposed methods includes only syntactic considerations.

Related Work. The family of pure sequent calculi was defined in [2]. The semantics for these calculi which lies in the basis of our proofs, is similar to the one in [10] (and takes its inspiration from [6]). Nevertheless, [10] investigates translations of derivability in analytic pure calculi to the classical satisfiability problem, leaving open the tasks of constructing analytic calculi, and checking whether a given calculus is analytic. In this paper we aim to fill this gap, by providing simple sufficient conditions (that hold in various known cases) for analyticity. Furthermore, we note that the notion of analyticity employed in [10] is generalized in the current paper: (1) here we also consider derivations from assumptions (also known as "non-logical axioms"); and (2) we use a more general parametrized notion of a subformula. A particular well-behaved subfamily of pure calculi, called *canonical calculi* was studied in [4]. For these calculi, it was shown that analyticity and cut-admissibility are equivalent, and both were precisely characterized by a simple and decidable *coherence* criterion. However, various useful pure calculi (some of which are included in examples below) are not canonical, and still their analyticity can be shown using the results of the current paper. Finally, the general framework of [11] allows one to encode all pure calculi in linear logic, and use linear logic to reason about them. Among the pure calculi, it is again only the canonical ones for which a decidable criterion for cut-admissibility is given in [11].

2 Pure Sequent Calculi

In what follows, we assume a propositional language for classical logic, that consists of a countably infinite set of atomic variables $At = \{p_1, p_2, \ldots\}$, the binary connectives $\wedge$, $\vee$ and $\supset$, the unary connective $\neg$, and the nullary connectives $\top$ and $\bot$. A *sequent* is a pair $\langle \Gamma, \Delta \rangle$ (denoted by $\Gamma \Rightarrow \Delta$) where Γ and Δ are finite sets of formulas. We employ the standard sequent notations, e.g. when writing expressions like $\Gamma, \psi \Rightarrow \Delta$ or $\Rightarrow \psi$. The union of sequents is defined by $(\Gamma_1 \Rightarrow \Delta_1) \cup (\Gamma_2 \Rightarrow \Delta_2) = \Gamma_1 \cup \Gamma_2 \Rightarrow \Delta_1 \cup \Delta_2$. For a sequent $\Gamma \Rightarrow \Delta$, $frm(\Gamma \Rightarrow \Delta) = \Gamma \cup \Delta$. This notation is naturally extended to sets of sequents. Given a set $\mathcal{F}$ of formulas, we say that a sequent s is an *$\mathcal{F}$-sequent* if $frm(s) \subseteq \mathcal{F}$. A *substitution* is a function from At to the set of formulas. A substitution σ is naturally extended to compound formulas by $\sigma(\diamond(\psi_1, \ldots, \psi_n)) = \diamond(\sigma(\psi_1), \ldots, \sigma(\psi_n))$ for every compound formula $\diamond(\psi_1, \ldots, \psi_n)$. Substitutions are also naturally extended to sets of formulas, sequents and sets of sequents.

We focus on a general family of relatively simple sequent calculi, called *pure sequent calculi*. Roughly speaking, these are propositional fully-structural calculi (calculi that include the structural rules: exchange, contraction and weakening),[1] whose derivation rules do not enforce any limitations on the context formulas (following [2], the adjective "pure" stands for this requirement). We note that additive applications are employed (i.e., all premises share one context sequent), rather than multiplicative ones. In the context of this paper, this is just a matter of taste, since the two options are obviously equivalent when all structural rules are available.

Definition 1. A *pure rule* is a pair $\langle S, s \rangle$ (denoted by S / s) where S is a set of sequents and s is a sequent. The elements of S are called the *premises* of the rule and s is called the *conclusion* of the rule. The set S of premises of a pure rule is usually written without set braces, and its elements are separated by ";".

Definition 2. An *application* of a pure rule $s_1, \ldots, s_n / s$ is a pair of the form $\langle \{\sigma(s_1) \cup c, \ldots, \sigma(s_n) \cup c\}, \sigma(s) \cup c \rangle$ (denoted by $\dfrac{\sigma(s_1) \cup c, \ldots, \sigma(s_n) \cup c}{\sigma(s) \cup c}$) where σ is a substitution and c is a sequent (called a *context sequent*). The sequents $\sigma(s_i) \cup c$ are called the *premises* of the application and $\sigma(s) \cup c$ is called the *conclusion* of the application.

Note that every application of a pure rule is itself a pure rule. Moreover, every pure rule is an application of itself, obtained by taking the identity substitution and the empty context sequent. This duality will be exploited when we expand analytic calculi with new rules in the form of applications of other rules.

---

[1] Exchange and contraction are implicitly included, since sequents are taken to be pairs of *sets*.

Definition 3. A *pure calculus* is a finite set of pure rules. A *proof* in a pure calculus $\mathbf{G}$ is defined as usual, where in addition to applications of the pure rules of $\mathbf{G}$, the following standard schemes may be used:

$$(weak) \; \frac{\Gamma \Rightarrow \Delta}{\Gamma, \Gamma' \Rightarrow \Delta', \Delta} \qquad (id) \; \frac{}{\Gamma, \psi \Rightarrow \psi, \Delta} \qquad (cut) \; \frac{\Gamma \Rightarrow \psi, \Delta \quad \Gamma, \psi \Rightarrow \Delta}{\Gamma \Rightarrow \Delta}$$

Given a pure calculus $\mathbf{G}$, a set $\mathcal{F}$ of formulas, a set of $\mathcal{F}$-sequents S and an $\mathcal{F}$-sequent s, we write $S \vdash_{\mathbf{G}}^{\mathcal{F}} s$ if there is a proof of s from S in $\mathbf{G}$ consisting only of $\mathcal{F}$-sequents. When $\mathcal{F}$ is the set of all formulas, we write $\vdash_{\mathbf{G}}$ instead of $\vdash_{\mathbf{G}}^{\mathcal{F}}$.

In what follows, all rules and calculi are pure. There are many sequent calculi for non-classical logics (admitting cut-elimination) that fall in this framework. These include calculi for three and four-valued logics, various calculi for paraconsistent logics, and all canonical sequent systems [3,4,6].

Example 1. The propositional fragment of Gentzen's fundamental calculus for classical logic can be directly presented as a pure calculus, denoted henceforth by $\mathbf{LK}$. It consists of the following rules:

$(\perp \Rightarrow)$	$\emptyset \,/\, \perp \Rightarrow$	$(\Rightarrow \top)$	$\emptyset \,/\Rightarrow \top$
$(\neg \Rightarrow)$	$\Rightarrow p_1 \,/\, \neg p_1 \Rightarrow$	$(\Rightarrow \neg)$	$p_1 \Rightarrow \,/\Rightarrow \neg p_1$
$(\wedge \Rightarrow)$	$p_1, p_2 \Rightarrow \,/\, p_1 \wedge p_2 \Rightarrow$	$(\Rightarrow \wedge)$	$\Rightarrow p_1; \Rightarrow p_2 \,/\Rightarrow p_1 \wedge p_2$
$(\vee \Rightarrow)$	$p_1 \Rightarrow; p_2 \Rightarrow \,/\, p_1 \vee p_2 \Rightarrow$	$(\Rightarrow \vee)$	$\Rightarrow p_1, p_2 \,/\Rightarrow p_1 \vee p_2$
$(\supset \Rightarrow)$	$\Rightarrow p_1; p_2 \Rightarrow \,/\, p_1 \supset p_2 \Rightarrow$	$(\Rightarrow \supset)$	$p_1 \Rightarrow p_2 \,/\Rightarrow p_1 \supset p_2$

Example 2. The calculus from [3] for da Costa's historical paraconsistent logic C_1 can be directly presented as a pure calculus, that we call $\mathbf{G}_{C_1}$. It consists of the rules of $\mathbf{LK}$ except for $(\neg \Rightarrow)$ that is replaced by the following rules:

$p_1 \Rightarrow \,/\, \neg\neg p_1 \Rightarrow$

$\Rightarrow p_1; \Rightarrow \neg p_1 \,/\, \neg(p_1 \wedge \neg p_1) \Rightarrow \qquad \neg p_1 \Rightarrow; \neg p_2 \Rightarrow \,/\, \neg(p_1 \wedge p_2) \Rightarrow$

$\neg p_1 \Rightarrow; p_2, \neg p_2 \Rightarrow \,/\, \neg(p_1 \vee p_2) \Rightarrow \qquad p_1, \neg p_1 \Rightarrow; \neg p_2 \Rightarrow \,/\, \neg(p_1 \vee p_2) \Rightarrow$

$p_1 \Rightarrow; p_2, \neg p_2 \Rightarrow \,/\, \neg(p_1 \supset p_2) \Rightarrow \qquad p_1, \neg p_1 \Rightarrow; \neg p_2 \Rightarrow \,/\, \neg(p_1 \supset p_2) \Rightarrow$

The following properties of pure calculi will be particularly useful below:

Proposition 1. *Let* $\mathbf{G}$ *be a calculus,* $\mathcal{F}$ *a set of formulas,* S *a set of* $\mathcal{F}$*-sequents, and* s *an* $\mathcal{F}$*-sequent. Suppose that* $S \vdash_{\mathbf{G}}^{\mathcal{F}} s$. *Then, the following hold:*

1. $\sigma(S) \vdash_{\mathbf{G}}^{\sigma(\mathcal{F})} \sigma(s)$ *for every substitution* σ.
2. $\{s' \cup c \mid s' \in S\} \vdash_{\mathbf{G}}^{\mathcal{F}} s \cup c$ *for every* $\mathcal{F}$*-sequent* c.

2.1 Analyticity

Analyticity is a crucial property of proof systems. In the case of fully-structural propositional sequent calculi it usually implies their decidability and consistency (the fact that the empty sequent is not derivable). Roughly speaking, a calculus is analytic if whenever a sequent s is provable in it from a set S of sequents, s can be proven using only the "syntactic material available inside s and S". Usually, this "material" is taken to consist of all subformulas occurring in s. Next, we introduce a generalized analyticity property, based on a parametrized notion of a subformula in which negation plays a special role.

Definition 4. Let $k \geq 0$. A formula φ is an *immediate k-subformula* of a formula ψ if either $\psi = \neg\varphi$, or $\psi = \varphi_1 \sharp \varphi_2$ and $\varphi = \neg^m \varphi_i$ for some formulas φ_1, φ_2, $\sharp \in \{\wedge, \vee, \supset\}$, $0 \leq m \leq k$ and $i \in \{1,2\}$.[2] The *k-subformula relation* is the reflexive transitive closure of the immediate k-subformula relation. A k-subformula φ of a formula ψ is called *proper* if $\varphi \neq \psi$. We denote the set of k-subformulas of a formula ψ by $sub^k(\psi)$. This notation is naturally extended to sets of formulas, sequents and sets of sequents.

Definition 5. A calculus $\mathbf{G}$ is called *k-analytic* if $S \vdash_{\mathbf{G}} s$ entails $S \vdash_{\mathbf{G}}^{sub^k(S \cup \{s\})} s$ for every set S of sequents and sequent s.

0-subformulas are usual subformulas, and thus 0-analyticity amounts to the usual (global) subformula property of sequent calculi. Note that k-analyticity (for any k) ensures the decidability and consistency of a calculus. The following propositions will be useful in the sequel.

Proposition 2. *If a formula φ is a (proper) k-subformula of a formula ψ, then $\sigma(\varphi)$ is a (proper) k-subformula of $\sigma(\psi)$ for every substitution σ. Consequently, $\sigma(sub^k(\psi)) \subseteq sub^k(\sigma(\psi))$ for every formula ψ and substitution σ.*

Proposition 3. *Suppose that a calculus $\mathbf{G}'$ is obtained from a calculus $\mathbf{G}$ by one of the following:*

1. *Replacing some rule $S \mathbin{/} \Gamma \Rightarrow \psi, \Delta$ by $S; \psi \Rightarrow \mathbin{/} \Gamma \Rightarrow \Delta$.*
2. *Replacing some rule $S \mathbin{/} \Gamma, \psi \Rightarrow \Delta$ by $S; \Rightarrow \psi \mathbin{/} \Gamma \Rightarrow \Delta$.*
3. *Replacing two rules of the form $S; \Gamma \Rightarrow \Delta \mathbin{/} s$ and $S; \Gamma' \Rightarrow \Delta' \mathbin{/} s$ by the rule $S; \Gamma \cup \Gamma' \Rightarrow \Delta \cup \Delta' \mathbin{/} s$, given that $\Gamma \cup \Gamma' \cup \Delta \cup \Delta' \subseteq sub^k(S \cup \{s\})$.*

Then $\vdash_{\mathbf{G}'} = \vdash_{\mathbf{G}}$ and $\mathbf{G}'$ is k-analytic iff $\mathbf{G}$ is k-analytic.

Proposition 4. *Let $\mathbf{G}'$ be a calculus obtained from a calculus $\mathbf{G}$ by adding a premise s' to some rule $r = S \mathbin{/} s$ of $\mathbf{G}$. Suppose that $S \vdash_{\mathbf{G}'}^{sub^k(frm(r))} s'$. Then, $\mathbf{G}'$ is k-analytic iff $\mathbf{G}$ is k-analytic.[3]*

Proof. Suppose that $S = \{s_1, \ldots, s_n\}$ and let $S' = S \cup \{s'\}$. To show that $\mathbf{G}'$ is k-analytic iff $\mathbf{G}$ is k-analytic, we prove that $S_0 \vdash_{\mathbf{G}'}^{\mathcal{F}} s_0$ iff $S_0 \vdash_{\mathbf{G}}^{\mathcal{F}} s_0$ for every set $\mathcal{F}$ of formulas that is closed under k-subformulas, set S_0 of $\mathcal{F}$-sequents, and $\mathcal{F}$-sequent s_0:

($\Rightarrow$): Trivially, a proof in $\mathbf{G}'$ is also a proof in $\mathbf{G}$ with some redundant sequents.

($\Leftarrow$): Suppose $S_0 \vdash_{\mathbf{G}}^{\mathcal{F}} s_0$. Let $\hat{r} = \dfrac{\sigma(s_1) \cup c, \ldots, \sigma(s_n) \cup c}{\sigma(s) \cup c}$ be an application of

---

[2] $\neg^m \varphi$ is inductively defined by: $\neg^0 \varphi = \varphi$, and $\neg^{m+1} \varphi = \neg\neg^m \varphi$.

[3] frm is extended to pure rules and their applications in the obvious way, e.g. $frm(S \mathbin{/} s) = frm(S) \cup frm(s)$.

r in the proof of s_0 from S_0, such that $frm(\hat{r}) \subseteq \mathcal{F}$. Since $S \vdash_{\mathbf{G}'}^{sub^k(frm(S\,/\,s))} s'$, by Proposition 1, we have $\sigma(S) \vdash_{\mathbf{G}'}^{\sigma(sub^k(frm(S\,/\,s)))} \sigma(s')$. Since $frm(\hat{r}) \subseteq \mathcal{F}$ and $\mathcal{F}$ is closed under k-subformulas, $sub^k(\sigma(frm(S\,/\,s))) \subseteq \mathcal{F}$. By Proposition 2, $\sigma(sub^k(frm(S\,/\,s))) \subseteq \mathcal{F}$. Hence $\sigma(S) \vdash_{\mathbf{G}'}^{\mathcal{F}} \sigma(s')$. In addition, $frm(c) \subseteq \mathcal{F}$, and hence by Proposition 1, $\{\sigma(s_i) \cup c \mid 1 \leq i \leq n\} \vdash_{\mathbf{G}'}^{\mathcal{F}} \sigma(s') \cup c$. Hence we may add $\sigma(s') \cup c$ to the premises of $\hat{r}$, and obtain an application of $S'\,/\,s$ that consists only of formulas in $\mathcal{F}$. This can be done for every application of $S\,/\,s$, and hence $S_0 \vdash_{\mathbf{G}'}^{\mathcal{F}} s_0$. □

3 Sufficient Criterion for Analyticity

In this section we generalize the coherence condition from [4], and show that the generalized condition entails analyticity.

Definition 6. A rule r is called k-*closed* if its conclusion has the form $\Rightarrow \varphi$ or $\varphi \Rightarrow$, and its premises consist only of proper k-subformulas of φ. A calculus is called k-*closed* if it consists only of k-closed rules.

Notation 1. Given a k-closed rule r, we denote by φ_r the formula that appears in the conclusion of r.

The calculus **LK** (Example 1) is 0-closed (and hence it is k-closed for any k). For example, the rule $r = (\Rightarrow \supset)$ of **LK** is 0-closed and $\varphi_r = p_1 \supset p_2$. The calculus $\mathbf{G}_{C_1}$ (Example 2) is 1-closed.

Definition 7. A k-closed calculus $\mathbf{G}$ is called (*cut*)-*guarded* if for every two rules of $\mathbf{G}$ of the forms $S_1\,/\,\Rightarrow \varphi_1$ and $S_2\,/\,\varphi_2 \Rightarrow$, and substitutions σ_1, σ_2 such that $\sigma_1(\varphi_1) = \sigma_2(\varphi_2)$, we have that the empty sequent is derivable from $\sigma_1(S_1) \cup \sigma_2(S_2)$ using only (*cut*).

Note that it is decidable whether a given calculus is (*cut*)-guarded or not. Indeed, for each pair of rules $S_1\,/\,\Rightarrow \varphi_1$ and $S_2\,/\,\varphi_2 \Rightarrow$, one can first rename the atomic variables so that no atomic variable occurs in both rules, and then it suffices to check the above condition for the most general unifier of φ_1 and φ_2.

Theorem 1. *Every (cut)-guarded k-closed calculus is k-analytic.*

This theorem is obtained as a corollary of Theorem 2 below. Next, we present some examples of applications of it.

Example 3. **LK** is (*cut*)-guarded and 0-closed, and hence it is 0-analytic. Similarly, every canonical system (as defined in [4]) in the language of classical logic is 0-closed, and hence every (*cut*)-guarded canonical system is 0-analytic.

Example 4. The quotations-free fragment of the calculus from [5] for primal infon logic (see [7]) can be directly presented as a pure calculus, that we call **P**. It consists of the rules $(\wedge \Rightarrow)$, $(\Rightarrow \wedge)$, $(\Rightarrow \vee)$, $(\supset \Rightarrow)$, $(\Rightarrow \top)$ and $(\bot \Rightarrow)$ of **LK**, together with the rule $\Rightarrow p_2\,/\,\Rightarrow p_1 \supset p_2$. Clearly, **P** is 0-closed and (*cut*)-guarded. By Theorem 1, it is 0-analytic.

Remark 1. For the intended application of primal infon logic as a logic for access control, it is necessary to extend the language with quotations (i.e., unary connectives of the form "q said"), and add appropriate inference rules for them. Following [10], we note that the 0-analyticity of any given pure calculus entails the subformula property for its extension with quotations.

Example 5. The paper [8] investigates a hierarchy of weak double negations, by presenting an infinite set $\{L2^{n+2} \mid n \in \mathbb{N}\}$ of calculi. For example, the calculus $L4$, that captures the relevance logic of first-degree entailment (see [1]), can be obtained by augmenting $\mathbf{LK} \setminus \{(\neg \Rightarrow), (\Rightarrow \neg)\}$ with the following rules:

$$p_1, \neg p_2 \Rightarrow / \neg(p_1 \supset p_2) \Rightarrow \qquad \Rightarrow p_1; \Rightarrow \neg p_2 / \Rightarrow \neg(p_1 \supset p_2)$$
$$\neg p_1 \Rightarrow; \neg p_2 \Rightarrow / \neg(p_1 \wedge p_2) \Rightarrow \qquad \Rightarrow \neg p_1, \neg p_2 / \Rightarrow \neg(p_1 \wedge p_2)$$
$$\neg p_1, \neg p_2 \Rightarrow / \neg(p_1 \vee p_2) \Rightarrow \qquad \Rightarrow \neg p_1; \Rightarrow \neg p_2 / \Rightarrow \neg(p_1 \vee p_2)$$
$$p_1 \Rightarrow / \neg\neg p_1 \Rightarrow \qquad \Rightarrow p_1 / \Rightarrow \neg\neg p_1$$

This calculus is (*cut*)-guarded and 1-closed, and hence, by Theorem 1, it is 1-analytic. Moreover, it can be easily observed that each $L2^{n+2}$ is (*cut*)-guarded and $(n+1)$-closed, and thus by Theorem 1, each $L2^{n+2}$ is $(n+1)$-analytic.

4   Constructing Analytic Calculi

Theorem 1 allows us to prove that many calculi are k-analytic, by observing that they are k-closed and (*cut*)-guarded. However, this criterion is not necessary. For example, $\mathbf{G}_{C_1}$ from Example 2 is 1-analytic (this can be shown as a consequence of cut-elimination), but it is not (*cut*)-guarded. Indeed, for the rules $p_1 \Rightarrow / \Rightarrow \neg p_1$ and $p_1 \Rightarrow / \neg\neg p_1 \Rightarrow$, and the substitutions σ_1, σ_2 with $\sigma_1(p_1) = \neg p_1$ and $\sigma_2(p_1) = p_1$, we have $\sigma_1(\neg p_1) = \sigma_2(\neg\neg p_1)$, but the empty sequent is not provable from $\{(\neg p_1 \Rightarrow), (p_1 \Rightarrow)\}$ only with (*cut*). In order to capture $\mathbf{G}_{C_1}$ and other useful calculi, we introduce a more general method to prove analyticity. More precisely, we present a method for constructing k-analytic calculi by joining applications of rules of a certain basic (*cut*)-guarded k-closed calculus.

In what follows $\mathbf{B}$ *denotes an arbitrary k-closed (cut)-guarded calculus, that serves as a basic calculus.*

Definition 8. An application of a rule $s_1, \ldots, s_n / s$ is called k-*safe* if it has the form $\dfrac{\sigma(s_1) \cup c, \ldots, \sigma(s_n) \cup c}{\sigma(s) \cup c}$ for some substitution σ and sequent c, such that c consists only of proper k-subformulas of formulas that occur in $\sigma(s)$.

Example 6. The following are 0-safe, 1-safe and 2-safe applications of the rule $(\supset \Rightarrow)$ of $\mathbf{LK}$ (respectively):

$$\frac{p_1 \Rightarrow p_1 \wedge p_2 \quad p_1, p_2 \Rightarrow}{p_1, (p_1 \wedge p_2) \supset p_2 \Rightarrow} \qquad \frac{\neg p_1 \Rightarrow p_1 \wedge p_2 \quad \neg p_1, p_2 \Rightarrow}{\neg p_1, (p_1 \wedge p_2) \supset p_2 \Rightarrow}$$

$$\frac{\neg\neg p_3 \Rightarrow p_1 \wedge p_2, \neg(p_1 \wedge p_2) \quad \neg\neg p_3, p_2 \supset p_3 \Rightarrow \neg(p_1 \wedge p_2)}{\neg\neg p_3, (p_1 \wedge p_2) \supset (p_2 \supset p_3) \Rightarrow \neg(p_1 \wedge p_2)}$$

Proposition 5. *Consider a k-closed rule $r = s_1, \ldots, s_n \,/\, s$, and a k-safe appli-*
cation of r, $\hat{r} = \dfrac{\sigma(s_1) \cup c, \ldots, \sigma(s_n) \cup c}{\sigma(s) \cup c}$. Then all formulas in $\mathrm{sub}^k(\sigma(s_i) \cup c)$
are proper k-subformulas of $\sigma(\varphi_r)$ (and thus $\mathrm{sub}^k(\mathrm{frm}(\hat{r})) \subseteq \mathrm{sub}^k(\sigma(\varphi_r)))$.

Proof. Let $\psi \in \mathrm{sub}^k(\sigma(s_i) \cup c)$ and let $\varphi \in \sigma(\mathrm{frm}(s_i)) \cup \mathrm{frm}(c)$ such that ψ
is a k-subformula of φ. We show that φ is a proper k-subformula of $\sigma(\varphi_r)$.
Since ψ is a k-subformula of φ, it follows that ψ is also a proper k-subformula
of $\sigma(\varphi_r)$. If $\varphi = \sigma(\varphi')$ for some $\varphi' \in \mathrm{frm}(s_i)$, then since r is k-closed, φ' is
a proper k-subformula of φ_r. By Proposition 2, φ is a proper k-subformula of
$\sigma(\varphi_r)$. Otherwise, $\varphi \in \mathrm{frm}(c)$, and since $\hat{r}$ is k-safe, φ is a proper k-subformula
of $\sigma(\varphi_r)$. $\square$

Theorem 2. *Every calculus that consists solely of k-safe applications of rules*
of $\mathbf{B}$ is k-analytic.

Theorem 2 will be proved in the next section. First, observe that Theorem 1
is obtained as a corollary:

Proof (of Theorem 1). Every rule of $\mathbf{B}$ is a trivial k-safe application of itself,
and hence by Theorem 2, $\mathbf{B}$ itself is k-analytic. $\square$

Before proving Theorem 2, we present some consequences and examples of it.
For these examples, we take the basic calculus $\mathbf{B}$ to be $\mathbf{LK}$ (that is (cut)-guarded
and k-closed for every k).

Example 7. A sequent calculus $\mathbf{G}_{P_1}$ for the *atomic* paraconsistent logic P_1 from
[12] can be constructed using Theorem 2.[4] Begin with $\mathbf{LK} \setminus \{(\neg \Rightarrow)\}$, and add
the following 0-safe applications of $(\neg \Rightarrow)$ to allow left-introduction of negation
for compound formulas:

$$\Rightarrow \neg p_1 \,/\, \neg\neg p_1 \Rightarrow \qquad\qquad \Rightarrow p_1 \wedge p_2 \,/\, \neg(p_1 \wedge p_2) \Rightarrow$$
$$\Rightarrow p_1 \vee p_2 \,/\, \neg(p_1 \vee p_2) \Rightarrow \qquad \Rightarrow p_1 \supset p_2 \,/\, \neg(p_1 \supset p_2) \Rightarrow$$

Note that the context sequent c is empty in each of these applications. By The-
orem 2, this calculus is 0-analytic. In $\mathbf{G}_{P_1}$ we have $\nvdash_{\mathbf{G}_{P_1}} p_1, \neg p_1 \Rightarrow p_2$, but
$\vdash_{\mathbf{G}_{P_1}} \varphi, \neg\varphi \Rightarrow \psi$ for every compound formula φ and formula ψ. Note that $\mathbf{G}_{P_1}$
is also 0-closed and (cut)-guarded, and hence its analyticity directly follows from
Theorem 1.

In some cases, k-safe applications of rules of $\mathbf{B}$ turn out to have premises
that are already derivable. For example, suppose we would like to augment the
calculus $\mathbf{P}$ from Example 4 with the rule $\bot \Rightarrow p_1 \,/\, \Rightarrow \bot \supset p_1$, which is a 0-safe
application of $(\Rightarrow \supset)$. Since the sequent $\bot \Rightarrow p_1$ is provable in $\mathbf{P}$, it is a redundant
premise. In this case, one can add the rule $\emptyset \,/\, \Rightarrow \bot \supset p_1$ directly. The following
proposition is used for omitting redundant premises in the examples below.

---

[4] $\mathbf{G}_{P_1}$ is equivalent to a sequent calculus for P_1 given by Arnon Avron in an unpublished manuscript.

Proposition 6. *Let* $\mathbf{G}_1$ *be a* k-*analytic calculus, and* $\mathbf{G}_2$ *be a calculus that consist solely of* k-*safe applications of rules of* $\mathbf{B}$. *Suppose that* $\mathbf{G}_1 \cup \mathbf{G}_2$ *is a* k-*analytic calculus, and that* $\vdash_{\mathbf{G}_1} s$ *for every premise* s *of a rule of* $\mathbf{G}_2$. *Let* $\mathbf{G}_3 = \{\emptyset\,/\,s \mid S\,/\,s \in \mathbf{G}_2\}$. *Then* $\mathbf{G}_1 \cup \mathbf{G}_3$ *is* k-*analytic.*

Proof. Let $S\,/\,s \in \mathbf{G}_2$ and $s' \in S$. Since $\mathbf{G}_1$ is k-analytic and $\vdash_{\mathbf{G}_1} s'$, we have $\vdash_{\mathbf{G}_1}^{sub^k(s')} s'$. By Proposition 5, $sub^k(s') \subseteq sub^k(s)$. Therefore, $\vdash_{\mathbf{G}_1}^{sub^k(s)} s'$, and so $\vdash_{\mathbf{G}_1 \cup \mathbf{G}_3}^{sub^k(s)} s'$. By repeatedly applying Proposition 4, since $\mathbf{G}_1 \cup \mathbf{G}_2$ is k-analytic, we obtain that $\mathbf{G}_1 \cup \mathbf{G}_3$ is k-analytic as well. □

Example 8. The calculus $\mathbf{G}_{C_1}$ from Example 2 is 1-analytic. Using Proposition 6, we construct a 1-analytic equivalent calculus that we call $\mathbf{G}'_{C_1}$. Let $\mathbf{G}_1 = \mathbf{LK} \setminus \{(\neg \Rightarrow)\}$. $\mathbf{G}'_{C_1}$ is obtained by augmenting $\mathbf{G}_1$ with the following rules:

$$\emptyset\,/\,\neg\neg p_1 \Rightarrow p_1$$
$$\emptyset\,/\,p_1, \neg p_1, \neg(p_1 \wedge \neg p_1) \Rightarrow \qquad \emptyset\,/\,\neg(p_1 \wedge p_2) \Rightarrow \neg p_1, \neg p_2$$
$$\emptyset\,/\,\neg(p_1 \vee p_2) \Rightarrow \neg p_1, p_2 \qquad \emptyset\,/\,\neg(p_1 \vee p_2) \Rightarrow \neg p_1, \neg p_2$$
$$\emptyset\,/\,\neg(p_1 \vee p_2) \Rightarrow p_1, \neg p_2 \qquad \emptyset\,/\,\neg(p_1 \supset p_2) \Rightarrow p_1, p_2$$
$$\emptyset\,/\,\neg(p_1 \supset p_2) \Rightarrow p_1, \neg p_2 \qquad \emptyset\,/\,\neg(p_1 \supset p_2) \Rightarrow \neg p_1, \neg p_2$$

Every rule in this list has the form $\emptyset\,/\,s$, where s is the conclusion of a 1-safe application of the rule $(\neg \Rightarrow)$ of $\mathbf{LK}$, whose premises are all provable in $\mathbf{G}_1$. For example, the sequent $\neg(p_1 \wedge p_2) \Rightarrow \neg p_1, \neg p_2$ is the conclusion of $\dfrac{\Rightarrow p_1 \wedge p_2, \neg p_1, \neg p_2}{\neg(p_1 \wedge p_2) \Rightarrow \neg p_1, \neg p_2}$, which is a 1-safe application of the rule $(\neg \Rightarrow)$ of $\mathbf{LK}$, and its premise $\Rightarrow p_1 \wedge p_2, \neg p_1, \neg p_2$ is derivable in $\mathbf{G}_1$. By Theorem 2, augmenting $\mathbf{G}_1$ with these applications results in a 1-analytic calculus. $\mathbf{G}'_{C_1}$ is obtained by discarding their premises, and its 1-analyticity is guaranteed by Proposition 6. Using Proposition 3, it is easy to see that $\mathbf{G}'_{C_1}$ is equivalent to $\mathbf{G}_{C_1}$, and furthermore, the 1-analyticity of $\mathbf{G}'_{C_1}$ entails the 1-analyticity of $\mathbf{G}_{C_1}$.

Example 9. The calculus $\mathbf{P}$ from Example 4 enjoys a linear time decision procedure (see, e.g., [7]). As shown in [10], it is possible to augment $\mathbf{P}$ with additional rules in order to make it somewhat closer to $\mathbf{LK}$, without compromising the linear time complexity.[5] Such extension is obtained as follows. Begin with a calculus $\mathbf{G}_0$ that consists of the rules $(\wedge \Rightarrow)$, $(\Rightarrow \wedge)$, $(\Rightarrow \vee)$, $(\supset\Rightarrow)$, $(\Rightarrow \top)$ and $(\bot \Rightarrow)$ of $\mathbf{LK}$. Add the rule $p_2, p_1 \Rightarrow p_2\,/\,p_2 \Rightarrow p_1 \supset p_2$ to obtain a calculus that we call $\mathbf{P}'$. This rule is a 0-safe application of the rule $(\Rightarrow\supset)$ of $\mathbf{LK}$. Now, add the following set of rules to recover some natural properties of the classical connectives (none of these rules is derivable in $\mathbf{P}'$):

$$\emptyset\,/\,\Rightarrow \bot \supset p_1 \qquad\qquad \emptyset\,/\,p_1 \vee p_1 \Rightarrow p_1 \qquad\qquad \emptyset\,/\,\Rightarrow p_1 \supset p_1$$
$$\emptyset\,/\,\bot \vee p_1 \Rightarrow p_1 \qquad\qquad \emptyset\,/\,p_1, \neg p_1 \Rightarrow \qquad\qquad \emptyset\,/\,\Rightarrow (p_1 \wedge p_2) \supset p_1$$
$$\emptyset\,/\,p_1 \vee \bot \Rightarrow p_1 \qquad\qquad \emptyset\,/\,p_1 \vee (p_1 \wedge p_2) \Rightarrow p_1 \qquad \emptyset\,/\,\Rightarrow (p_1 \wedge p_2) \supset p_2$$
$$\emptyset\,/\,(p_1 \wedge p_2) \vee p_1 \Rightarrow p_1 \qquad \emptyset\,/\,\Rightarrow p_2 \supset (p_1 \supset p_2)$$

Every rule in this list has the form $\emptyset\,/\,s$, where s is the conclusion of a 0-safe

---

[5] A manual ad-hoc proof of analyticity of the extended calculus was needed in [10].

application of a rule of **LK**, whose premises are all derivable in $\mathbf{P}'$. For example, the sequent $\Rightarrow p_2 \supset (p_1 \supset p_2)$ is the conclusion of the 0-safe application
$$\frac{p_2 \Rightarrow p_1 \supset p_2}{\Rightarrow p_2 \supset (p_1 \supset p_2)}$$
of $(\Rightarrow\supset)$, and its premise $p_2 \Rightarrow p_1 \supset p_2$ is derivable in $\mathbf{P}'$. By Theorem 2, augmenting $\mathbf{P}'$ with these applications results in a 0-analytic calculus. By Proposition 6, 0-analyticity is preserved when discarding their premises. Using Proposition 6 again, we may also discard the premise $p_2, p_1 \Rightarrow p_2$ of the rule $p_2, p_1 \Rightarrow p_2 \,/\, p_2 \Rightarrow p_1 \supset p_2$. Using Proposition 3, it is easy to see that we may replace the new rule $\emptyset \,/\, p_2 \Rightarrow p_1 \supset p_2$ by $\Rightarrow p_2 \,/\, \Rightarrow p_1 \supset p_2$, which is the original right introduction rule of implication in $\mathbf{P}$.

5 Proof of Theorem 2

This section is devoted to prove Theorem 2. Our proof relies on a semantic interpretation of pure calculi, that gives rise to a semantic characterization of analyticity, as was shown in [10]. Note that we have to slightly strengthen the soundness and completeness theorem given in [10] in order to cover derivations with assumptions (i.e. $S \vdash_{\mathbf{G}}^{\mathcal{F}} s$ for non-empty set S).

Definition 9. A *bivaluation* is a function v from some set $dom(v)$ of formulas to $\{0, 1\}$. A bivaluation v is extended to $dom(v)$-sequents by: $v(\Gamma \Rightarrow \Delta) = 1$ iff $v(\varphi) = 0$ for some $\varphi \in \Gamma$ or $v(\varphi) = 1$ for some $\varphi \in \Delta$. v is extended to sets of $dom(v)$-sequents by: $v(S) = min\{v(s) \mid s \in S\}$, where $min\,\emptyset = 1$. Given a set $\mathcal{F}$ of formulas, by an $\mathcal{F}$-*bivaluation* we refer to a bivaluation v with $dom(v) = \mathcal{F}$. A bivaluation v whose domain $dom(v)$ is the set of all formulas is called *full*.

Definition 10. A bivaluation v *respects* a rule $S \,/\, s$ if $v(\sigma(S)) \leq v(\sigma(s))$ for every substitution σ such that $\sigma(frm(S \,/\, s)) \subseteq dom(v)$. v is called $\mathbf{G}$-*legal* for a calculus $\mathbf{G}$ if it respects all rules of $\mathbf{G}$.

Example 10. A $\{p_1, \neg\neg p_1\}$-bivaluation v respects the rule $p_1 \Rightarrow \,/\, \neg\neg p_1 \Rightarrow$ iff either $v(p_1) = v(\neg\neg p_1) = 0$ or $v(p_1) = 1$. Note that **LK**-legal bivaluations are exactly usual classical valuation functions.

Theorem 3 (Soundness and Completeness). *Let $\mathbf{G}$ be a calculus, $\mathcal{F}$ be a set of formulas, S be a set of $\mathcal{F}$-sequents, and s be an $\mathcal{F}$-sequent. Then, $S \vdash_{\mathbf{G}}^{\mathcal{F}} s$ iff $v(S) \leq v(s)$ for every $\mathbf{G}$-legal $\mathcal{F}$-bivaluation v.*

Proof. **Soundness** Assume $S \vdash_{\mathbf{G}}^{\mathcal{F}} s$ and let v be a $\mathbf{G}$-legal $\mathcal{F}$-bivaluation such that $v(S) = 1$. We prove that $v(s) = 1$ by induction on the length of the proof of s from S in $\mathbf{G}$:

1. If $s \in S$ or s is a conclusion of an application of (cut), $(weak)$ or (id), then this is obvious.
2. If s is the conclusion of an application of a rule of $\mathbf{G}$, then there exist $s_1, \ldots, s_n \,/\, s_0 \in \mathbf{G}$, an $\mathcal{F}$-sequent c and a substitution σ such that $\sigma(frm(\{s_1, \ldots, s_n, s_0\})) \subseteq \mathcal{F}$, $s = \sigma(s_0) \cup c$, and $S \vdash_{\mathbf{G}}^{\mathcal{F}} \sigma(s_i) \cup c$ for every $1 \leq i \leq n$. If $v(c) = 1$, then $v(\sigma(s_0) \cup c) = 1$. Otherwise, by the induction hypothesis, $v(\sigma(s_i)) = 1$ for every $1 \leq i \leq n$. Since v is $\mathbf{G}$-legal, $v(\sigma(s_0)) = 1$, and hence $v(s) = v(\sigma(s_0) \cup c) = 1$.

Completeness Assume $S \not\vdash^{\mathcal{F}}_{\mathbf{G}} s$. We prove that there exists a $\mathbf{G}$-legal $\mathcal{F}$-bivaluation v such that $v(S) = 1$ and $v(s) = 0$. Define an ω-$\mathcal{F}$-sequent to be a pair $\langle L, R \rangle$ (denoted by $L \Rightarrow R$) such that L and R are (possibly infinite) subsets of $\mathcal{F}$. We write $S \vdash^{\mathcal{F}}_{\mathbf{G}} L \Rightarrow R$ if there exist finite $\Gamma \subseteq L$ and $\Delta \subseteq R$ such that $S \vdash^{\mathcal{F}}_{\mathbf{G}} \Gamma \Rightarrow \Delta$. All other definitions for sequents are naturally extended to ω-sequents. It is straightforward to extend s to an ω-$\mathcal{F}$-sequent $L^* \Rightarrow R^*$ that has the following properties:

- $\Gamma' \subseteq L^*$ and $\Delta' \subseteq R^*$ where $s = \Gamma' \Rightarrow \Delta'$.
- $S \not\vdash^{\mathcal{F}}_{\mathbf{G}} L^* \Rightarrow R^*$.
- $S \vdash^{\mathcal{F}}_{\mathbf{G}} L^* \Rightarrow R^*, \psi$ for every $\psi \in \mathcal{F} \setminus R^*$.
- $S \vdash^{\mathcal{F}}_{\mathbf{G}} L^*, \psi \Rightarrow R^*$ for every $\psi \in \mathcal{F} \setminus L^*$.

Since the identity axiom (id) is available, we obviously have $L^* \cap R^* = \emptyset$. Similarly, using (cut), it can be shown that $\mathcal{F} = frm(L^* \Rightarrow R^*)$. Hence L^* and R^* partition $\mathcal{F}$. Define an $\mathcal{F}$-bivaluation v by: $v(\psi) = 1$ if $\psi \in L^*$, and $v(\psi) = 0$ if $\psi \in R^*$. Clearly, $v(L^* \Rightarrow R^*) = 0$ and therefore $v(s) = 0$. We prove that $v(S) = 1$ and that v is $\mathbf{G}$-legal. Let $\Gamma \Rightarrow \Delta \in S$. Obviously, $S \vdash^{\mathcal{F}}_{\mathbf{G}} \Gamma \Rightarrow \Delta$. Since $S \not\vdash^{\mathcal{F}}_{\mathbf{G}} L^* \Rightarrow R^*$, we have either $\Gamma \not\subseteq L^*$ or $\Delta \not\subseteq R^*$. If there exists $\varphi \in \Gamma \setminus L^*$, then $\varphi \in R^*$ and hence $v(\varphi) = 0$. Otherwise, there exists $\varphi \in \Delta \setminus R^*$, and hence $\varphi \in L^*$, which means that $v(\varphi) = 1$. Either way, $v(\Gamma \Rightarrow \Delta) = 1$.

Let $S_0 / s_0 \in \mathbf{G}$ and σ be a substitution such that $\sigma(frm(S_0 / s_0)) \subseteq \mathcal{F}$. We assume that $v(\sigma(S_0)) = 1$, and prove that $v(\sigma(s_0)) = 1$. Suppose that $S_0 = \{\Gamma_1 \Rightarrow \Delta_1, \ldots, \Gamma_n \Rightarrow \Delta_n\}$. We construct the following sequent $\Gamma \Rightarrow \Delta$: For every $1 \le i \le n$, there exists either $\psi_i \in \Gamma_i$ such that $v(\sigma(\psi_i)) = 0$ or $\psi_i \in \Delta_i$ such that $v(\sigma(\psi_i)) = 1$. If the first option holds, we add $\sigma(\psi_i)$ to Δ. If the second option holds, we add $\sigma(\psi_i)$ to Γ. Clearly, $v(\sigma(\Gamma \Rightarrow \Delta)) = 0$. In addition, $\Gamma \subseteq L^*$ and $\Delta \subseteq R^*$. Now, for every $1 \le i \le n$, using (id) and $(weak)$, we get that $S \vdash^{\mathcal{F}}_{\mathbf{G}} \sigma(\Gamma_i \Rightarrow \Delta_i) \cup (\Gamma \Rightarrow \Delta)$. Applying S_0 / s_0 with $\Gamma \Rightarrow \Delta$ as a context sequent, we get that $S \vdash^{\mathcal{F}}_{\mathbf{G}} \sigma(s_0) \cup (\Gamma \Rightarrow \Delta)$. Since $\Gamma \subseteq L^*$ and $\Delta \subseteq R^*$, $S \vdash^{\mathcal{F}}_{\mathbf{G}} \sigma(s_0) \cup (L^* \Rightarrow R^*)$. Let $\sigma(s_0) = \Gamma_0 \Rightarrow \Delta_0$. It follows that either $\Gamma_0 \not\subseteq L^*$ or $\Delta_0 \not\subseteq R^*$. Hence, $v(\psi) = 0$ for some $\psi \in \Gamma_0$ or $v(\psi) = 1$ for some $\psi \in \Delta_0$. Therefore, $v(\sigma(s_0)) = 1$. $\qquad \square$

Using the theorem above, we formulate a semantic property of calculi that is equivalent to k-analyticity.

Definition 11. A calculus $\mathbf{G}$ is called *semantically k-analytic* if every $\mathbf{G}$-legal bivaluation v can be extended to a $\mathbf{G}$-legal full bivaluation, provided that $dom(v)$ is finite and closed under k-subformulas.

Theorem 4. *A calculus $\mathbf{G}$ is k-analytic iff it is semantically k-analytic.*

Proof. If $\mathbf{G}$ is not k-analytic, then there is a set S of sequents and a sequent s such that $S \vdash_{\mathbf{G}} s$ and $S \not\vdash^{sub^k(S \cup \{s\})}_{\mathbf{G}} s$. Hence, there exists finite $S' \subseteq S$ such that $S' \vdash_{\mathbf{G}} s$, and $S' \not\vdash^{sub^k(S' \cup \{s\})}_{\mathbf{G}} s$. According to Theorem 3, there exists a $\mathbf{G}$-legal $sub^k(S' \cup \{s\})$-bivaluation v such that $v(S') = 1$ and $v(s) = 0$, and $u(S') \le u(s)$

for every **G**-legal full bivaluation u. Therefore, v cannot be extended to a **G**-legal full bivaluation. In addition, $dom(v) = sub^k(S' \cup \{s\})$ is finite and closed under k-subformulas.

For the converse, suppose that v is a **G**-legal bivaluation, $dom(v)$ is finite and closed under k-subformulas, and v cannot be extended to a **G**-legal full bivaluation. Let $\Gamma = \{\psi \in dom(v) \mid v(\psi) = 1\}$, $\Delta = \{\psi \in dom(v) \mid v(\psi) = 0\}$, and $s = \Gamma \Rightarrow \Delta$. Then $dom(v) = sub^k(s)$ and $v(s) = 0$. We show that $u(s) = 1$ for every **G**-legal full bivaluation u. Indeed, every such u does not extend v. Hence there is some $\psi \in dom(v)$ such that $u(\psi) \neq v(\psi)$. Then, $u(\psi) = 0$ if $\psi \in \Gamma$, and $u(\psi) = 1$ if $\psi \in \Delta$. In either case, $u(s) = 1$. By Theorem 3, $\vdash_{\mathbf{G}}^{sub^k(s)} s$ and $\vdash_{\mathbf{G}} s$. □

We use the semantic characterization of analyticity given in Theorem 4 to prove Theorem 2. Thus, we provide a method for extending bivaluations whose domains are finite and closed under k-subformulas.

This method is iterative: in each step we extend a given bivaluation v with a truth value for a single formula ψ, such that $dom(v) \cup \{\psi\}$ is closed under k-subformulas. We call such formulas k-*addable*:

Definition 12. A formula ψ is called k-*addable* to a bivaluation v if $dom(v)$ contains all proper k-subformulas of ψ.

The extension of partial bivaluations is determined according to the basic calculus **B**, as given in the following definition:

Definition 13. Let v be a bivaluation and ψ be a formula. The $dom(v) \cup \{\psi\}$-bivaluation $v_{\mathbf{B}}^{\psi}$ is defined as follows: 1) $v_{\mathbf{B}}^{\psi}(\varphi) = v(\varphi)$ for every $\varphi \in dom(v)$. 2) If $\psi \notin dom(v)$: $v_{\mathbf{B}}^{\psi}(\psi) = 1$ iff there exist a rule of the form $S \mathbin{/} \Rightarrow \varphi$ in **B** and a substitution σ such that $\sigma(frm(S)) \subseteq dom(v)$, $\sigma(\varphi) = \psi$ and $v(\sigma(S)) = 1$.

If the above extension method "works" for a given calculus **G**, we say that **G** is **B**-k-*analytic*. Formally, this is defined as follows.

Definition 14. A calculus **G** is called **B**-k-*analytic* if $v_{\mathbf{B}}^{\psi}$ is **G**-legal for every **G**-legal bivaluation v whose domain is finite and closed under k-subformulas and formula ψ that is k-addable to v.

Proposition 7. *Every* **B**-k-*analytic calculus is* k-*analytic.*

Proof. Let **G** be a **B**-k analytic calculus. By Theorem 4, it suffices to prove that **G** is semantically k-analytic. Let v be a **G**-legal bivaluation whose domain is finite and closed under k-subformulas. We extend v to a **G**-legal full bivaluation v'. It is a routine matter to enumerate all formulas and obtain an infinite sequence $\psi_1, \psi_2, \ldots$ such that: a) If $\psi_i \in dom(v)$ and $\psi_j \notin dom(v)$ then $i < j$. b) If ψ_i is a k-subformula of ψ_j then $i \leq j$. We define a sequence of bivaluations $v_0, v_1, \ldots$ as follows: $v_0 = v$, and $v_i = v_{i-1}{}_{\mathbf{B}}^{\psi_i}$ for every $i > 0$. $dom(v_i) = dom(v) \cup \{\psi_1, \ldots, \psi_i\}$ for every i, and therefore each ψ_i is k-addable to v_{i-1}. Since **G** is **B**-k-analytic, each v_i is **G**-legal. The full bivaluation v' is defined by $v'(\psi_i) = v^i(\psi_i)$ for every $i > 0$. In order to see that v' is **G**-legal, let $S\,/\,s \in \mathbf{G}$ and let σ be a substitution. Let $j = \max\{i \mid \psi_i \in \sigma(frm(S\,/\,s))\}$. Then $v'(\psi) = v_j(\psi)$ for every $\psi \in \sigma(frm(S\,/\,s))$. Recall that v_j is **G**-legal, and therefore we have that $v'(\sigma(S)) = v_j(\sigma(S)) \leq v_j(\sigma(s)) = v'(\sigma(s))$. □

Next, we prove that **B**-k-analyticity is preserved when a calculus is augmented with one k-safe application of a rule of **B**.

Theorem 5. *Let* **G** *be a* **B**-k-*analytic calculus, and* **G**′ *be a calculus obtained by augmenting* **G** *with a* k-*safe application* $\hat{r}$ *of a rule* r *of* **B**. *Then* **G**′ *is* **B**-k-*analytic.*

Proof. Suppose $r = S\,/\,s$ with $S = \{s_1, \ldots, s_n\}$, and $\hat{r} = \hat{S}\,/\,\hat{s}$. Let α be a substitution and c be a sequent such that $\hat{S} = \{\alpha(s_1) \cup c, \ldots, \alpha(s_n) \cup c\}$ and $\hat{s} = \alpha(s) \cup c$. Now, let v be a **G**′-legal bivaluation whose domain is finite and closed under k-subformulas, and ψ be a formula that is k-addable to v. We prove that the bivaluation $v_{\mathbf{B}}^{\psi}$ is **G**′-legal. Let $S_0\,/\,s_0 \in \mathbf{G}'$ and σ be a substitution such that $\sigma(frm(S_0\,/\,s_0)) \subseteq dom(v_{\mathbf{B}}^{\psi})$. We show that $v_{\mathbf{B}}^{\psi}(\sigma(S_0)) \leq v_{\mathbf{B}}^{\psi}(\sigma(s_0))$. If $S_0\,/\,s_0 \in \mathbf{G}$ then this holds since **G** is **B**-k-analytic. If $\psi \notin \sigma(frm(S_0\,/\,s_0))$ or $\psi \in dom(v)$ then this holds since v is **G**′-legal. Assume now that $S_0\,/\,s_0 = \hat{r}$, $\psi \in \sigma(frm(S_0\,/\,s_0))$ and $\psi \notin dom(v)$.

We first prove that $\psi = \sigma(\alpha(\varphi_r))$. Otherwise, $\sigma(\alpha(\varphi_r)) \in dom(v)$. By Proposition 5, $frm(\hat{r}) \subseteq sub^k(\alpha(\varphi_r))$, and by Proposition 2, we also have that $\sigma(sub^k(\alpha(\varphi_r))) \subseteq sub^k(\sigma(\alpha(\varphi_r)))$, and hence $\sigma(frm(\hat{r})) \subseteq sub^k(\sigma(\alpha(\varphi_r)))$. Since $dom(v)$ is closed under k-subformulas and $\sigma(\alpha(\varphi_r)) \in dom(v)$, we have that $sub^k(\sigma(\alpha(\varphi_r))) \subseteq dom(v)$, and hence $\sigma(frm(\hat{r})) \subseteq dom(v)$. Since $\psi \in \sigma(frm(\hat{r}))$, it follows that $\psi \in dom(v)$, which is a contradiction.

Similarly, we show that $\sigma(frm(\hat{S})) \subseteq dom(v)$. Indeed, let $\varphi \in \sigma(frm(\hat{S}))$ and let $\varphi' \in frm(\hat{S})$ such that $\varphi = \sigma(\varphi')$. By Proposition 5, φ' is a proper k-subformula of $\alpha(\varphi_r)$, and hence by Proposition 2, φ is a proper k-subformula of $\psi = \sigma(\alpha(\varphi_r))$. In particular, $\varphi \neq \psi$. Since $\sigma(frm(\hat{S})) \subseteq dom(v_{\mathbf{B}}^{\psi})$, it follows that $\varphi \in dom(v)$.

Now suppose $v_{\mathbf{B}}^{\psi}(\sigma(\alpha(s_i) \cup c)) = 1$ for every $1 \leq i \leq n$. We prove that $v_{\mathbf{B}}^{\psi}(\sigma(\alpha(s) \cup c)) = 1$. If $v_{\mathbf{B}}^{\psi}(\sigma(c)) = 1$ then we are clearly done. Suppose otherwise. Then we have $v_{\mathbf{B}}^{\psi}(\sigma(\alpha(S))) = 1$. We prove that $v_{\mathbf{B}}^{\psi}(\sigma(\alpha(s))) = 1$ (it would then follow that $v_{\mathbf{B}}^{\psi}(\sigma(\alpha(s) \cup c)) = 1$). Since $\sigma(frm(\hat{S})) \subseteq dom(v)$, we also have $\sigma(\alpha(frm(S))) \subseteq dom(v)$. Hence, $v(\sigma(\alpha(S))) = 1$. We distinguish two cases. If s is $\Rightarrow \varphi_r$ then since $\sigma(\alpha(frm(S))) \subseteq dom(v)$, $\sigma(\alpha(\varphi_r)) = \psi$ and $v(\sigma(\alpha(S))) = 1$, by Definition 13, we have $v_{\mathbf{B}}^{\psi}(\psi) = 1$, and so $v_{\mathbf{B}}^{\psi}(\sigma(\alpha(s))) = 1$. Otherwise s is $\varphi_r \Rightarrow$. To prove that $v_{\mathbf{B}}^{\psi}(\sigma(\alpha(s))) = 1$, we show that $v_{\mathbf{B}}^{\psi}(\psi) = 0$. By Definition 13, it suffices to prove that for every rule of the form $S'\,/ \Rightarrow \varphi'$ in **B** and substitution σ' such that $\sigma'(frm(S')) \subseteq dom(v)$ and $\sigma'(\varphi') = \psi$, we have

$v(\sigma'(S')) = 0$. Let $S' / \Rightarrow \varphi'$ and σ' as above. Since $\mathbf{B}$ is (cut)-guarded, the empty sequent is derivable from $\sigma(\alpha(S)) \cup \sigma'(S')$ using only (cut). It easily follows that $\sigma(\alpha(S)), \sigma'(S') \vdash_{\mathbf{G'}}^{dom(v)} \Rightarrow$. By Theorem 3, since v is $\mathbf{G'}$-legal and $v(\sigma(\alpha(S))) = 1$, we must have $v(\sigma'(S')) = 0$.   □

Finally, we obtain Theorem 2 as a corollary:

Proof (of Theorem 2). Let $\mathbf{G}$ be a calculus that consists solely of k-safe applications of rules of $\mathbf{B}$. Begin with the empty calculus and add the rules of $\mathbf{G}$ one by one. The empty calculus is clearly $\mathbf{B}$-k-analytic, and by Theorem 5, in each step we obtain a $\mathbf{B}$-k-analytic calculus. By Proposition 7, $\mathbf{G}$ is k-analytic.   □

6   Further Research

While we focused on the language of classical logic for the sake of simplicity and clarity, the definitions and results of this paper can be straightforwardly adapted for arbitrary propositional languages. In addition, the following extensions and questions naturally arise and are left for a future work. First, unlike the case of canonical calculi [4], the relations between cut-elimination and analyticity in pure calculi are still unclear. We plan to apply semantic methods (see, e.g., [9]) to investigate cut-elimination in pure calculi. Second, while this paper studies only pure calculi, that have a simple semantic interpretation, we believe that a similar approach can be useful for more complicated families of sequent calculi. In particular, the family of basic sequent calculi that was studied in [9], and has a Kripke-style semantic interpretation, is an interesting subject for a similar investigation of analyticity. Lastly, it will be interesting and useful to extend the current method also for many-sided sequents (using many-valued valuation functions), as well as for calculi for first-order logics.

References

1. Anderson, A.R., Belnap, N.D.: Entailment: The Logic of Relevance and Neccessity, vol. I. Princeton University Press (1975)
2. Avron, A.: Simple consequence relations. Inf. Comput. 92(1), 105–139 (1991)
3. Avron, A., Konikowska, B., Zamansky, A.: Modular construction of cut-free sequent calculi for paraconsistent logics. In: Proceedings of the 27th Annual IEEE Symposium on Logic in Computer science (LICS), pp. 85–94 (2012)
4. Avron, A., Lev, I.: Non-deterministic multiple-valued structures. Journal of Logic and Computation 15(3), 241–261 (2005)
5. Beklemishev, L., Gurevich, Y.: Propositional primal logic with disjunction. Journal of Logic and Computation 24(1), 257–282 (2014)
6. Béziau, J.-Y.: Sequents and bivaluations. Logique et Analyse 44(176), 373–394 (2001)
7. Cotrini, C., Gurevich, Y.: Basic primal infon logic. Journal of Logic and Computation (2013)

8. Kamide, N.: A hierarchy of weak double negations. Studia Logica 101(6), 1277–1297 (2013)
9. Lahav, O., Avron, A.: A unified semantic framework for fully structural propositional sequent systems. ACM Trans. Comput. Logic 27, 1–33 (2013)
10. Lahav, O., Zohar, Y.: Sat-based decision procedure for analytic pure sequent calculi. To appear in Proceedings of the 7th International Joint Conference on Automated Reasoning, IJCAR (2014)
11. Miller, D., Pimentel, E.: A formal framework for specifying sequent calculus proof systems. Theoretical Computer Science 474(0), 98–116 (2013)
12. Sette, A.M.: On the propositional calculus P1. Mathematica Japonicae 18(13), 173–180 (1973)

On the Broader Epistemological Significance of Self-Justifying Axiom Systems

Dan E. Willard

University at Albany Computer Science and Mathematics Departments

Abstract. This article will be a continuation of our research into self-justifying systems. It will introduce several new theorems (one of which will transform our previous infinite-sized self-verifying logics into formalisms or purely finite size). It will explain how self-justification is useful, even when the Incompleteness Theorem clearly limits its scope.

1 Introduction

Gödel's Incompleteness Theorem has two parts. Its first half indicates no decision procedure can identify arithmetic's true statements. Its "Second Incompleteness" result specifies sufficiently strong logics *cannot* verify their own consistency. Gödel was careful to insert a caveat into his historic paper [11], indicating a *diluted* form of Hilbert's Consistency Program might have some success:

> * *"It must be expressly noted Proposition XI represents no contradiction of the formalistic standpoint of Hilbert. For this standpoint presupposes only the existence of a consistency proof by finite means, and there might conceivably be finite proofs which cannot be stated in P or in ..."*

Some scholars have interpreted * as, possibly, anticipating attempts to confirm Peano Arithmetic's consistency, via either Gentzen's formalism or Gödel's Dialetica interpretation. On the other hand, the Stanford's Encyclopedia's entry about Gödel quotes him, in its Section 2.2.4, stating he was hesitant to view the Second Incompleteness Theorem as fully ubiquitous, until learning of Turing's work. Moreover, Yourgrau [45] states von Neumann *"argued against Gödel himself"* in the early 1930's, about the definitive termination of Hilbert's consistency program, which *"for several years"* after [11]'s publication, Gödel *"was cautious not to prejudge"*. Also, it is known [6,13,45] that Gödel did initially presume the second theorem was false, before proving its stunning result.

In any case several year after he wrote *'s initial statement, Gödel gave a 1933 lecture [12], where he told his audience that Hilbert's initial 1926 objectives, summarized formally by ** below, had *"unfortunately"* no *"hope of succeeding along"* its originally intended plans.

> ** (Hilbert [17] 1926): *"Where else would reliability and truth be found if even mathematical thinking fails? The definitive nature of the infinite has become necessary, not merely for the special interests of individual sciences, but rather for the honor of human understanding itself."*

U. Kohlenbach et al. (Eds.): WoLLIC 2014, LNCS 8652, pp. 221–236, 2014.
© Springer-Verlag Berlin Heidelberg 2014

Our research, in both the current article and prior papers [35–44], was stimulated by the prospect that we find ∗∗ enticing, even though the Second Incompleteness Theorem *unequivocally* demonstrates that logics *cannot* recognize their own consistency *in a robust sense*. Accordingly, we have studied *both* generalizations and boundary-case exceptions for the Second Incompleteness Theorem in [35–44]. The current article will seek to *both* strengthen these prior results, in the context of axiom systems with *strictly finite cardinalities*, and to also provide a more intuitive explanation of the meaning behind [35–44]'s results.

The thesis of this article will be delicate because there can be no doubt that the Second Incompleteness Theorem is sharply robust, when viewed from a conventional purist mathematical perspective. On the other hand, we will argue that there are certain facets of a "Self-Justifying Logics", that are tempting under a hard-nosed engineering perspective, contemplating sharply *curtailed forms* of Hilbert's goals. These results will be fragile *but not fully immaterial*.

2   Background Setting

Let (α, d) denote any axiom system and deduction method satisfying the simple "**Split Rule**" below [1]. This pair will be called "**Self Justifying**" when:

i   one of α's theorems will state that the deduction method d, applied to the system α, will produce a consistent set of theorems, and

ii   the axiom system α is in fact consistent.

For any (α, d), it is easy to construct a second $\alpha^d \supseteq \alpha$ that satisfies the Part-i requirement. For instance, α^d could consist of all of α's axioms plus an added "**SelfRef**(α, d)" sentence, defined as stating:

- There is no proof (using d's deduction method) of $0 = 1$ from the *union* of the system α with *this* sentence "SelfRef(α, d)" (looking at itself).

Kleene [20] noted how to encode rough analogs of "SelfRef(α, d)". Each of Kleene, Rogers and Jeroslow [19, 20, 29] noted α^d may, however, be inconsistent (despite SelfRef(α, d)'s assertion), thus causing it to violate Part-ii's requirement.

This problem arises in many contexts besides Gödel's paradigm, where α was an extension of Peano Arithmetic (see [1–5, 7, 9, 11, 14–16, 18, 21–23, 25–34, 38, 39, 43]). Such formalize paradigms where self-justification is infeasible, due to diagonalization issues. (It should, perhaps, be added that among this lengthy list of articles, it was especially [1, 4, 11, 23, 27, 31, 34]'s incompleteness results that influenced our work in [35–44].) In any case, the main point is that most logicians have hesitated to employ an analog of a SelfRef(α, d) axiom because $\alpha^d = \alpha + \text{SelfRef}(\alpha, d)$ is typically inconsistent.

---

[1] Our "Split Rule" is the trivial requirement that all the axiom sentences in α are technically *proper axioms*, and that deduction method d is required to include **BOTH** a finite number of rules of inference and whatever "logical axioms" are needed *(if any ?)* by d's methodology. (This trivial notation convention is helpful.)

Our research in [35, 37, 40–42] focused on paradigms where self-justification is feasible. It involved weakening the properties a logic can prove about addition and/or multiplication (to avoid potential difficulties). To be more precise, let $Add(x, y, z)$ and $Mult(x, y, z)$ denote 3-way predicates specifying $x + y = z$ and $x * y = z$. Then a logic will be said to **recognize** successor, addition and multiplication as **Total Functions** iff it includes sentences 1-3 as axioms.

$$\forall x \; \exists z \quad Add(x, 1, z) \tag{1}$$
$$\forall x \; \forall y \; \exists z \quad Add(x, y, z) \tag{2}$$
$$\forall x \; \forall y \; \exists z \quad Mult(x, y, z) \tag{3}$$

A logic α will be called **Type-M** iff it contains 1-3 as axioms, **Type-A** iff it contains only (1) and (2) as axioms, **Type-S** iff it contains only (1) as an axiom, and **Type-NS** iff it contains none of these axioms. The relationship of these constructs to self-justification is explained by items (a) and (b):

a. The existence of Type-A systems that can recognize their own consistency under semantic tableaux deduction, while proving analogs of all Peano Arithmetic's Π_1 theorems (in a slightly different language), was demonstrated in [40]. Also, [37, 41] noted that some specialized forms of Type-NS systems can likewise recognize their own Hilbert consistency.

b. The above evasions of the Second Incompleteness Theorem are known to be near-maximal in a mathematical sense. This is because the combined work of Pudlák, Solovay, Nelson and Wilkie-Paris [24, 27, 31, 34] implied no natural Type-S system can recognize its Hilbert consistency, and Willard subsequently [38, 43, 44] hybridized their formalisms with some techniques of Adamowicz-Zbierski [1, 2] to establish that most Type-M systems cannot recognize their own semantic tableaux consistency.

Other fascinating efforts to evade the Second Incompleteness Theorem have used the Kreisel-Takeuti "CFA" system [22] or the the *interpretational framework* of Friedman, Nelson, Pudlák and Visser [10, 24, 27, 33]. These systems are unrelated to our approach because they do not use Kleene-like *"I am consistent"* axiom-sentences. Instead, CFA uses the special properties of "second order" generalizations of Gentzen's *cut-free* Sequent Calculus, and the interpretational approach formalizes how some systems recognize their Herbrand consistency on localized sets of integers, which unbeknown to themselves, includes all integers. (These alternate results are interesting but unrelated to our approach.)

3   Defining Notation and Earlier Results

A function F will be called **Non-Growth** iff $F(a_1...a_j) \leq Maximum(a_1...a_j)$ holds. Six examples of non-growth functions are *Integer Subtraction* (where $x - y$ is defined to equal zero when $x \leq y$), *Integer Division* (where $x \div y$ equals x when $y = 0$, and it equals $\lfloor x/y \rfloor$ otherwise), $Maximum(x, y)$, $Logarithm(x)$ $Root(x, y) = \lceil x^{1/y} \rceil$ and $Count(x, j)$ designating the number of "1" bits among x's rightmost j bits. The term **U-Grounding Function** referred in [40] to a set of primitives, which included the preceding functions plus the *growth operations* of addition and $Double(x) = x + x$. Our language L^* was built out of these symbols, plus the primitives of "0", "1", "$=$" and "$\leq$".

In a context where t is any term in [40]'s language L^*, the quantifiers in the wffs $\forall\, v \leq t\ \Psi(v)$ and $\exists\, v \leq t\ \Psi(v)$ were called *bounded quantifiers*. Any formula in L^*, all of whose quantifiers are bounded, was called a Δ_0^* formula. The Π_n^* and Σ_n^* formulae were then defined by the usual rules that:

1. Every Δ_0^* formula is considered to be "Π_0^*" and also "Σ_0^*".
2. A wff is called Π_n^* when it is encoded as $\forall v_1 \ldots \forall v_k\ \Phi$ with Φ being Σ_{n-1}^*
3. Also, a wff is called Σ_n^* when it is encoded as $\exists v_1..\exists v_k\ \Phi$, where Φ is Π_{n-1}^*.

Our article [40] used the symbol D to denote a deduction method. In addition to using Fitting's version of semantic tableaux methodology [8], it defined an alternative, called **Tab-k deduction**, that consisted of a speeded-up version of a tableaux, which permitted a *limited analog* of Gentzen-style deductive cuts for Π_k^* and Σ_k^* formulae.

Thus, if H denotes a sequence of ordered pairs (t_1, p_1), (t_2, p_2), ... (t_n, p_n), where p_i is a Semantic Tableaux proof of the theorem t_i, then H was called a **"Tab-k Proof"** of a theorem T from α's axioms iff $T = t_n$ and also:

1. Each of the "intermediately derived theorems" $t_1, t_2, \ldots, t_{n-1}$ have a complexity no greater than that of either a Π_k^* or Σ_k^* sentence.
2. Each axiom in p_i's proof either comes from α or is one of $t_1, t_2, \ldots, t_{i-1}$.

Let us say an axiom system α has a **Level-J Understanding** of its own consistency under a deduction method D iff α can prove that there exists no proofs using its axioms and D's deduction of both a Π_j^* theorem and its negation. In this notation, items A and B summarize [36,38–40,43]'s main results:

A. For any axiom system A using L^*'s U-Grounding language, [40] showed its $\mathrm{IS}_D(A)$ formalism could prove all A's Π_1^* theorems and simultaneously verify its Level-1 consistency under Tab-1 deduction.

B. Two negative results, tightly complementing item A's positive result, were exhibited in [36, 38, 39, 43]. The first was that [36, 38, 43] showed most systems are unable to verify their Level-0 consistency under semantic tableaux deduction, when they included statement (3)'s "Type-M" axiom that multiplication is a total function. Moreover, [39] offered an alternate form of this incompleteness result, showing statement (2)'s *far weaker* Type-A systems cannot verify their Level-0 consistency under Tab-2 deduction.

The contrast between these positive and negative results had led to our conjecture that automated theorem provers are likely to eventually achieve a fragmentary part of the ambitions that were suggested by Hilbert in $**$. This is because the question of whether a formalism can support an *idealized Utopian* conception of its own consistency is *different* from exploring the degrees to which theorem-provers can possess a *fragmentary knowledge* of their own consistency. The Incompleteness Theorem has demonstrated an Utopian idealized form of self-justification is unobtainable, but our research has found some diluted cousins of this construct that are feasible

4   The $IS_D(A)$ Axiom System

In a context where A denotes any axiom system using L^*'s U-Grounding language, $IS_D(A)$ was defined in [40] to be an axiomatic formalism capable of recognizing all of A's Π_1^* theorems and corroborating its own Level-1 consistency under D's deductive method. It consisted of the following four groups of axioms:

Group-Zero: Two of the Group-zero axioms will define the constant-symbols, $\bar{c}_0$ and $\bar{c}_1$, designating the integers of 0 and 1. The Group-zero axioms will also define the growth functions of addition and $Double(x) = x + x$. The net effect of these axioms will be to set up a machinery to define any integer $n \geq 2$ using fewer than $3 \cdot \lceil \text{Log } n \rceil$ logic symbols.

Group-1: This axiom group will consist of a finite set of Π_1^* sentences, denoted as F, which can prove any Δ_0^* sentence that holds true under the standard model of the natural numbers. (Any finite set of Π_1^* sentences F with this property may be used to define Group-1, as [40] noted.)

Group-2: Let $\ulcorner \Phi \urcorner$ denote Φ's Gödel Number, and $\text{HilbPrf}_A(\ulcorner \Phi \urcorner, p)$ denote a Δ_0^* formula indicating p is a Hilbert-styled proof of theorem Φ from axiom system A. For each Π_1^* sentence Φ, the Group-2 schema will contain an axiom of form (4). (Thus $IS_D(A)$ can trivially prove all A's Π_1^* theorems.)

$$\forall p \ \{ \ \text{HilbPrf}_A(\ulcorner \Phi \urcorner, p) \ \Rightarrow \ \Phi \ \} \tag{4}$$

Group-3: The final part of the $IS_D(A)$ will be a self-referencing Π_1^* axiom, indicating $IS_D(A)$ meets §3's criteria of being "Level-1 consistent" under deductive method D. It is, thus, the following declaration:

> \# *No two proofs exist for a Π_1^* sentence and its negation, when D's deductive method is applied to an axiom system, consisting of the union of Groups 0, 1 and 2 with* **this sentence** *(looking at itself).*

One encoding of \#, as a self-referencing Π_1^* axiom, appears in [40]. Thus, (5) is a Π_1^* styled encoding for \# when: 1) $\text{Prf}_{IS_D(A)}(a, b)$ is a Δ_0^* formula indicating that b is a proof of a theorem a under $IS_D(A)$'s axiom system and D's deduction method, and 2) $Pair(x, y)$ is a Δ_0^* formula indicating that x is a Π_1^* sentence and that y represents x's negation.

$$\forall x \, \forall y \, \forall p \, \forall q \quad \neg \ [\ Pair(x, y) \ \wedge \ \text{Prf}_{IS_D(A)}(x, p) \ \wedge \ \text{Prf}_{IS_D(A)}(y, q) \ ] \tag{5}$$

Notation. An operation $I(\ \bullet\ )$ that maps an initial axiom system A onto an alternate system $I(A)$ will be called **Consistency Preserving** iff $I(A)$ is consistent whenever all of A's axioms hold true under the standard model of the natural numbers. In this context, [40] demonstrated:

Theorem 1. *Suppose the symbol D denotes either semantic tableaux deduction or its Tab-1 generalization. Then the $IS_D(\ \bullet\ )$ mapping operation is consistency preserving (e.g. $IS_D(A)$ will be consistent whenever all of A's axioms hold true under the standard model of the natural numbers).*

We emphasize the most difficult part of [40]'s result was neither the definition of its $IS_D(A)$'s axiom system nor the Π_1^* fixed-point encoding of (5)'s Group-3 axiom. Instead, the key challenge was the confirming of Theorem 1's "Consistency Preservation" property.

The confirming of this property is subtle because its invariant breaks down when D is a deduction method only slightly stronger than either semantic tableaux or Tab-1 deduction. Thus, Pudlák's and Solovay's work [27,31] implies Theorem 1's analog fails when D represents Hilbert deduction, and [39] showed its generalization fails even when D represents Tab-2 deduction.

5   A Finitized Generalization of Theorem 1's Methodology

One difficulty with $IS_D(A)$ is that it employs an infinite number of different incarnations of sentence (4) in its Group-2 scheme (since it contains one incarnation of this sentence for each Π_1^* sentence Φ in L^*'s language). Such a Group-2 schema is awkward because it simulates A's Π_1^* knowledge almost via a brute-force enumeration.

Our Definition 1 and Theorems 2 and 3 will show how to mostly overcome this problem by compressing the infinite number of instances of sentence (4) in $IS_D(A)$'s Group-2 schema into a purely finite structure.

Definition 1. Let β denote any finite set of axioms that have Π_1^* encodings. Then $IS_D^{\#}(\beta)$ will denote an axiom system, similar to $IS_D(A)$, except its Group-2 scheme will employ β's set of axioms, instead of using an infinite number of applications of statement (4)'s scheme. (Thus, the *"I am consistent"* statement in $IS_D^{\#}(\beta)$'s Group-3 axiom will be the same as before, except that the *"I am"* fragment of its self-referencing statement will reflect these changes in Group-2 in the obvious manner.)

Theorem 2. *Let D again denote either semantic tableaux or Tab-1 deduction, and β again denote a set of Π_1^* axioms. Then $IS_D^{\#}(\beta)$ will be consistent whenever all β's axioms hold true under the standard model. (In other words, $IS_D^{\#}(\beta)$ will satisfy an analog of Theorem 1's consistency preservation property for $IS_D(A)$.)*

Theorem 2's proof is almost identical to [40]'s proof of Theorem 1. It will. not be repeated in this extended abstract. Instead, this section will apply Theorem 2 to show how **finite-sized** self-justifying logics can provide an **infinite amount** of "kernelized" Π_1^* information.

Definition 2. Let $\text{Test}_i(t, x)$ denote any Δ_0^* formula, and $\ulcorner \Psi \urcorner$ denote Ψ's Gödel number. Then $\text{Test}_i(t, x)$ will be called a **Kernelized Formula** iff Peano Arithmetic can prove every Π_1^* sentence Ψ satisfies (6)'s identity:

$$\Psi \quad \Longleftrightarrow \quad \forall x \; \text{Test}_i(\ulcorner \Psi \urcorner, x) \tag{6}$$

There are infinitely many Δ_0^* predicates $\text{Test}_1(t, x)$, $\text{Test}_2(t, x)$, $\text{Test}_3(t, x)$ ... satisfying this kernelized condition (one of which is illustrated by Example 1). An enumerated list of all the available kernels is called a **Kernel-List**.

Example 1. The set of true Σ_1^* sentences is r.e. This implies there exists a Δ_0^* formula, called say $\mathrm{Probe}(g, x)$, such that g is the Gödel number of a Σ_1^* statement that holds true in the Standard Model iff (7) is true:

$$\exists\, x \quad \mathrm{Probe}(g, x) \,\wedge\, x \geq g \tag{7}$$

Now, let $\mathrm{Pair}(t, g)$ denote a Δ_0^* formula that specifies t is the Gödel number of a Π_1^* statement and g is the Σ_1^* formula which is its negation. Then our notation implies that t is a true Π_1^* statement if and only if (8) holds true:

$$\forall\, x \quad \neg\,[\,\exists\, g \leq x \quad \mathrm{Pair}(t, g) \,\wedge\, \mathrm{Probe}(g, x)\,] \tag{8}$$

Thus if $\mathrm{Test}_0(t, x)$ denotes the Δ_0^* formula of $\neg\,[\,\exists\, g \leq x\ \mathrm{Pair}(t, g) \wedge \mathrm{Probe}(g, x)]$, it is one example of what Definition 2 would call a "Kernelized Formula".

Definition 3. Let us recall Definition 2 defined **Kernel-List** to be an enumeration of all the kernelized formulae $\mathrm{Test}_1(t, x)$, $\mathrm{Test}_2(t, x)$, $\mathrm{Test}_3(t, x)$.... . Assuming $\mathrm{Test}_i(t, x)$ is the i–th element in this list and Ψ is an arbitrary Π_1^* sentence, Ψ's **i-th Kernel Image** will be defined as the following Π_1^* sentence:

$$\forall\, x \quad \mathrm{Test}_i(\,\ulcorner \Psi \urcorner, x\,) \tag{9}$$

Example 2. The Definitions 2 and 3 suggest that there is a subtle relationship between a sentence Ψ and its i–th Kernel Image. This is because Definition 2 indicates that Peano Arithmetic can prove the invariant (6), indicating that Ψ is equivalent to its i–th Kernel Image. However, a weak axiom system can be plausibly uncertain about whether this equivalence holds.

Thus if a weak axiom system proves statement (9) (rather than Ψ), it may not be able to equate these results. This problem will apply to Theorem 3's formalism. However, Theorem 3 will be still of much interest because §6 will illustrate a methodology that overcomes many of Theorem 3's limitations.

Theorem 3. *Let A denote any system, whose axioms hold true in arithmetic's standard model, and i denote the index of any of Definition 2's kernelized formulae $\mathrm{Test}_i(t, x)$. Then it is possible to construct a finite-sized collection of Π_1^* sentences, called say $\beta_{A,i}$, where $IS_D^{\#}(\beta_{A,i})$ satisfies the following invariant:*

If Ψ is one of the Π_1^* theorems of A then $IS_D^{\#}(\beta_{A,i})$ can prove (9)'s statement (e.g. it will prove the "the i–th kernelized image" of Ψ).

Proof Sketch: Our justification of Theorem 3 will use the following notation:

1. $\mathrm{Check}(t)$ will denote a Δ_0^* formula that produces a Boolean value of "True" when t represents the Gödel number of a Π_1^* sentence.
2. $\mathrm{HilbPrf}_A(t, q)$ will denote a Δ_0^* formula that indicates q is a Hilbert-style proof of the theorem t from axiom system A.
3. For any kernelized $\mathrm{Test}_i(t, x)$ formula, $\mathrm{GlobSim}_i$ will denote (10)'s Π_1^* sentence. (It will be called A's i–th **"Global Simulation Sentence"**.)

$$\forall t \ \forall q \ \forall x \ \{ \ [\ \text{HilbPrf}_A (t,q) \ \wedge \ \text{Check}(t) \ ] \ \implies \ \text{Test}_i(t,x) \ \} \ (10)$$

In this notation, the requirements of Theorem 3 will be satisfied by any version of the axiom system $\text{IS}_D^{\#}(\beta)$, whose Group-2 schema β is a finite sized consistent set of Π_1^* sentences that has (10) as an axiom. (This includes the minimal sized such system, that has only (10) as an axiom.) This is because if Ψ is any Π_1^* theorem of A, whose proof is denoted as $\bar{p}$, then both the Δ_0^* predicates of $\text{HilbPrf}_A (\ulcorner \Psi \urcorner, \bar{p})$ and $\text{Check}(\ulcorner \Psi \urcorner)$ are true. Moreover, $\text{IS}_D^{\#}(\beta)$'s Group-1 axiom subgroup was defined so that it can automatically prove all Δ_0^* sentences that are true. Thus, $\text{IS}_D^{\#}(\beta)$ will prove these two statements and hence corroborate (via axiom (10)) the further statement:

$$\forall x \ \text{Test}_i (\ulcorner \Psi \urcorner , x) \tag{11}$$

Hence for each of the infinite number of Π_1^* theorems that A proves, the above defined formalism will prove a matching statement that corresponds to the i−th kernelized image of each such proven theorem.   □

6   L-Fold Generalizations of Theorem 3

Theorem 3 is of interest because every axiom system A will have its formalism $\text{IS}_D^{\#}(\beta_{A,i})$ prove the i−th kernelized image of every Π_1^* theorem that A proves. This fact is helpful because (6)'s invariance holds for all Π_1^* sentences. Moreover, our "U-Grounded" Π_1^* sentences capture all Conventional Arithmetic's *crucial* Π_1 information because they can view multiplication as a 3-way Δ_0^* predicate $\text{Mult}(x,y,z)$ via (12)'s encoding of this predicate.

$$[\ (x = 0 \vee y = 0) \Rightarrow z = 0] \ \wedge \ [\ (x \neq 0 \wedge y \neq 0) \ \Rightarrow \ (\frac{z}{x} = y \wedge \frac{z-1}{x} < y \ )] \ (12)$$

One difficulty with $\text{IS}_D^{\#}(\beta)$ and $\text{IS}_D^{\#}(\beta_{A,i})$ was mentioned by Example 2. It was that while Peano Arithmetic can corroborate (6)'s invariance for every Π_1^* sentence Ψ , these latter systems cannot also do so.

While there will probably never be a perfect method for fully resolving this challenge, there is a pragmatic engineering-style solution that is often available. This is essentially because our proof of Theorem 3 employed a formalism β that used essentially only one axiom sentence (e.g. (10)'s Π_1^* declaration).

Since the $\text{IS}_D^{\#}(\beta)$ formalism was intended for use by any finite-sized system β, it is clearly possible to include any finite number of formally true Π_1^* sentences in β. Thus for some fixed constant L, one can easily let β include L copies of (10)'s axiom framework for a finite number of different $\text{Test}_1, \text{Test}_2 \ldots \text{Test}_L$ predicates, each of which satisfy Definition 2's criteria for being kernelized formulae. In this case, $\text{IS}_D^{\#}(\beta)$ will formally map each initial Π_1^* theorem Ψ of some axiom system A onto L resulting different Π_1^* theorems of the form (9).

Remark 1. Our basic conjecture is, essentially, that a goodly number of issues, concerning logic-based engineering applications called say E, may have convenient solutions via self-justifying logics, that follow the preceding outlined L-fold strategy. Thus, we are suggesting that if β is a large-but-finite set of axioms, that consists of L copies of (10)'s axiom framework for different $\text{Test}_1 ... \text{Test}_L$ predicates, then some future engineering applications E may possibly have their needs met by an $\text{IS}_D^{\#}(\beta)$ formalisms, when a software engineer meticulously chooses an appropriately constructed finite-sized β.

Remark 2. The preceding was not meant to overlook that the Second Incompleteness Theorem is a robust result, applying to all logics of sufficient strength. Our suggestion, however, is that computers are becoming so powerful, in both speed and memory size as the 21st century is progressing, that there will likely emerge engineering-style applications E that will benefit from $\text{IS}_D^{\#}(\beta)$'s self-referencing formalisms when a *large-but-finite-sized* β is delicately chosen. Moreover, it is of interest to speculate whether such computers can partially imitate a human being's approximate instinctive conjectures about his own consistency (that, as common colloquially held conjectures, seem to serve as *essential prerequisites* for humans to gain their motivation to cogitate).

Sections 7-9 will examine the preceding issues in further detail. One of their themes will be that our exceptions to Gödel's second theorem, while sometimes nontrivial, clearly do not narrow the main intentions of Gödel's result.

7   Comparing Type-M and Type-A Formalisms

Let us recall axioms (1)-(3) indicated Type-A systems differ from Type–M formalisms by treating Multiplication as a 3-way relation (rather than as a total function). For the sake of accurately characterizing what our systems can and cannot do, we have described our results as being fringe-like exceptions to the Second Incompleteness Theorem, from the perspective of an Utopian view of Mathematics, while perhaps being more significant results from an engineering-style perspective of knowledge. Our goal in this section will be to amplify upon this perspective by taking a closer look at Type-A and Type-M formalisms.

Let us assume that $x_0 = 2 = y_0$ and that $x_1, x_2, x_3, ...$ and $y_1, y_2, y_3, ...$ are defined by the recurrence rules of:

$$x_{i+1} = x_i + x_i \qquad \text{AND} \qquad y_{i+1} = y_i * y_i \qquad (13)$$

The sequences $x_0, x_1, x_2, ...$ and $y_0, y_1, y_2, ...$ will thus represent the growth rates associated with the addition and multiplication primitives, lying in the statements (2) and (3)'s **"Type-A"** and **"Type-M"** axioms.

Since $x_0 = 2 = y_0$, the rule (13) implies $y_n = 2^{2^n}$ and $x_n = 2^{n+1}$. The $y_0, y_1, y_2, ...$ sequence will, thus, grow much more quickly than the $x_0, x_1, x_2, ...$ sequence (since y_n's binary encoding will have an $\text{Log}(y_n) = 2^n$ length while x_n's binary encoding will have a shorter length $= \text{Log}(x_n) = n + 1$).

Our prior papers noted that the difference between these growth rates was the reason that [36, 38, 43] showed all natural Type-M systems, recognizing integer-multiplication as a total function, were unable to recognize their tableaux-styled consistency — while [35, 37, 40] showed some Type-A systems could simultaneously prove all Peano Arithmetic Π_1^* theorems and corroborate their own tableaux consistency. Their gist was that a Gödel-like diagonalization argument, which causes an axiom system to become inconsistent as soon as it proves a theorem affirming its own tableaux consistency, stems, ultimately, from the exponential growth in the series $y_0, y_1, y_2, \cdots$.

This growth, thus, facilitates an intense amount of self-referencing, using the identity $\text{Log}(y_n) \cong 2^n$, that will, ultimately, invoke the force of Gödel's seminal diagonalization machinery. It thus raises raises the following question:

> *** How natural are exponentially growing sequences, such as $y_0, y_1, y_2 \cdots$, whose n-th member needs 2^n bits for its encoding, when such lengths are greater than the number of atoms in the universe when merely $n > 100$? Is such a sequence's use, for corroborating the Second Incompleteness Effect, an inherently artificial construct ?

We will not attempt to derive a Yes-or-No answer to Question $***$ because it is one of those epistemological questions that can be debated endlessly. Our point is that $***$ probably does not require a definitive positive or negative answer because both perspectives are useful. Thus, the theoretical existence of a sequence integers of $y_0, y_1, y_2, \ldots$, whose binary encodings are doubling in length, is tempting from the perspective of an Utopian view of mathematics, while awkward from an engineering styled perspective. We therefore ask: *"Why not be tolerant of both perspectives? "*

One virtue of this tolerance is it ushers in a greater understanding for the statements $*$ and $**$ that Gödel and Hilbert made during 1926 and 1931. This is because the Incompleteness Theorem demonstrates no formalism can display an understanding of its own consistency in an idealized Utopian sense. On the other hand, §6 suggested these remarks by Gödel and Hilbert might receive more sympathetic interpretations, if one sought to explore such questions from a less ambitious almost engineering-style perspective.

Our main thesis is supported by a theorem from [42]. It indicated that tableaux variations of self-justifying systems have no difficulty in recognizing that an infinitized generalization of a computer's floating point multiplication (with rounding) is a total function. The latter differs from integer-multiplication, by not having its output become double the length of its input when a number is multiplied by itself. Thus, the intuitive reason [42]'s multiplication-with-rounding operation is compatible with self-justification is because it avoids the inexorable exponential growth under rule (13)'s sequence $y_0, y_1, y_2 \cdots$.

Also, Theorem 4 indicates self-justifying logics can view a double-precision form of integer multiplication as likewise a total function. Its proof, exactly analogous to [42]'s methodology, will appear in a longer version of this paper.

Theorem 4. *Let us assume the* A *in* $IS_D(A)$ *and* $IS_D^\#(\beta_{A,i})$ *represents Peano Arithmetic. Then* $IS_D(A)$ *and* $IS_D^\#(\beta_{A,i})$ *can formalize two total functions, called* $Left(a, b)$ *and* $Right(a, b)$, *where any pair of integers* (a, b) *is mapped onto the left and right halves of* a *and* b's *multiplicative product.*

Remark 3. One slightly tricky aspect is that our positive results, involving [42]'s floating point multiplication and Theorem 4's double precision multiplication, *should not be confused* with a different examination of Herbrandized consistency in [44]. The latter took advantage of the fact that our Herbrand-styled proofs, in [44]'s paradigm, are exponentially longer than their tableaux counterparts, thus allowing [44] to formalize a limited use of multiplication (because its deductive methods was exponentially less efficient). Thus [44]'s results, while perhaps theoretically interesting, are basically irrelevant to engineering environments, e.g. the main concern of Theorems 1–4 (especially in regards to their particular interpretations given in Remark 2).

8   A Different Type of Evidence Supporting Our Thesis

Let us recall Pudlák and Solovay [27, 31] observed that essentially all Type-S systems, containing merely statement (1)'s axiom that successor is a total function, cannot verify their own consistency under Hilbert deduction. (See also related work by Buss-Ignjatovic [5], Švejdar [32] and the Appendix A of [37])

It turns out that [39] generalized these results to show that Equation (2)'s Type-A systems are unable to verify their own consistency, under the Tab−2 deduction (defined in §3). At the same time, the IS_D and $IS_D^\#$ frameworks, from Sections 4 and 5, can verify their own consistency under Tab−1 deduction. Our goal in this section will be to illustrate how the tight contrast between these positive and negative results is analogous to the differing growth rates of the sequences $x_0, x_1, x_2, \ldots$ and $y_0, y_1, y_2, \ldots$ from rule (13).

During our discussion $G_i(v)$ will denote the scalar-multiplication operation that maps an integer v onto $2^{2^i} \cdot v$. Also, Υ_i will denote the statement, in the U-Grounding language, that declares that G_i is a total function. Our paper [39] proved that Υ_i has a Π_2^* encoding. It is also implied that G_i satisfied:

$$G_{i+1}(v) \quad = \quad G_i(\; G_i(v) \;) \tag{14}$$

It was noted in [39] that this identity implies one can construct an axiom system β, comprised of solely Π_1^* sentences, where a semantic tableaux proof can establish Υ_{i+1} from $\beta + \Upsilon_i$ in a constant number of steps. This implies, in turn, that a Tab−2 proof from β will require no more that $O(n)$ steps to prove Υ_n (when it uses the obvious n-step process to confirm in chronological order Υ_1, Υ_2, $\ldots \Upsilon_n$.)

These observations are significant because $G_n(1) = 2^{2^n}$. Thus, [39] showed a Tab−2 proof from β can verify in $O(n)$ steps that this integer exists.

This example is helpful because it illustrates the difference between the growth speeds under Tab−1 and Tab−2 deduction, is analogous to the differing growth rates of the sequences $x_0, x_1, x_2, \ldots$ and $y_0, y_1, y_2, \ldots$ from rule (13). Hence once again, a faster growth-rate will usher in the Second Incompleteness Theorem's power (e.g. see [39]).

This analogy suggests that the Second Incompleteness Theorem has different implications from the perspectives of Utopian and engineering theories about the intended applications of mathematics. Thus, a Utopian may possibly be comfortable with a perspective, that contemplates sequences $y_0, y_1, y_2, \ldots$ with elements growing in length at an exponential speed, but many engineers may be suspicious of such growths.

A hard-core engineer, in contrast, might surmise that the inability of self-justifying formalisms to be compatible with Tab-2 deduction is not as disturbing as it might initially appear to be. This is because Tab-2 differs from Tab-1 deduction by producing exponential growths that are so sharp that their material realization has no analog in the everyday mechanical reality that is the focus of an engineer's interest.

Our personal preference is for a perspective lying half-way between that of an Utopian mathematician and a hard-nosed engineer. Its dualistic approach suggests some form of diluted partial agreement with Hilbert's goals in $**$.

9   Related Reflection Principles

An added point is that there are many types of self-justifying systems available, with some better suited for engineering environments than others.

Ideally, one would like to develop self-justifying systems S that could corroborate the validity of (15)'s reflection principle for all sentences Φ.

$$\forall p \ [\ Prf_S^D(\ulcorner \Phi \urcorner, p) \ \Rightarrow \ \Phi \ ] \tag{15}$$

Löb's Theorem establishes, however, that all systems S, containing Peano Arithmetic's strength, are able to prove (15)'s invariant *only in the degenerate case* where they prove Φ itself. Also, the Theorem 7.2 from [37] showed essentially all axiom systems, *weaker* than Peano Arithmetic, are unable to prove (15) for all Π_1^* sentences Φ simultaneously. Thus, Theorem 5 will be near optimal:

Theorem 5. *For any input axiom system A, it is possible to extend the self-justifying $IS_D(A)$ and $IS_D^\#(\beta_{A,i})$ systems, from Theorems 1 and 3, so that the resulting self-justifying logics S can also:*

1. *Verify that Tab-1 deduction supports the following analog of (15)'s self-reflection principle under S for any Δ_0^* and Σ_1^* sentences Φ :*

$$\forall p \ [\ Prf_S^{\text{Tab}-1}(\ulcorner \Phi \urcorner, p) \ \Rightarrow \ \Phi \ ] \tag{16}$$

2. *Verify (17)'s more general* **"root-diluted"** *reflection principle for S whenever θ is Σ_1^* and Φ is a Π_2^* sentence of the form "$\forall u_1 \ldots \forall u_n \ \theta(u_1 \ldots u_n)$ ".*

$$\forall p \ [\ Prf_S^{\text{Tab}-1}(\ulcorner \Phi \urcorner, p) \ \Longrightarrow \ \forall x \ \forall u_1 < \sqrt{x} \ \ldots \ \forall u_n < \sqrt{x} \ \theta(u_1 \ldots u_n) \ ] \tag{17}$$

Theorem 5's proof will rest upon hybridizing the techniques from [37]'s tangibility reflection principle with Theorem 3's methodologies, in a natural manner, as will be demonstrated in a longer version of this article. Analogous to our other results, Theorem 5 reinforces the theme about how exceptions to the Second Incompleteness Theorem may appear to be *quite minor* from the perspective of an Utopian view of mathematics, while being significant from an engineering standpoint. In Theorem 5's particular case, this is because:

A. The ability of Theorem 5's system S to support (16)'s self-reflection principle under Tab-1 proofs for any Δ_0^* and Σ_1^* sentence, as well as to support (17)'s root reflection principle for Π_2^* sentences, is clearly significant.

B. The incompleteness result of [37]'s Theorem 7.2 imposes, however, sharp limitations upon Item A's generality (in that it cannot be extended to fully all Π_1^* sentences, *in an undiluted sense*).

Thus, the tight fit between A and B is reminiscent of other slender borderlines, that separated generalizations and boundary-case exceptions for the Incompleteness Theorem, explored earlier. Once again, the Second Incompleteness Theorem is seen as robust, from an idealized Utopian perspective on mathematics, while permitting caveats from engineering styled perspectives.

This dualistic viewpoint allows one to nicely share *partial (and not full)* agreement with Hilbert's main aspirations in **, while also appreciating the stunning achievement of the Second Incompleteness Theorem.

10 Concluding Remarks

At a purely technical level, this article has reached beyond our prior papers in several respects, including §5's demonstration that any initial system A can have a kernelized image of its Π_1^* knowledge duplicated by IS$_D^{\#}(\beta_{A,i})$'s **strictly finite sized** self-justifying system, and also by Section 6's and Remark 2's quite pragmatic L-fold generalizations of this result.

These results help resolve the mystery that has enshrouded the Second Incompleteness Theorem and the statements * and ** of Gödel and Hilbert. This is because we have *meticulously separated* the goals of a pristine theoretical study of mathematical logic from those of a *finite-sized* axiomatic subset of mathematics, intended for modeling mostly an engineering environment.

There is no question that Gödel's Second Theorem is ideally robust, relative to a purely pristine approach to mathematics. On the other hand, we suspect Hilbert was *half-way correct* by speculating in ** about humans possessing a knowledge about their own consistency, *in at least some strikingly weak and tender sense*, as essentially a fundamental prerequisite for *psychologically motivating* their cogitations.

Thus in a context where the limitations of axiom systems, that fail to recognize multiplication as a total function, are manifestly obvious, even when such systems duplicate Peano Arithmetic's central Π_1^* knowledge, it is legitimate to inquire whether posterity might find some *partial-albeit-not-full* redeeming value in formalisms having *weak-style* knowledges of their Tab-1 consistency?

More precisely, Sections 5-9 were intended to provide a unified broad-scale interpretation of our diverse earlier results from [35–44]. In a context where the Incompleteness Theorem is firmly understood to be sufficiently ubiquitous to preclude Hilbert's aspirations in ** from ever being fully realized, they show how some *fragmentary portion* of Hilbert's conjectures can be corroborated by *judiciously weakened* logics, using a formalism, that is *much less* than ideally robust, *although not fully immaterial.*

Acknowledgments. I thank Bradley Armour-Garb and Seth Chaiken for many useful suggestions about how to improve the presentation of our results. I also thank the anonymous referees for their comments. This research was partially supported by NSF Grant CCR 0956495.

References

1. Adamowicz, Z.: Herbrand consistency and bounded arithmetic. Fundamenta Mathematicae 171(3), 279–292 (2002)
2. Adamowicz, Z., Zbierski, P.: On Herbrand consistency in weak theories. Archive for Mathematical Logic 40(6), 399–413 (2001)
3. Bezboruah, A., Shepherdson, J.C.: Gödel's second incompleteness theorem for Q. Jour. Symb. Logic 41(2), 503–512 (1976)
4. Buss, S.R.: Bounded Arithmetic. Studies in Proof Theory, Lecture Notes 3. Published by Bibliopolis (1986) (Revised version of Ph. D. Thesis)
5. Buss, S.R., Ignjatovic, A.: Unprovability of consistency statements in fragments of bounded arithmetic. Annals Pure and Applied Logic 74(3), 221–244 (1995)
6. Dawson, J.W.: Logical Dilemmas the life and work of Kurt Gödel. AKPeters (1997)
7. Feferman, S.: Arithmetization of metamathematics in a general setting. Fundamenta Mathematicae 49, 35–92 (1960)
8. Fitting, M.: First Order Logic and Automated Theorem Proving. Springer (1996)
9. Friedman, H.M.: On the consistency, completeness and correctness problems. Technical report, Ohio State Univ. (1979), See Pudlák [28]'s summary of this result
10. Friedman, H.M.: Translatability and relative consistency. Technical report, Ohio State Univ. (1979), See Pudlák [28]'s summary of this result
11. Gödel, K.: Über formal unentscheidbare Sätze der Principia Mathematica und verwandter Systeme I. Monatshefte für Math. Phys. 38, 173–198 (1931)
12. Gödel, K.: The present situation in the foundations of mathematics. In: Feferman, S., Dawson, J.W., Goldfarb, W., Parson, C., Solovay, R. (eds.) Collected Works Volume III: Unpublished Essays and Lectures, pp. 45–53. Oxford University Press (2004), Our quotes from this 1933 lecture come from its page 52
13. Goldstein, R.: Incompleteness The Proof and Paradox of Kurt Gödel. Norton (2005)
14. Hájek, P.: Mathematical fuzzy logic and natural numbers. Fundamenta Mathematicae 81, 155–163 (2007)
15. Hájek, P.: Towards metamathematics of weak arithmetics over fuzzy. Logic Journal of the IPL 19(3), 467–475 (2011)

16. Hájek, P., Pudlák, P.: Metamathematics of First Order Arithmetic. Springer (1991)
17. Hilbert, D.: Über das Unendliche. Mathematische Annalen 95, 161–191 (1926)
18. Hilbert, D., Bernays, P.: Grundlagen der Mathematik, vol. II. Springer (1939)
19. Jeroslow, R.G.: Consistency statements in formal theories. Fundamenta Mathematicae 72, 17–40 (1971)
20. Kleene, S.C.: On notation for ordinal numbers. Jour. Symb. Logic 3(1), 150–156 (1938)
21. Kołodziejczyk, L.A.: On the Herbrand notion of consistency for finitely axiomatizable fragments of bounded arithmetic theories. Jour. Symb. Logic 71(2), 624–638 (2006)
22. Kreisel, G., Takeuti, G.: Formally self-referential propositions in cut-free classical analysis and related systems. Dissertationes Mathematicae 118, 1–50 (1974)
23. Löb, M.H.: A solution of a problem of Leon Henkin. Jour. Symb. Logic 20(2), 115–118 (1955)
24. Nelson, E.: Predicative Arithmetic. Math. Notes. Princeton Univ. Press (1986)
25. Parikh, R.: Existence and feasibility in arithmetic. Jour. Symb. Logic 36(3), 494–508 (1971)
26. Parsons, C.H.: On $n-$quantifier induction. Jour. Symb. Logic 37(3), 466–482 (1972)
27. Pudlák, P.: Cuts, consistency statements and interpretations. Jour. Symb. Logic 50(2), 423–442 (1985)
28. Pudlák, P.: On the lengths of proofs of consistency. In: Collegium Logicum: Annals of the Kurt Gödel, vol. 2, pp. 65–86. Springer-Wien (1996)
29. Rogers, H.A.: Theory of Recursive Functions and Effective Compatibility. McGraw Hill (1967)
30. Salehi, S.: Herbrand consistency of some arithmetical theories. Jour. Symb. Logic 77(3), 807–827 (2012)
31. Solovay, R.M.: Telephone conversation in 1994 describing Solovay's generalization of one of Pudlák's theorems [27], using some methods of Nelson and Wilkie-Paris [24, 34] (The Appendix A of [37] offers a 4-page summary of this conversation) (1994)
32. Svejdar, V.: An interpretation of Robinson arithmetic in its Grzegorczjk's weaker variant. Fundamenta Mathematicae 81, 347–354 (2007)
33. Visser, A.: Faith and falsity. Annals Pure and Applied Logic 131(1), 103–131 (2005)
34. Wilkie, A.J., Paris, J.B.: On the scheme of induction for bounded arithmetic. Annals Pure and Applied Logic 35, 261–302 (1987)
35. Willard, D.E.: Self-verifying axiom systems. In: Mundici, D., Gottlob, G., Leitsch, A. (eds.) KGC 1993. LNCS, vol. 713, pp. 325–336. Springer, Heidelberg (1993)
36. Willard, D.E.: The semantic tableaux version of the second incompleteness theorem extends almost to Robinson's arithmetic Q. In: Dyckhoff, R. (ed.) TABLEAUX 2000. LNCS, vol. 1847, pp. 415–430. Springer, Heidelberg (2000)
37. Willard, D.E.: Self-verifying systems, the incompleteness theorem and the tangibility reflection principle. Jour. Symb. Logic 66(2), 536–596 (2001)
38. Willard, D.E.: How to extend the semantic tableaux and cut-free versions of the second incompleteness theorem almost to Robinson's arithmetic Q. Jour. Symb. Logic 67(1), 465–496 (2002)

39. Willard, D.E.: A version of the second incompleteness theorem for axiom systems that recognize addition but not multiplication as a total function. In: Hendricks, V., Neuhaus, F., Pederson, S.A., Scheffler, U., Wansing, H. (eds.) First Order Logic Revisited, pp. 337–368. Logos Verlag, Berlin (2004)
40. Willard, D.E.: An exploration of the partial respects in which an axiom system recognizing solely addition as a total function can verify its own consistency. Jour. Symb. Logic 70(4), 1171–1209 (2005)
41. Willard, D.E.: A generalization of the second incompleteness theorem and some exceptions to it. Annals Pure and Applied Logic 141(3), 472–496 (2006)
42. Willard, D.E.: On the available partial respects in which an axiomatization for real valued arithmetic can recognize its consistency. Jour. Symb. Logic 71(4), 1189–1199 (2006)
43. Willard, D.E.: Passive induction and a solution to a Paris-Wilkie open question. Annals Pure and Applied Logic 146(2), 124–149 (2007)
44. Willard, D.E.: Some specially formulated axiomizations for $I\Sigma_0$ manage to evade the Herbrandized version of the second incompleteness theorem. Information and Computation 207(10), 1078–1093 (2009)
45. Yourgrau, P.: A World Without Time: The Forgotten Legacy of Gödel and Einstein. Basic Books (2005), See page 58 for the passages we have quoted

Author Index